URBAN SOILS

URBAN SOILS

Applications and Practices

PHILLIP J. CRAUL
Senior Lecturer in Landscape Architecture
The Graduate School of Design
Harvard University
Cambridge, Massachusetts

JOHN WILEY & SONS, INC.
New York • Chichester • Weinheim • Brisbane • Singapore • Toronto

This book is printed on acid-free paper. ♾

Published simultaneously in Canada.

Library of Congress Cataloging-in-Publication Data:

Craul, Phillip J., 1932–
Urban soils : applications and practices / Phillip J. Craul and evapotranspiration estimation
p. cm.
Includes bibliographical references (p.) and index.
ISBN 0-471-18903-0 (cloth : alk. paper)
1. Urban soils. I. Title.
S592.17.U73C735 1999 98-38197
631.4′7173′2--DC21

Printed in the United States of America.

10 9 8 7 6 5 4 3 2 1

For

Joanne

". . . promising with Divine Assistance
to be unto Thee
a loving and faithful wife
so long as we both may live."

. . . and so it is for 40 years.

". . . the stuff that lies
beneath grass isn't dirt.
It's the elixir of life."

Phillip J. Craul,
as quoted in the *Geraldine R. Dodge Foundation Report 1996,* p. 69.

". . . close living to the good earth
is a source of physical and moral
and spiritual strength. This is the
verdict of history as far back as it goes,
and those civilizations who have forgotten it,
have not fared well in times of crises.
Being too far removed from the soil has often
caused loss of security, a warping of outlook,
weakening of moral force, distortion of the
sense of values, a weaning away of self-dependence,
and even a failure to survive."

Richard L. Evans, *This Day . . . And Always,* Harper & Bros., New York, 1942. p. 46.

CONTENTS

PREFACE

The purpose of this book is to provide detailed information on some of the more commonly used procedures and applications of urban soils. It extends the efforts of the author in his book *Urban Soil in Landscape Design* (John Wiley and Sons, 1992) and the more recent need to be more definitive in several aspects of urban soil and its design applications. This information provides perhaps new knowledge to the landscape architect, horticulturist, and urban forester. Much of the information can be used to develop specifications or preliminary criteria for a design application.

The author needs to acknowledge the people who have had significant impact on the ideas given here. Dr. Jim Patterson, formerly of the National Park Service at Washington, DC, now retired; Dr. Tom Perry, Natural Systems Associates, Raleigh, North Carolina; Jim Urban of Jim Urban Associates, Annapolis, Maryland; and Dr. Gary Watson and Dr. Pat Kelsey, Morton Arboretum, Lisle, Illinois. These scientists have all contributed much knowledge, many practical ideas and much experience. To them I give my heartfelt thanks for their help and also their longtime friendship. Further, thanks to those clients I have served over the years who have provided the opportunity to practice these applications to one degree or another, and to associates on the various projects who have made the execution of them successful. Included are: Susan Child and Douglas Reed (formerly) of Child Associates, Boston; Clarissa Rowe of Brown and Rowe, Boston; Laura Solano, Sally Coyle, and Michael Van Valkenburgh of Michael Van Valkenburgh Associates, Cambridge, Massachusetts; Don Vitters, Henry Ricciuti, George Burr, and Lucy Chamberlain of Sasaki Associates, Watertown, Massachusetts; Bob Bodell, Dennis McGlade, and Cindy Saunders of Olin Partnership (formerly Hanna/Olin, Ltd.), Philadelphia; Frank Clare, DPCO, Boston; Ralph DeGregorio of Massachusetts Highway Department, and Kelly Duke of Valley Crest Landscaping, Los Angeles. Without this practical experience as proof to the success of the applications discussed here, the book would not have its validity. To them I offer my appreciation for their support and confidence in improving our approach to the solution of urban soil challenges. It was only through their conscientious efforts that made the case studies as successful as they are.

Especially I thank Dr. Philip Manor, Senior Editor, Professional and Trade Division, for his guidance and assistance, and friendship, over the years involving the publication of two books. Our dinner conversations are always a delight. Thanks also to Rose Leo Kish, his assistant, for her help.

As always, the author must take full responsibility for any oversights and errors of omission or commission in the book. *However, the author makes no warranty, expressed or implied, as to the success or failure of any application given herein since he does not exercise any control over its final form and execution. The user assumes all liability.*

PHILLIP J. CRAUL

And to the inspiration of Sec. 31, T24N, R10W,
Montana Prime Meridian, on "Hannibal."
Manlius, New York
December 1998

1

INTRODUCTION

THE NEED FOR PRACTICAL URBAN SOIL APPLICATIONS

Review of scientific publications, personal experience, and contact with many landscape architects, scientists, and contractors over the years indicate that new views and approaches must be taken toward the treatment of soil in urban landscape design. Much unused technology and original experience exist that provide some of the more successful approaches. Also, the context in which much landscape design is accomplished is changed. No longer are there many "virgin" undisturbed sites for development; most are now restoration sites that require new approaches to handling the soil question. Pollution and contamination elimination have become watchwords of the process. Suggestions for the new approaches are made with practical examples as illustrations. Other concerns become apparent as the discussion proceeds.

McHarg (1969) indirectly suggests that we may be creating a moon landscape here on planet earth (p. 25). What he means is that the urban sprawl of our metropolitan areas spilling over the landscape with little attention to design and leaving inappropriate open space is creating a spartan environment. The central business districts have towering high-rise buildings with "wind-tunnel" streets and little vegetation. These areas are surrounded by urban blight and ghettos requiring deliberate restoration. The eventual effect could be transformation of earth's natural landscape of biodiveristy and "greenness" into a very sterile, monotonous viewscape of steel, glass, asphalt, and concrete.

According to Van der Ryn and Cowan (1996), the environmental crisis is a design crisis, stating that "it is a consequence of how things are made, buildings are constructed, and landscapes are used. . . ." They define *design* "as the intentional shaping of matter, energy, and process to meet a perceived need or desire." According to this definition, many kinds of people, including farmers and foresters, can be designers besides those professionals who we usually think of as designers.

The same authors contend that standard templates have been applied to development across the U.S. continent. The result is *dumb design:* "design that fails to consider the health of human communities or of ecosystems, let alone the prerequisites of creating an actual place." The definition tends to follow the same thought McHarg was suggesting earlier. One needs to look no further than a nearby shopping mall or the local international airport to see the emphasis on vehicular transportation and the lack of infrastructure for pedestrians.

Old versus New Methods

Several factors particularly influence the need for new methods in urban soil applications. These include the demand for soil specifications that are testable and enforceable, the use of revised planting techniques for trees, the potential for improved technology for the biophysical analysis of planting sites, and the trend toward the design and specification of sustainable soil.

Soil Specifications. Soil specifications in the past have been superficial or incomplete at best. The attitude commonly was, and still is in some areas, "dirt is dirt and it's cheap." Because many of the past landscape planting failures can be attributed to poor soil conditions, it is necessary to make soil specifications just as detailed and specific, and technically correct, as are the planting stock specifications, the engineering and the architectural specifications.

Further, the greater sensitivity to and requirements for sustainability in design for many projects suggest that the specified soil has recyclable components rather than composed entirely of nonrenewable resources. The incorporation of recyclable components requires a high level of detail in specification to prevent the use of unsuitable or unsafe materials on the project. Governmental regulations also require specification of limits on various potential contaminants that may be present in the recyclable materials. These recyclable materials may include composted sewage sludge, municipal solid wastes, agricultural wastes, and food-processing wastes. They may contain contaminants such as heavy metals, PCBs, glass, plastic, metal, and so on.

Thus, the soil specification becomes several pages or more in length rather than a simple paragraph or one page. A definition of "an upland, well-drained, friable, natural loam topsoil" is totally inadequate for present and future projects. Any specification that simple should be rejected out of hand. The writing of soil specifications is a major focus of this manual and is covered in detail in Chapter 5. Elements of specifications are discussed in many parts of other chapters.

Planting Techniques. There is a tendency to follow "old tried-and-true" planting techniques. In some instances, where the site conditions are ideally suited to tree planting, this attitude is appropriate. However, there are few sites in urban areas where the conditions are ideal for planting. As a result, it is necessary to adopt new and innovative planting techniques to meet the adverse planting conditions found in many urban or newly developed sites (Urban, 1989). These are discussed in Chapter 9.

The innovative planting techniques require greater appreciation of the plant needs for water and nutrients, aeration and a quality rooting medium for the roots, and adequate rooting volume to provide these amenities. Greater technical knowledge by the landscape architect is a prerequisite. Greater detail in planting design is required, and installation and its supervision is more costly. But the greater potential for long-term success of the overall project design justifies the added requirements compared to the old way of doing business.

Biophysical Analyses. Many plantings are made with little appreciation and attention to the above-ground environmental conditions presented by the site. Involved here are the microclimatic factors of solar radiation and reflected heat, wind, air temperature, and humidity (Bassuk and Whitlow, 1985). Evaluation of these factors needs to become a more routine part of site analysis to make plantings more successful. Rooftop applications should have contained these analyses on a routine basis from the beginning of this technology.

Analysis of the microclimatic factors may be called *biophysical analysis,* since it is these factors that influence the actual and potential evapotranspiration of plants. The internal plant physiological processes are affected by the degree of transpirational stress experienced by the plant. Solar radiation, reflected heat, wind speed, air temperature and humidity, and, thus, the temperature of the plant, can be evaluated for a given site by using the right data and tools. Some of the appropriate tools and procedures are discussed in Chapter 3. The user needs to supply the necessary data for a given project site. In certain instances, the user must obtain some of the data. Measurements with simple instruments sometimes is most helpful when data are lacking. Obviously, a complete picture of the environmental conditions is not always possible, but a few critical measurements are better than nothing. A description of the critical measurements with their value and use are also given.

Reliability and Liability

There comes a time for landscape architects to say, "No!" or use the fourth "R"—*refuse!* And drugs and sex have nothing to do with it.

In a talk before the Boston Society of Landscape Architects in October 1995, Jim Urban, a landscape architect from Annapolis, Maryland, used this analogy to illustrate the point of saying, "No!" A client and owner goes to an engineer to say that there is serious cost overrun. Therefore, to save money, we want you as the engineer to scale down the size of the steel beams and girders in the building. I think we all know what the reply from the engineer would be. He would refuse on the spot because a reduced safety margin is involved, and if the building failed, he is the one who would be liable because his professional stamp is on the drawings. He would lose everything. Landscape architects need to adopt the same attitude anytime a request is made to reduce or modify the design in such a way that it will not produce its full potential for survival and development. Soil specifications are one of the easiest documents to attack by clients and contractors. Many people claim that "dirt is dirt" and shouldn't cost anything, and, besides, public safety is not involved! Any reduction in the quality, and especially the quantity (volume, depth, or whatever form of measurement used) of the soil, should be vigorously resisted. The results of modification or reduction of soil specifications nearly always have a detrimental effect on the growth success of the design and may even doom it to failure. As Jim suggests, a few successful liability cases against landscape architects over plant design failure due to excessive compromise will develop some backbone in the profession.

Many bases for defending oneself from "cost-cutting intimidation" will be provided in this book. The other points of discussion from earlier in this chapter are addressed in the following chapters.

Landscape Restoration

Baltimore is preparing to tear down 800 vacant rowhouses in two neighborhoods of the city according to the *Baltimore Sun* of March 23, 1996. There are more than 9000 vacant buildings scattered across the city. The action is part of a vast, multimillion dollar undertaking to restore abandoned and deteriorated neighborhoods of the inner city. Most of the land would be transformed into open greenspace. Only a quarter of the properties would be redeveloped for housing, and a few others would become neighborhood shopping centers. In Sandtown, as many as 870 homes stand vacant and boarded in the 72 square blocks. Half are to be renovated and the other half razed. Across town, another 2300 rowhouses have been abandoned in the 250 blocks around the Johns Hopkins Hospital, the Hopkins medical school, and other medical organizations. Some 400 of these vacant houses will be raised. Unlike Sandtown, where modern townhouses or shops will be build on demolished sites, the East Baltimore group wants to use the land strictly for neighborhood parks.

Most large cities of the United States and the world have the similar problem of abandoned or deteriorated inner city areas that need restoration. In order to ensure the success of the small- and large-scale efforts of restoration, it behooves the planners, landscape architects, horticulturists, and managers to appreciate the site and soil problems presented by these areas. These concerns have been expressed earlier by Hardin (1968), McHarg (1992), and Spirn (1984, 1992). Plans must be made to overcome the problems and provide an alternative set of solutions during the very initial phase of the planning process, even before the drafting table is approached.

As McHarg (1969) states, we need nature as much in the city as in the countryside. New York City has Central and Riverside parks and others, Philadelphia has Fairmont and Wissahickon Parks, Chicago has Grant and Lincoln Parks, and Washington has Rock Creek Park, the C and O Canal, and the Mall. But these are concentrated, and many blighted, ungreen areas exist where open greenspace is badly needed. When asked what they would like most for their school yard, the pupils of an elementary school in St. Helens, England (suburb of Liverpool), chose to have a green lawn surrounded with raised planters rather then swing sets and sandboxes, and so on (Fig. 1.1).

Emphasis on Sustainability

There has been much written, discussed, and reported about the term *sustainability*. To reduce confusion for our purposes I define *sustainability* as

> the ecological and technological planning, design, implementation, management, and maintenance of a landscape project that reduces environmental impact both on the project site and off the site.

This concept includes the following considerations:

- The project is located on land that does not reduce production from prime agricultural land; or already has been so detrimentally impacted that its return to agricultural production is not feasible

Fig. 1.1 School yard in St. Helens, England, with greenery established over asphalt paving, 1987.

- The construction of the project and its final form have minimal or no direct or indirect environmental impact on adjacent and other off-site locations
- The materials and products used in the project are of recycled or renewable resources
- The production and byproducts of these materials and products have minimal environmental impact
- The final project requires minimal or no maintenance needing prolonged input of energy-consuming products or nonrenewable materials.

The former solution of covering the site material with topsoil and planting in that layer is no longer appropriate for two major reasons. The first is that true topsoil has become a scarce and expensive material with a high level of off-site environmental impact. Topsoil from other projects is no longer an acceptable practice. In most cases, the topsoil should remain for use on the original site rather than playing musical chairs with the topsoil among several different projects. The second and more important reason is spreading topsoil or any material as a layer over unknown subbase material is an unwise practice. The entire soil profile should be designed to accommodate the existing conditions and the planting plan and it palette. Opportunities for design alternatives may arise from information derived from the intensive sampling and analyses. The use of existing site materials as a component of a designed mix should be explored together with sources of recycled materials such as composted organic wastes and mineral materials. All of these actions would significantly reduce the environmental impact of restoration activities.

The technology of recyclable material use in constructing soil has progressed to a sufficient degree that permits the serious consideration of a totally sustainable soil. That is, a soil composed entirely of recyclable materials and containing no nonrenewable resources. This is the topic of Chapter 4.

Soil Costs in the Project Budget

The soil material design and specification many times in the past have been looked on as a place to save money, leaving more of the budget allocated to the plant palette (more plants per square yard) or more lighting and fancier structures, or a bigger fountain. Often, this scheme worked because the project "lucked out" in obtaining some fine natural topsoil (not necessarily well specified) that was not mishandled and probably was in the way of another project, so it was cheap. The planting was a rousing success, leading to the invalid conclusion "Why waste money on soil!"

Experience has indicated that we need to spend more money on soil. Some of this impact is caused by design considerations and some is caused by factors outside of the design and project realm and in the realm of global economics. The ratio of plant palette costs to soil requires elucidation. In one notable planting failure, the ratio was about 7.5:1. The cost of specifying, obtaining, testing, and inspecting of soil should amount to a greater proportion of the project budget. The future ratio may approach 3:1, with hope that the plant budget is not cut to reduce the ratio to that level.

THE INTERPLAY AMONG THE SOIL SCIENTIST, LANDSCAPE ARCHITECT, AND OTHER PROFESSIONALS

It is imperative that the landscape architect become knowledgeable in soil science and the terms used in the field. It is the soil that provides the medium for rooting of the stock in the plant palette, and success or failure depends on how well the landscape architect designs and specifies the soil. This task cannot be accomplished without attention to the assessment of the site and the soil design details. The details need to be done by a landscape architect knowledgeable in soil design, or lacking that knowledge, a soil scientist specializing in urban soils should be hired to do the job. There is no hesitation within the profession for individual landscape architects to study the engineering involved in structure, lighting, and so on, or to hire an engineer to do the design. The same approach should be applied to soil design.

By being knowledgeable in soil, the landscape architect can do some of the soil design, converse intelligently with a soil scientist who will do the work, and to defend the design in interplay with the project architect and engineer. Many of the terms used by a soil scientist in designed soil are similar, if not the same, as used by soil engineers. Thus, the engineer can understand and appreciate what the landscape architect wishes to achieve in the landscape design. Unfortunately, the soil characteristics desired by the engineer and architect are nearly diametrically opposite of those desired by the landscape architect. If the landscape architect is knowledgeable in these concepts and terms, greater success will be achieved in gaining the desired results.

It has been the author's experience over 25 years of consulting on landscape projects that it is imperative for the landscape architect to appreciate soil and to become knowledgeable in its properties and characteristics and how it can be designed if the practitioner wishes to be successful at all levels of project planning, design, installation, and maintenance. It is hoped, the material in this book will aid in the quest.

2

SITE SOIL ASSESSMENT AND INTERPRETATION DEVELOPMENT

Foreknowledge is power.

INTRODUCTION

Site assessment involves the investigation of site conditions to provide data and information so that planting design can take advantage of existing opportunities and avoid problems that will influence the outcome of the planting operation.

Site assessment requires an initial overview of the existing conditions in order to develop an appropriate approach to the design and methodology of planting. The level of effort and detail may range from a "dashboard" survey to intensive and detailed soil sampling with multiple laboratory soil analyses. The effort for many planting projects will fall near the middle of this range: some soil sampling with some laboratory analyses.

A preliminary checksheet for site assessment is given in Table 2.1. The checksheet may be applied to all types of sites. After it is established that the site is undeveloped or developed, the process of assessment differs because of the disparity in the types of resource data available, which in turn, has significant influence on the type and intensity of on-site sampling and analyses required to produce the data for making a valid assessment.

Making the decision on a site being undeveloped or developed is straightforward and easy. Undeveloped (and undisturbed except for agricultural, forestry, or rangeland activities) sites are usually outside urban limits and have natural soil and vegetation in place. The topography is little disturbed except for accelerated erosion in some instances. Developed land has disturbed topography, soil, and vegetation. The land potentially is polluted or contaminated in some way. Thus, the undeveloped site has a natural history of evolution, whereas the developed site has an anthropic constructional history

TABLE 2.1 Site Assessment Overview Checklist

A. Site Conditions
  1. Natural or urban
  2. Exposure:
    (a) open (360°)
    (b) partial:
      (1) orientation/aspect
      (2) degree of sun-shade
      (3) wind
  3. Infrastructure on-site:
    (a) existing curbs, foundation walls, sidewalks, streets, pipe traces, roofs, culverts
  4. Surface:
    (a) covering, if not bare
    (b) topography:
      (1) slope gradient and shape (convex, concave, complex)
  5. Vegetation:
    (a) none
    (b) if present, type:
      (1) to be preserved?
      (2) not to be preserved?
  6. Drainage:
    (a) surface:
      (1) direction, etc.
    (b) watertable depth
    (c) streams, rivers, floodbasins present and distance?

B. Soil Conditions
  1. Type:
    (a) natural: soil series if known (otherwise, see what follows):
  2. Urban "made":
    (a) profile type:
      (1) fill:
      (2) cut:
      (3) mound or berm
    (b) fill:
      (1) type
        —construction debris
        —"soil" fill
        —dredgings
        —waste products (ash, tailings, slag, foundry sand, mine spoil)
      (2) depth of fill (to lithologic discontinuity of rock, etc., or natural soil):
    (c) cut:
      (1) depth to bedrock or other discontinuity
      (2) type of material below discontinuity
    (d) mound or berm:
      (1) height of mound above discontinuity
      (2) type of material below discontinuity:
  3. texture: (textural class with coarse fragment content, including anthropeic materials):
  4. drainage class or permeability
  5. density estimate (compaction?)

C. Sketch of Site Area

Source: Craul, 1992.

that is far more complex than the undeveloped site. In many instances, the complete constructional history may not be known.

Urban (1992) suggests a three-step protocol for determining needed site modification. Step one involves the determination of soil quality. The determination is based on four classes of quality:

1. Not graded and not compacted
2. Not graded but compacted
3. Graded but not compacted
4. Graded and compacted.

He defines the term *graded* as a soil that has had its topsoil (A horizon) disturbed, removed, or replaced. *Compacted* is defined as a soil that has been compressed to a bulk density greater than 1.6 Mg/m^3. The second step is to determine the level of urbanization. To be more specific, it is the area of open soil available for plant roots as opposed to paved area and that occupied by structures on the project site, coupled with the amount of resources available per tree to modify the planting site. Thus, a small area beside a tall office or apartment building would have a high degree of intensity of use and therefore large resources per tree. The percentage of impervious surface after construction falls into five classes: (1) less than 15%, (2) 15 to 50%, (3) 50 to 75%, (4) 75 to 90%, and (5) 90% or greater. Step three consists of finding the site minimum design criteria. Urban provides a soil quality/urbanization surface matrix containing the design criteria of the size of the planting hole for each tree, the drainage modification required for given soil percolation rates, and the aeration modification techniques that include mechanical aeration and mulching, planting on berms, and installation of wicking. The entire process is excellent for a preliminary planning design that would lead to detailed design documents. The details for the latter documents are obtained by greater intensity of sampling, profile description, and evaluation, and, in some cases of undeveloped land, construction of an interpretation matrix for the project objective. Discussion of these details follow.

RESOURCE DATA

There are much data and information usually available for undeveloped sites if the investigator knows the sources (Way, 1973). Geologic, hydrologic, and topographic data and information are derived from publications and maps of the U.S. Geologic Survey (USGS). In fact, much of the data have now been digitized for input to geographic information systems (GISs). Topography is in the form of digital line graphs (DLGs) or digital elevation models (DEMs). However, the data points may be too broad for adequate accuracy and precision as applied to most landscape architecture projects with limited area. Actual topography may be available from the project engineering survey. The contour intervals may be as narrow as 1 to 2 feet (30 to 60 cm; here, English units are given first since most U.S. surveyors and engineers still use English rather than metric land units).

The geologic reports will provide information on the type of bedrock, its depth and structure, and infer the surficial geology (the geologic material at the surface in areas of transported materials, i.e., glaciated areas) if actual data are not provided. Newer reports may have direct information on surficial geology. This information is useful for hydrologic and pedologic interpretations.

On-site assessment will reveal a large amount of information (Table 2.1), especially if other sources are scarce or lacking.

Soil data are obtained from the soil survey report for the given county. These are published throughout the United States by the USDA-Natural Resources Conservation Service (NRCS), formerly the Soil Conservation Service (SCS). Some areas of the country have yet to be surveyed.

The county soils are mapped at a scale ranging from 1:15,840 to 1:24,000, with the most recent and all future maps to be in the latter scale. This is the same scale as the USGS topographic quadrangle sheets and allows the overlay of data from one map to the other, especially when the soil map is based on an orthophoto of the area. The soil survey maps are useful mainly for projects that are about 25 acres (62 ha) or larger in area. At the 1:24,000 scale, the smallest delineated soil unit is about 6 acres. The potential maximum number of units shown would be a little over four units. Thus, the scale produces too little detail to be useful to the landscape architect and soil scientist in planning a detailed soil and planting plan for an area smaller than 25 acres. For these projects, a special detailed soil survey is strongly recommended and should be included in the project budget. The scale should be at least 1 inch equal to 400 feet (1:4800) or larger.

SAMPLING DESIGN

A degree of variability in soil conditions exists across the landscape. The gradient in this variability may be *continuous, discontinuous,* or *discrete,* as illustrated in Fig. 2.1. Continuous defines a variability trend consistently in the same direction, as on a uniform topographic slope. Discontinuous defines

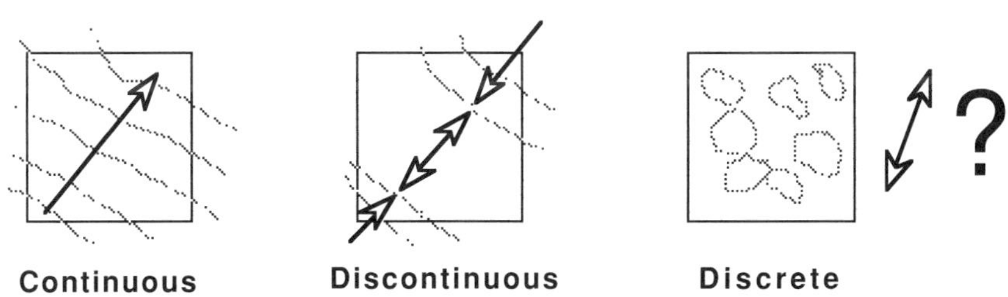

Fig. 2.1 The types of soil variability across the landscape.

a variability whose gradient in a given direction is interrupted, such as a concave slope between two convex slopes. Discrete (sometimes termed *clustered*) variability has no gradient in any direction. A topographic analog is the pit and mound relief found on glacial moraines or on many shallow fragipan soils of the northern hardwood forest. In the latter case, the microrelief is caused by a long history of tree windthrow. The discrete variability is typical of many developed urban sites as well.

Observed and measured internal soil properties and characteristics tend to follow the same types of variability. To obtain these data, a sampling method must be used that best "fits" the variability. The methods are illustrated in Fig. 2.2.

The *random* method (A) is useful when variability is thought to be random or the variability is unknown and a preliminary sampling is made to gain knowledge of the variability. The *systematic,* or *grid,* method (B and C) is also valid when the variability is unknown. Samples are taken at line intersects or on the intervals. The latter generally provides a larger number of samples. *Transecting* is also a systematic method. The method has two variations, the line intercept (D) and the point intercept (E). Both require knowledge of boundary locations. All the sampling methods discussed so far provide a measure of overall variability on the site. When knowledge of variability within units is required, the *stratified* method (F) is necessary.

The question arises as to the length of the interval (density) of sampling, that is how far apart should be the sampling lines? Theoretically, sample

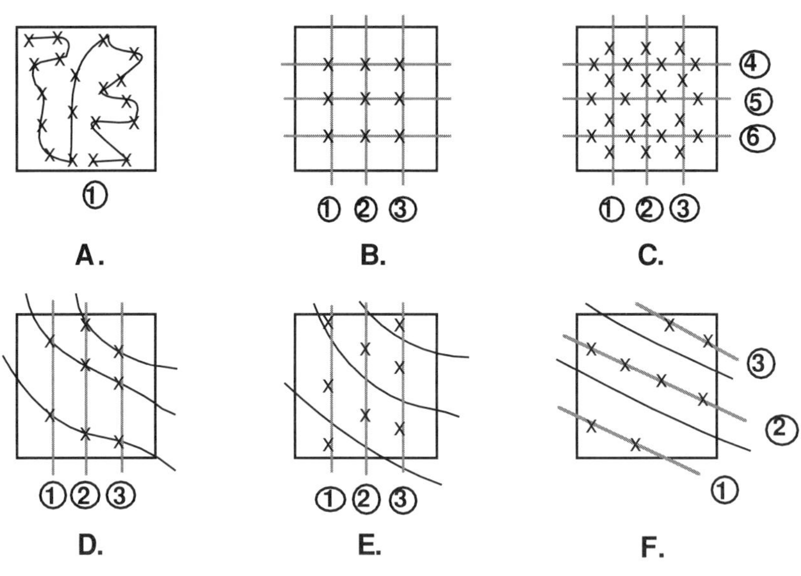

Fig. 2.2 Various sampling methods for soil analysis and assessment. A, random; B, systematic line intercept; C, systematic interval intercept; D, line intercept transect; E, point intercept transect; F, stratified transect. The dashed lines represent the sampling survey lines.

points should be within the half-cycle of spatial variation. Soil properties tend to follow a *sine-wave* pattern with varying amplitude of maxima and minima over space. In complex soil areas, it is difficult to predict the wavelength of the properties without intensive excavation and detailed soil description. Warrick and colleagues (1986) give sample spacings based on results on a list of research studies (Table 2.2). Obviously, the sampling density (interval between sample points) depends on the soil property to be measured. The variability in the interval length suggests that compromise is necessary. Also, there is the question of costs versus number of samples collected. One method to reduce the number of samples is by compositing. A number of samples is obtained within a relatively homogeneous area, then mixed together, and a subsample extracted for laboratory testing and analysis. The sampling area must be divided into at least perceived homogeneous areas prior to compositing. For areas less than 25 acres in size, an interval of 50 to 100 feet (15 to 30 m) is satisfactory.

SOIL PROFILE EVALUATION

The four physical soil properties of texture, structure, color, and density may be used to evaluate the planting potential of an urban site according to the characteristics of permeability, drainage, and waterholding capacity. Structure is a difficult property to describe and assess in a field soil profile by nonsoil scientists. It is extremely difficult to evaluate in urban soils since it is highly modified or destroyed (nonexistent) in many cases. Thus, the process is limited to the more obvious properties of texture, color as it indicates drainage and aeration, and density in that order (Urban and Craul, 1996).

The Field Process

1. **Texture** is first determined in the field, as illustrated in Table 2.3. A more extensive field process is described in Jones and colleagues (1983). Laboratory determination of particle size distribution using ASTM D-422-63 (Appendix) is appropriate for conformance purposes.

TABLE 2.2 Sample Spacings for Selected Soil Properties

Soil Property	Sample Spacing, m (ft)
Soil reaction (pH)	1.5 to 21 (5 to 69)
Sand content	> 5 to 40 (16 to 131)
Wilting point	0.6 to 32 (2 to 105)
Depth of A horizon	< 2 (6.5)

Source: Warrick et al., 1986.

TABLE 2.3 Field Determination of Soil Texture

Texture can be roughly approximated in the field by the sense of feel in the fingers. A handful of moist soil is worked to make casts or to form ribbons between the forefinger and thumb, as described in what follows. This method does not discriminate all soil textural classes equally well, but it is adequate for many field purposes.

Sand: Loose and single-grained. The individual grains of sand can be readily seen or felt. Squeezed in the hand when dry, it will fall apart when the pressure is released. Squeezed when moist it will form a cast, but crumble when touched.

Sandy loam: A soil containing much sand, but which has enough silt and clay to make it somewhat coherent. The individual sand grains can be readily seen and felt. Squeezed when dry, it will form a cast that will readily fall apart, but if squeezed when moist, a cast can be formed that will bear careful handling without breaking.

Loam: A soil having a relatively even mixture of different grades of sand and of silt and clay. It is mellow with a somewhat gritty feel, yet fairly smooth and slightly plastic. Squeezed when dry, it will form a cast that will bear careful handling, and the cast formed by squeezing the moist soil can be handled quite freely without breaking.

Silt loam: A soil having a moderate amount of the fine fractions of sand and only a small amount of clay, over half of the particles being of silt size. When dry, it may appear cloddy, but the lumps can be readily broken, and when pulverized, it feels soft and floury. When wet, the soil readily runs together and puddles. Either dry or moist, it will form casts that can be freely handled without breaking, but when moistened and squeezed between the thumb and forefinger it will not "ribbon" but will give a broken appearance.

Clay loam: A fine-textured soil that usually breaks into clods or lumps that are hard when dry. When the moist soil is pinched between the thumb and forefinger, it will form a thin "ribbon" that will break readily, barely sustaining its own weight. The moist soil is plastic and will form a cast that will bear much handling. When kneaded in the hand, it does not crumble readily but tends to work into a heavy compact mass.

Clay: A fine-textured soil that usually forms very hard lumps or clods when dry and is quite plastic and usually sticky when wet. When the moist soil is pinched out between the thumb and fingers, it will form a long, flexible "ribbon." Some fine clays very high in certain colloids are friable and lack plasticity in all conditions of moisture.

Source: Craul, 1992.

After texture is established, the drainage is determined by using the soil and the redoximorphic features, formerly called *mottling,* as described in the second step.

2. **Drainage** of the soil profile is determined by the examination of color with the use of a Munsell color chart, as described in Craul (1992), Olson (1976, 1981), and the *National Soils Handbook* (Soil Survey Staff, 1992). The flowchart illustrated in Fig. 2.3 links soil mottling to soil drainage class.

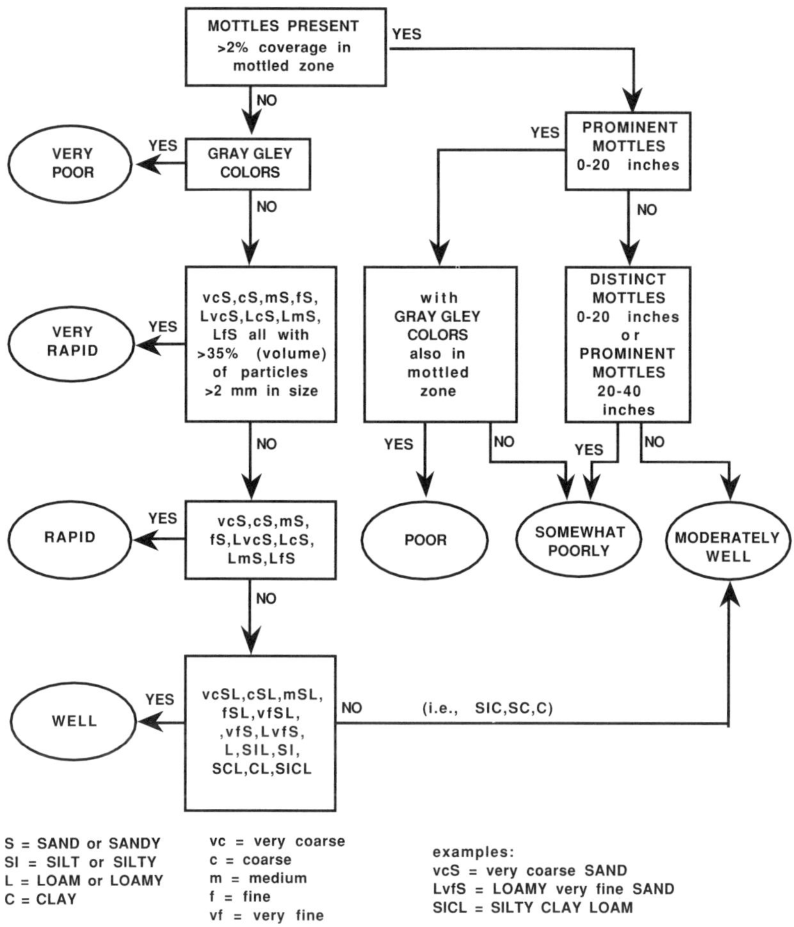

Fig. 2.3 The soil drainage determination flowchart. (Adapted from Jones et al., 1983).

(a) Mottling (redoximorphic features): Soil color is described on the basis of the Munsell color system (Olson, 1981). It is comprised of the *hue, value,* and *chroma* of the color (Fig. 2.4) illustrated on each page by a series of rows and columns of colored chips. The page represents the hue, the row represents the values, and the column represents the chromas for that hue. The ground, or matrix, color of the soil is first determined by comparison with the color chart. The mottle color(s) is then determined in the same way. After the mottle color(s) is identified, the evaluation of the mottling on the basis of coverage, contrast, and size is made.

(1) Area coverage (*quantity*) of mottling:

(a) Few: <2%

(b) Common: 2 to 20%

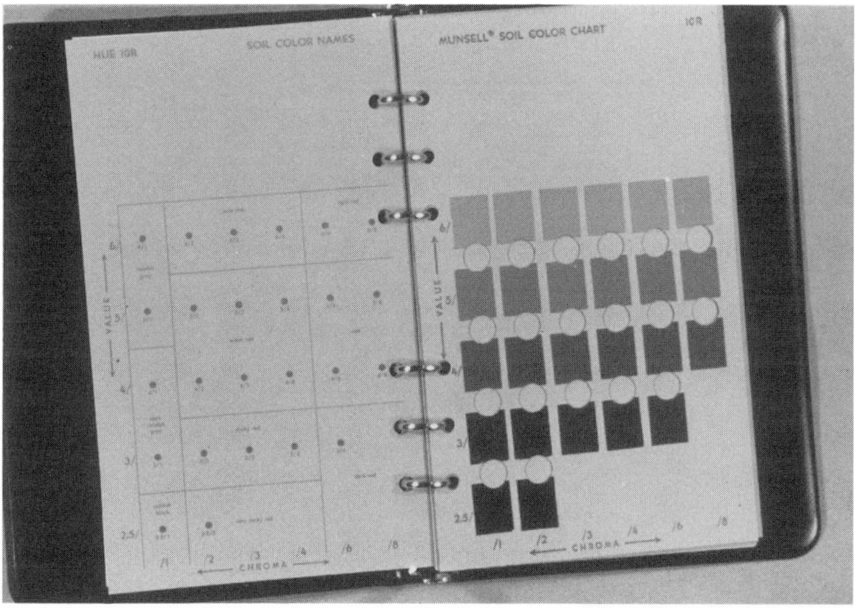

Fig. 2.4 The Munsell soil color book.

(c) Many: >20%

To illustrate coverage degree, some examples are given in Fig. 2.5.

(2) Degree (*contrast*) of mottling:

(a) Prominent: 5 units (two pages) of hue if chroma and value are the same; at least 4 units of value or 1 unit chroma or 2 units of value if hue differs by 2.5 units (one page).

(b) Distinct: same *hue* and differ 2 to 4 *chroma* units, or 3 to 4 units of *value* **or** differ 2.5 units of *hue* (one page) but no more than 1 unit of *chroma* or 2 units of *value.*

(c) Faint: same *hue* and no more than 1 unit of *chroma* or 2 units of *value.*

(3) Size:

(a) Fine: <5 mm

(b) Medium: 5 to 15 mm

(c) Coarse: >15 mm

The greater the coverage, contrast, and size of mottling, the greater the restriction in drainage (aeration). For horizons saturated for most of the year, they usually have a solid olive gray to gray color. Thus, in a soil profile with somewhat restricted drainage, the upper hori-

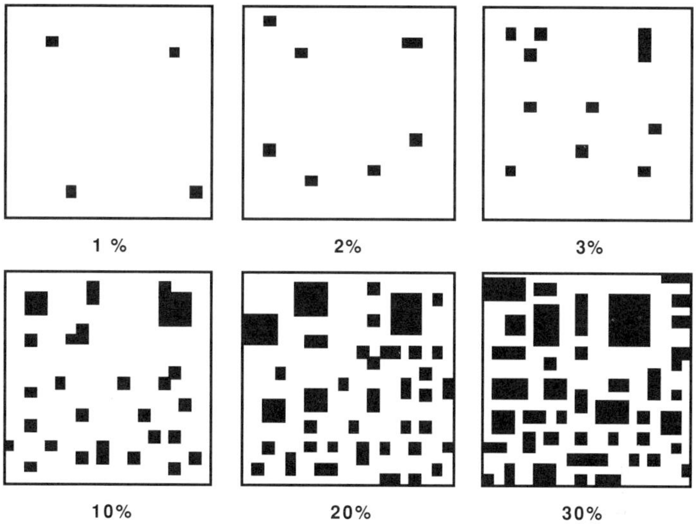

Fig. 2.5 Coverage charts for estimating proportion of mottles. (Adapted from Olson, 1976, 1981.)

zon(s) may have *few, faint, fine* mottles and increases to *many, prominent, coarse* mottles in the lower horizon(s). A well-drained subsoil would have reddish- to yellowish-brown or olive-brown matrix color without mottles above 107 cm (42 in.). Very poorly drained soils would have mottling to the surface. Topsoil (A horizon) matrix colors are darker due to the presence of organic matter.

(b) USDA drainage class:
Determination of the drainage class is based on the average depth to mottling in the soil profile. The flowchart as adapted from Jones and colleagues (1983) is a good starting point. The classes recognized by the NRCS are given in Table 2.4.

TABLE 2.4 USDA-NRCS Soil Drainage Class Based on Mottling Depth

Depth to Mottling, cm (in.)	Soil Drainage Class
15 (6)	Very poorly drained
30 (12)	Poorly drained
51 (20)	Somewhat poorly drained
81 (32)	Moderately well drained
107 (42)	Well drained
132 (52)	Somewhat excessively drained
> 152 (> 60)	Excessively drained

3. **Density** may be used to infer drainage (and aeration) if evidence of mottling is not present or obvious. Thus, it is another method for indicating the drainage of a soil profile. Generally, the denser the soil, the slower the movement of water through the soil, especially in a saturated condition, or *saturated flow.* Soil in a saturated condition usually is detrimental to most tree species other than the hydrophytes. However, if permeability or saturated flow through the profile is rapid, the soil is not saturated long enough to cause problems. *It must be kept continuously in mind that the soil profile drains only as well as the least permeable horizon.* Variation in density may indicate which horizon is least permeable and also may indicate the relative degree of permeability. The information is organized by three degrees of density.

   Permeability may be determined by the standard percolation test if the evaluator is inclined to make the effort to determine actual values. The data of Table 2.5 connect the two groups of information for use.

   The "High" class as defined by the NRCS (Soil Survey Staff, 1993) has a range of nearly 13 inches per hour. This class is subdivided here to better exhibit the differences in drainage between 1 and 3 inches per hour, 3 and 6 inches per hour, and 6 to 14 inches per hour. The subdivision is given in Table 2.6. The 1–3-inch range is commonly used in soil specifications.

   Completion of the process should give the evaluator a better idea of soil conditions than just punching some holes. However, if a planting zone does not drain or shows poor drainage through common sense, why go to all the trouble? *Sometimes you have to prove the condition to somebody else who may not have common sense!*

Planting Potential

Once the depth to mottling and its drainage class is determined, a rough guide to species planting may be inferred. Obviously, there are many factors influencing the final choice including the amplitude of adaptability of the various species (and their genotypes) and their cultivars. For example, Red maple (*Acer rubrum,* L.) has a wide amplitude in nature for moisture and drainage requirements, being found on wet, lowland soils and high dry sites. But to plant a "wet" Red maple on a dry site is asking for failure. Further, a soil most desirable for planting is one with medium density, moderately low to high permeability, being moderately well drained to well drained (see Table 2.7). Other combinations can be useful if appropriately modified. The modifications are discussed in Chapter 5, "Soil Specifications."

SOIL TESTING AND ANALYSIS

It is not necessary to make all soil analyses for all projects. The planned use of the project, its history, and the designed plant palette all determine the soil

TABLE 2.5 The Relation Between Density, Texture and Permeability

Bulk Density	Texture Class[1]	Permeability
High: 1.46 to 1.72 Mg/m^3	C, SIC, SICL, CL, SC	Low
	SI, SIL, L, SCL	Moderately low
	SL, LS, S	Moderately high
Medium: 1.19 to 1.59 Mg/m^3	SIC, C	Low
	SICL, CL, SC	Moderately low
	SI, SIL, L, SCL, 1/2 SL	Moderately high
	1/2 SL, LS, S	High
Low: 0.93 to 1.19 Mg/m^3	1/2 SICL, SIC, C	Moderately low
	SI, SIL, 1/2 SICL, CL, L, SC, 1/2 SCL	Moderately high
	SL, SCL, LS, S	High

[1]For definitions, see footnotes for Table 2.1.

Source: Compiled from data in Soil Survey Staff, 1993.

analyses required. The most useful analyses are ranked in Table 2.8. For example, soil reaction (pH) is one of the most useful soil analyses; iron and manganese are needed only under special circumstances.

Knowledge of variability and the proper method of sampling can greatly reduce the costs involved in soil analysis. The costs include the field work to obtain the soil samples, the number of samples to provide sufficient information for soil assessment, and the laboratory costs for the various analyses. Only a few samples are required for relative uniform soil properties, especially some of the physical properties, whereas many of the chemical and biological properties exhibited high degrees of variability. In some instances, the compositing (combining) of samples may be a method by which the number of laboratory samples are reduced. Some degree of the measure of variability is lost by compositing.

TABLE 2.6 Saturated Flow Rates for Permeability Classes

	Saturated Flow Rates	
Permeability Class	cm/sec	in./hr
Very low	$<1.55 \times 10^{-7}$	< 0.0014
Low	1.55×10^{-6}	0.0014 to 0.01417
Moderately low	1.55×10^{-5}	0.01417 to 0.1417
Moderately high	1.55×10^{-4}	0.1417 to 1.417
High	1.55×10^{-4} to 3.74×10^{-4}	1.417 to 3.417
Very high	3.74×10^{-4} to 7.02×10^{-4}	3.417 to 6.417
Excessively high	7.02×10^{-4} to 1.55×10^{-3}	6.417 to 14.17
Extreme	$>1.55 \times 10^{-3}$	> 14.17

Source: Data from Soil Survey Staff, 1993, with modifications.

TABLE 2.7 Planting Guide to Drainage Classes for Urban Trees

Drainage Class	Planting Recommendation
Very poorly drained	DON'T PLANT—unless deliberately drained!
Poorly drained	Only wetland or hydrophilic species
Somewhat poorly drained	Wetland and some mesophytic species
Moderately well drained	More moist mesophytic species
Well drained	Any mesophytic species for the locality
Somewhat excessively drained	Drier mesophytic, some xerophytic species
Excessively drained	Only xerophytic species, unless precipitation and/or irrigation permits otherwise

Testing and analysis of contaminated sites require much laboratory work. It is expensive and time-consuming. If a site is suspected to be contaminated, testing and analysis should be made a major budget item. In addition, provision should be made for cleanup responsibility. Generally, federal law holds that the owner is responsible even if ownership is assumed from the owner who caused the contamination. These and other questions must be clarified and agreed upon before any project work is begun. The potential for contamination on restoration sites is very strong; *in fact, restoration sites should be assumed to be contaminated unless proven otherwise.* Bridges (1987) provides guidelines for the survey and testing of contaminated land.

TABLE 2.8 Ranking of Common Soil Analyses for Derivation of Useful Information

Soil Analysis	Usefulness Rating
pH (1 soil:1 distilled water)	100 (most usefull)
Phosphorus	85
Potassium	85
Sodium adsorption ratio (SAR)	85
Exchangeable sodium percentage (ESP)	85
Organic matter	75
Sulfur	30
Magnesium	35
Manganese (above pH 7.5)	40
Manganese (below pH 7.5)	15
Iron (above pH 7.5)	30
Iron (below pH 7.5)	15
Nitrogen (estimated from organic matter)	40
Nitrogen (total nitrogen)	50
Nitrogen (nitrate nitrogen)	50
Cation-exchange capacity	60
Aluminum	40
Soluble salts (electrical conductivity)	40

Source: Peck, 1981, with additions and modifications.

The other soil characteristics requiring examination and measurement listed in Table 2.1 are part of the suite of data to be analyzed in the site assessment.

THE CHARACTER AND SOILS OF RESTORATION SITES

Abandoned sites or those that have structures demolished and cleared present several site and soil problems when a landscape project is proposed. The following definitions and descriptions are provided as a framework for the character of restoration sites.

Bridges (1987) gives a definition of derelict land that is most useful for application to restoration sites:

> Land which has been so damaged by extractive or other industrial processes or by any form of urban development that in default of special attention it is unlikely to be effectively used again within reasonable time and may well be a public nuisance in the meanwhile.

He lists the following features of derelict land:

(a) Unusual or complex topography, even holes in the ground next to piles of waste
(b) Varying degrees of unstable conditions caused by piled material, exposed, bare soil, and so on, subsidence, and erosion
(c) Presence of toxic substances
(d) Interrupted, disturbed drainage, and possible flooding
(e) Fragmented or absentee ownership
(f) May be in areas with limited access by normal infrastructure
(g) Abandoned buildings and other structures in derelict condition creating hazards
(h) Unsightliness; ruined buildings, bare and eroding soil, or coarse vegetation
(i) Absence of infrastructure; probably removed from the site earlier
(j) Surrounding areas in depressed state; poor housing, and so on

Smith (1985) adds a definition of contaminated land:

> Land that contains substances that when present in sufficient quantities or concentrations are likely to cause harm, directly or indirectly to man, to the environment, or on occasions to other targets.

It should be obvious that the terms can be applied with equal validity to the soil. In addition, derelict land and soil can be contaminated but not necessarily so, and that natural soil, not necessarily derelict, can also be contaminated. However, much derelict land is contaminated, and preliminary site exploration and assessment for any restoration project should be initially conducted as if contamination is expected. The constructional history of the site is most useful in the preliminary exploration.

Haines (1981) places contaminated land into eight categories:

1. Definitely contaminated
2. Probably contaminated
3. Potentially contaminated
4. Early housing where lead paint might have been exposed
5. Mixed land use with many small potentially contaminated sites
6. Definitely uncontaminated
7. Unknown
8. Previously contaminated sites that are now in "sensitive" uses

A protocol exists that can be followed in the assessment of land and soil that may have the potential to be contaminated. Bridges provides a very workable flowchart to aid the process (Fig. 2.6).

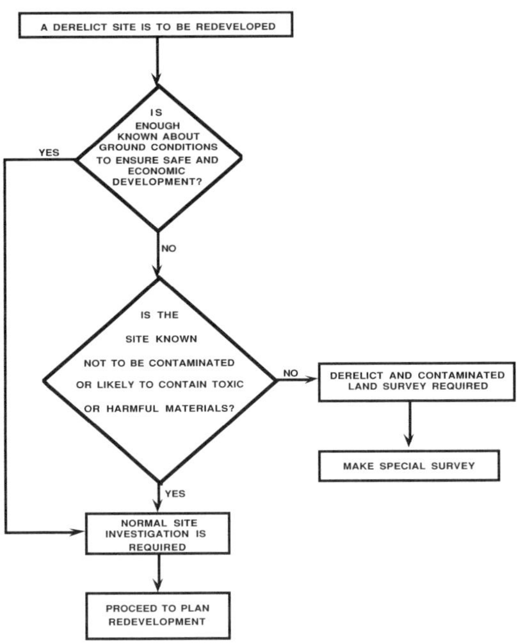

Fig. 2.6 Bridges (1987) flow chart for investigation and assessment of contaminated lands.

Obviously, the concern for contaminated conditions should arise very early in the planning process. Presence or absence of contamination and its type will have significant influence on the subsequent planning process.

THE SOIL INTERPRETATION PROCESS

Because soil maps do not directly supply meaning of usefulness to the user, a process called interpretation is necessary. Interpretation is defined as (Aandahl, 1958)

> Soil survey interpretation comprises the organization and presentation of knowledge about characteristics, qualities and behavior of soils as they are classified and outlined on maps.

1. A clear definition of the specific use of the soils to be interpreted. How will the information be used for what purpose?
2. Assignment of a quantitative value for each applicable soil property to evaluate its influence on each soil mapping unit as it relates to the specific use defined by the first principle. This is the weakest point in the interpretation process because quantitative values don't exist for each soil property as it relates to a specific use.
3. Soil properties to be evaluated are determined as based on existing empirical data, previous relevant research, and the knowledge and judgment of the interpreter.
4. The interpretation should result in a relative classification of soils in a given area, including a simple expression of their limitations and capabilities.
5. The soil properties used in the interpretation should be mappable.

Types of Soil Maps

In addition to the conventional soil maps discussed earlier in this chapter and elsewhere (Craul, 1992), other detailed and generalized maps may be constructed. Schematic maps are other products. These are graphic representations derived from information concerning the factors of soil formation. They are not usually derived from detailed soil survey maps or field mapping.

Generalized maps are produced for the purposes of interpretation. They are produced from the more detailed conventional soil survey maps of Order 2, 1:24,000 scale (Fig. 2.7, which is an old scale of 1:20,000) by reorganization and omission of some information. The maps are generalized because it is generally not necessary to carry the detailed information into the maps used for interpretation. They are more comprehensible and more readily useful. Thus, they are simplications of known mapping units and their distribution. These generalized maps fall into several categories. Graphic, or

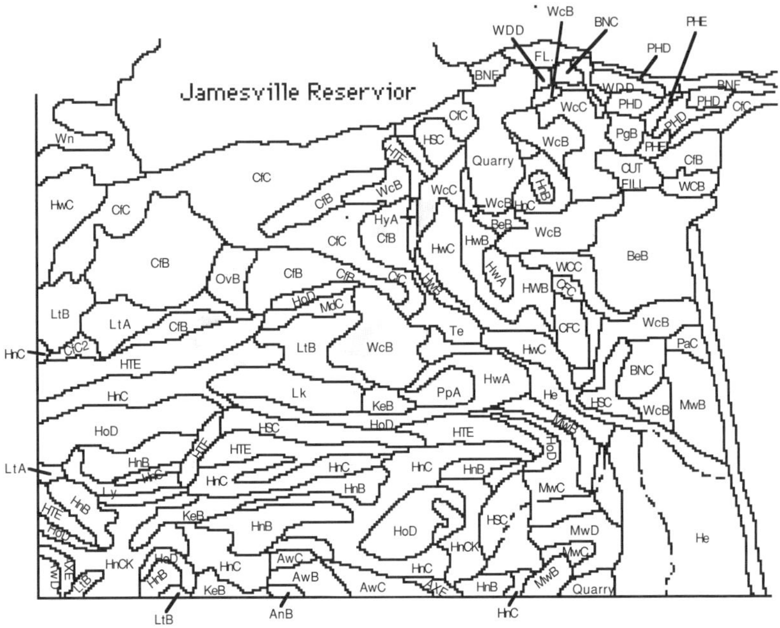

Fig. 2.7 A detailed, conventional soil survey map, Order 2, 1:20,000 scale. Scale at 1:24,000 would not significantly change the details.

cartographic generalization, is the omission of soil boundaries or the adjustment of the shape of line segments while maintaining the original mapping legend. Taxonomic, or *categorical* generalization, is the omission of boundaries and combining adjacent taxonomically similar mapping units. These two generalizations are mechanical in nature and do not require technical soil science knowledge. *Spatial* generalization involves combining consistently adjacent mapping units, whether or not they are taxonomically similar. This technique is used for land management or technically similar uses. Technical knowledge of soil patterns of the region is needed because significant changes in the original legend of the primary source map is required; thus, a soil scientist should be employed to develop these maps. Adjacent soils may have entirely different capabilities and limitations (i.e., interpretations) for a given land use. Spatial generalization may create soil units with a wide range in variation of properties and characteristics. This limitation should be recognized before a spatial generalization is developed. Both of the maps in Figs. 2.8 and 2.9 have been generalized "upward"; that is, the boundaries remaining are those from the original conventional Order 2 map. No new lines have been added and no original lines have been shifted. These are "real" boundaries.

The resulting generalized map for interpretive purposes is referred to as a *technical* map to distinguish it from the more detailed conventional soil survey maps as produced by the NRCS from which the technical map is derived.

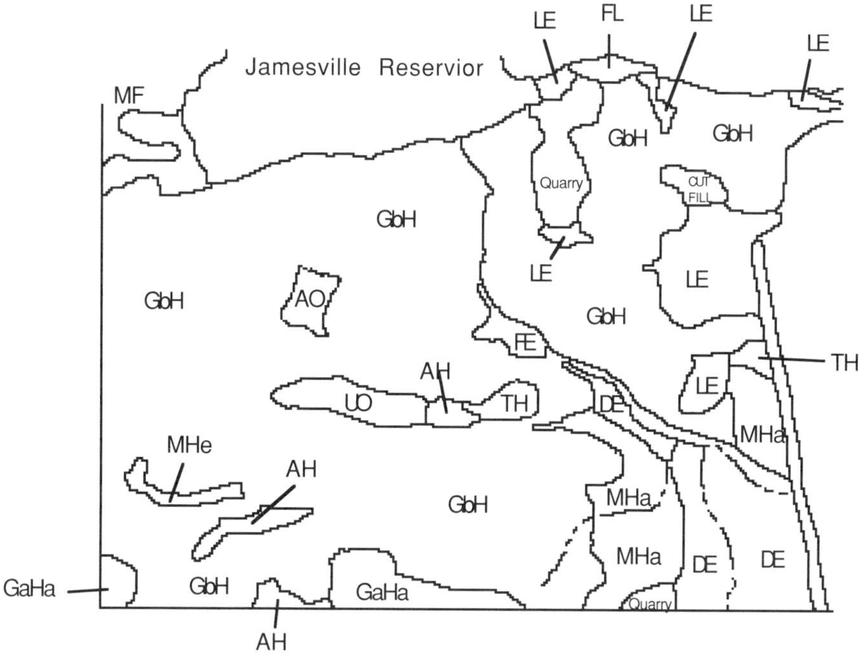

Fig. 2.8 A *taxonomic* generalized map of Fig. 2.7. The generalization basis is the Soil Subgroup category. The legend is AH = Aeric Haplaqupts; AO = Aeric Ochraqualfs; DE = Dystric Eutrochrepts; FE = Fluvic Eutrochrepts; GaHa = Glossaquic Hapludalfs; GbH = Glossaboric Hapludalfs; LE = Lithic Eutrochrepts; MF = Mollic Fluvaquents; MHa = Mollic Hapludalfs; MHe = Mollic Haplaquepts; TH = Typic Hapludolls; and UO = Udollic Ochraqualfs.

If a map is to be generalized, it is imperative that the information required for the objective of the interpretation not be lost or partially omitted. Thus, it is important to have knowledge of those soil and site factors that influence the soil behavior sought in the interpretation. A review of the factors that influence soil behavior for several land uses follows.

Though some of the direct factors for interpretation are not mapped, the information is related to the mapping units and found in accompanying tables in the county soil survey report.

Significant Interpretive Factors for Several Land Use Examples

Several interpretation guideline examples have been chosen to illustrate those soil factors that are significant to interpretation for soil behavior under various land uses. There are guidelines available for other interpretations, but space does not permit covering all of them. The *National Soils Handbook,* Soil Survey Staff, (1992) of the NRCS is the appropriate source for the guidelines. Those interpretations having the greatest usefulness to landscape architects

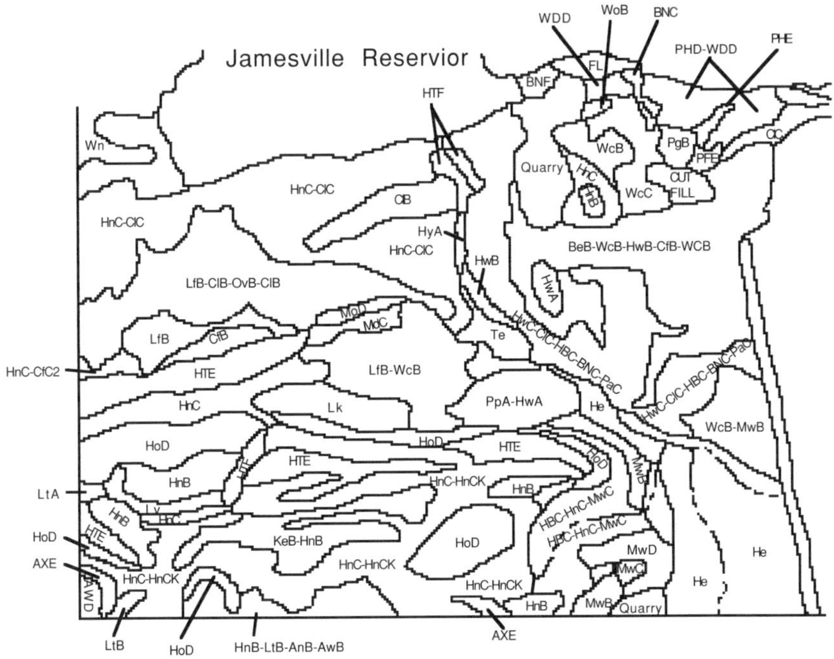

Fig. 2.9 A *spatially* generalized map of Fig. 2.7. Slope class (A, B, C, D) is the spatial parameter.

have been chosen. These guidelines provide the criteria for applying an interpretation to a soil mapping unit of a soil survey map.

On-Site Waste Disposal. Many home sites are located in the urban fringe and beyond, out of the reach of sewer service. On-site waste disposal capability becomes a determining factor for building permits by local governments. The significant factors are permeability class, hydraulic conductivity rate, percolation rate, depth to the watertable, flooding potential, slope gradient, depth to hard rock, stoniness and rockiness classes, and so on. The full list is given with their ratings and restrictive features in Table 2.9.

The capacity of the soil to absorb water without interference from a high watertable, shallow bedrock, or dense soil on a level to gently sloping soil are the major considerations for this land use. Stones, boulders, and shallow depth interfere with installation. Subsidence interferes with installation and maintenance. Excessive slope can cause lateral seepage and surfacing of the effluent on downslope areas.

Agriculture. Agricultural land, especially crop land, is typified by level or gently sloping land with well-drained, structurally stable, and fertile soils. Land for hay crops and pasture have less favorable characteristics for culti-

TABLE 2.9 Soil Interpretation Ratings for On-Site Waste Disposal

Soil Property	Limits			Restrictive Feature
	Slight	Moderate	Severe	
USDA texture	—	—	Ice	Permafrost
Total subsidence, in.	< 12	12 to 24	> 24	Subsides
Flooding	None	Rare	Frequently to occasional	Flooding
Depth to bedrock, in.	> 72	40 to 72	< 40	Depth to rock
Depth to cemented pan, in.	> 72	40 to 72	< 40	Cemented pan
Ponding	—	—	+	Ponding
Depth to high watertable, ft	> 6	4 to 6	< 4	Wetness
Permeability (24–60″), in./hr	2.0 to 6.0	0.6 to 2.0	< 0.6	Percs slowly
Permeability (24–60″), in./hr	—	—	> 6.0	Poor filter
Percolation, min/in.	< 45	45 to 60	> 60	Percs slowly
Slope, %	< 8	8 to 15	> 15	Slope
Weight percent 3–10″ (weighted ave. to 40″)	< 25	25 to 50	> 50	Too cobbly
Weight percent >10″ (weighted ave. to 40″)	< 5	5 to 15	> 15	Too stony

vation and land management. Unfortunately, good agricultural land is "good" for most other land uses and leads to strong competition among other land uses and the eventual conversion to nonagricultural uses due to overwhelming economic considerations rather than wise land husbandry and land diversity as proposed by McHarg (1969) and others.

The factors influencing the agricultural behavior of soil are texture with surface modifiers, drainage class, slope gradient, reaction (pH), stoniness and rockiness classes, depth to watertable, depth to bedrock, and so on. The full list is given with their ratings and restrictive features in Table 2.10.

Those soils with extreme texture would have severe limitations for crop production, having either low fertility and waterholding capacity or too heavy tilth for cultivation and high compaction potential. Available water, organic matter content, and clay activity also indicate fertility or productivity level.

Landfill Siting. The siting of landfills is still a major land use, though their importance is decreasing as recycling and the number of alternatives to landfilling increase. However, some degree of landfilling will be required into the future.

The factors that influence landfilling as a land use include depth to seasonal watertable, soil drainage class, flooding potential, permeability, slope gradient, soil texture, depth to bedrock, stoniness and rockiness classes, and so on. The full list is given with their ratings and restrictive features in Table 2.11.

Recreation. Recreation implies use and foot traffic throughout the year in most cases. Therefore, the soil must be compaction-resistant and well drained. The significant factors are texture with surface modifiers, slope, ponding potential, depth to watertable, flooding potential, stoniness class, depth to bedrock, and so on. The full list is given with their ratings and restrictive features in Table 2.12.

Drastically Disturbed Lands. Some proposed projects may occur in areas that have been drastically disturbed. These may include former mine spoil banks, steel mill slag piles, limestone quarry floors, gravel and sand pits, highway borrow pits, and other industrial processing waste sites. Besides the potential pollution problems encountered with abandoned sites, the physical characteristics of the site materials may be very undesirable for a planting project. The ratings and restrictions are given in Table 2.13. Design of a sustainable soil for a project on this type of site is almost a foregone conclusion. Chapter 5 discusses the design of sustainable soils.

The Output or Interpretation Map

After the interpretation guideline criteria are applied to each mapping unit, resulting in a rating for the unit, a generalized interpretation or technical map is created. The usual process is to apply colors to the rating classes for the

TABLE 2.10 Soil Interpretation Ratings for Agricultural Cropland

Soil Property	Limits: Slight	Moderate	Severe	Restrictive Feature
USDA texture	—	—	Ice	Permafrost
Texture (surface layer)[1]	—	—	COS, S, LCOS	Too sandy
Texture (surface layer)[1]	—	—	SC, SIC, C	Too clayey
Stoniness class[2]	0, 1	2, 3	4, 5	Too stony
Slope, %	< 8	8 to 15	> 15	Slope
Depth to high watertable, ft	> 2.5	1.5 to 2.5	< 1.5	Wetness
Depth to bedrock, in.	> 40	20 to 40	< 20	Depth to rock
Depth to cemented pan, in.	> 40	20 to 40	< 20	Cemented pan
Drainage class[3]	WD, MWD	SWEXD, SWD	EXD, PD, VPD	Drainage
Organic matter, %	> 1	0.5 to 1.0	< 0.5	Low fertility
Clay activity, CEC/clay[4]	> 0.24	0.16 to 0.24	< 0.16	Low fertility
Sodium adsorption ratio[5]	< 4	4 to 13	> 13	Excess sodium
Salinity, dS/m	< 8	8 to 16	> 16	Excess salt
Soil reaction, pH (0–40″)	5.0 to 8.5	4.0 to 5.0	< 4.0	Too acid
Soil reaction, pH (> 40″)	—	< 4.0	—	Too acid
Soil reaction, pH (> 40″)	—	—	> 8.5	Too alkaline
Available water capacity, in./in.	> 0.10	0.05 to 0.10	< 0.05	Droughty

[1]Texture classes: C = clay; COS = coarse sand; LCOS = loamy coarse sand; SC = sandy clay; SIC = silty clay; LFS = loamy fine sand; LVFS = loamy very fine sand; S = sand; FS = fine sand; VFS = very fine sand; SIL = silt loam; SI = silt; L = loam.

[2]Stoniness classes:

Class 0: No stones or too few to interfere with tillage. Stones cover less than 0.01% of the surface.

Class 1: Sufficient stones to interfere with tillage, but does not make intertilled crops impracticable. Stones 1 ft in diameter occur 30 to 100 ft apart and cover 0.01 to 0.1% of the surface.

Class 2: Sufficient stones to make intertilled crops impracticable; capable for hay or pasture. If stones are 1 ft in diameter, they occur about 5 to 30 ft apart and occupy about 0.1 to 3% of the surface.

Class 3: Sufficient stones to make use of all machinery impracticable. Used for wild pasture or forest. Stones 1 ft in diameter occur 2.5 to 5 ft apart and occupy 3 to 15% of the surface.

Class 4: Sufficient stones to make use of all machinery impracticable. Use restricted to poor pasture or forestry. Stones 1 ft in diameter are 2.5 ft or less apart and they occupy 15 to 90% of the surface.

Class 5: Land essentially paved with stones that occupy more than 90% of the exposed surface (rubble).

[3]Drainage classes: EXD = excessively drained; SWEXD = somewhat excessively drained; WD = well drained; MWD = moderately well drained; SWD = somewhat poorly drained; PD = poorly drained; VPD = very poorly drained.

[4]Clay activity: CEC/clay ratio = cation-exchange capacity, $cmol^+/100$ g divided by the weight of clay in 100-g soil.

[5]Sodium adsoprtion ratio: $SAR = Na^+/([Ca^{2+} + Mg^{2+}]/2)^{1/2}$

Source: The author's ratings.

TABLE 2.11 Soil Interpretation Ratings for Trench Landfill Siting[1]

Soil Property	Limits			Restrictive Feature
	Slight	Moderate	Severe	
Flooding	None	Rare	Frequently, occasional	Flooding
Depth to bedrock, hard, in.	—	—	< 72	Depth to rock
Depth to bedrock, soft, in.	—	—	< 72	Depth to rock
Depth to cemented pan, thin, in.	—	< 72	—	Cemented pan
Depth to cemented pan, thick, in.	—	—	< 72	Cemented pan
Permeability (bottom layer), in./hr	—	—	> 2.0	Seepage
Permeability (bottom layer) if Aridisol and not Salorthid or Aquic subgroups, all Aridic subgroups, all Torric great groups except Aquic subgroups	Any entry	—	—	Seepage
Ponding	—	—	Any entry	Ponding
Depth to high watertable, apparent, ft	—	—	< 6	Wetness
Depth to high watertable, perched, ft	> 4	2 to 4	< 2	Wetness
Slope, %	< 8	8 to 15	> 15	Slope
USDA texture	—	—	Ice	Permafrost
Texture (thickest layer 10–72″)	—	CL, SC, SICL	SIC, C	Too clayey
Texture (thickest layer 10–72″ Oxisols, Oxic Subgroups, and kaolinitc minerals)	—	SIC, C	—	Too clayey
Texture (thickest layer 10–72″)	—	LCOS, LS, LFS, LVFS	COS, S, FS, VFS	Too sandy
Unified (thickest layer 10–72″)	—	—	OL, OH, PT	Excess humus
Weight percent 3–10″ (weighted ave. to 72″)	< 20	20 to 35	> 35	Too cobbly
Weighted percent > 10″ (weighted ave. to 72″)	< 5	5 to 15	> 15	Too stony
Sodium absorption ratio	< 4	4 to 13	> 13	Excess sodium
Soil reaction, pH (any depth)	—	—	< 3.5	Too acid
Salinity, dS/m (any depth)	< 2	2 to 4	> 4	Excess salt

[1]For definitions and classes, see footnotes to Table 2.10.

Source: Soil Survey Staff, 1992.

TABLE 2.12 Soil Interpretation Ratings for Recreation[1]

Soil Property	Limits: Slight	Limits: Moderate	Limits: Severe	Restrictive Feature
USDA texture	—	—	Ice	Permafrost
Texture modifier (surface)[2]	—	STV, BYV, CB, FL	STX, BYX, CBX, FLX, CBV, FLV, CNX	Too cobbly
Texture (surface layer)	—	—	SC, SIC, C	Too clayey
Texture (surface layer; Tor, Ust, Arid, Xer, Oxi Suborders, Great Groups, and Subgroups Oxisols)	—	SC, SIC, C	—	Too clayey
Texture (surface layer; others)	—	LCOS, VFS, LFS, LS	COS, S, FS	Too sandy
Texture (surface layer finer material < 20″)	—	LCOS, VFS	—	Too sandy
Texture (surface layer; Tor, Arid, or Xer subgroups and great Groups)	—	SIL, SI, VFSL, L, ashy	—	Dusty
Slope, %	< 2	2 to 6	> 6	Slope
Ponding	—	—	Any entry	Ponding
Depth to high watertable, ft	> 2.5	1.5 to 2.5	< 1.5	Wetness
Flooding	None, rare	Occasional	Frequent	Flooding
Stoniness class	0, 1	2	3, 4, 5	Too stony
Weight percent 2 mm to 3″ (surface layer)	< 10	10 to 25	> 25	Too gravelly
Unified (surface)	—	—	PT	Excess humus
Depth to bedrock, ft (> 2% slopes, in.)	> 40	20 to 40	< 20	Depth to rock
Depth to bedrock, ft (≤ 2% slopes, in.)	≥ 20	—	< 20	Depth to rock
Depth to cemented pan (> 2% slope, in.)	> 40	20 to 40	< 20	Cemented pan
Depth to cemented pan (≤ 2% slope, in.)	> 20	—	< 20	Cemented pan
Permeability (0–40″, in./hr)	> 0.6	0.06 to 0.6	< 0.06	Percs slowly
Permeability (0–40″; Tor, Ust, Arid, Xer, Oxi suborders, Great Groups, and Subgroups Oxisols)	≥ 0.2	< 0.2	—	Percs slowly
Sodium adsorption ratio	< 4	4 to 13	> 13	Excess sodium
Salinity (surface layer), dS/m	< 4	4 to 8	> 8	Excess salt
Soil reaction (pH)	—	—	< 3.5	Too acid

[1]For definitions, see footnotes for Table 2.10.

[2]Texture modifiers: BYV = very bouldery; BYX = extremely bouldery; CB = cobbly; CBV = very cobbly; CBX = extremely cobbly; CNV = very channery; FL = flaggy; FLX = extremely flaggy; FLV = very flaggy; STV = very stony; STX = extremely stony.

Source: Soil Survey Staff, 1992.

TABLE 2.13 Soil Interpretation Ratings for Drastically Disturbed Land[1]

Soil Property	Limits: Slight	Limits: Moderate	Limits: Severe	Restrictive Feature
Sodium adsorption ratio	< 4	4 to 13	> 13	Excess sodium
Salinity, dS/m	< 8	8 to 16	> 16	Excess salt
Soil reaction, pH (0–40″)	5.0 to 8.5	4.0 to 5.0	< 4.0	Too acid
Soil reaction, pH (> 40″)	—	< 4.0	—	Too acid
Soil reaction, pH (> 40″)	—	—	> 8.5	Too alkaline
Available water capacity, in./in.	> 0.10	0.05 to 0.10	< 0.05	Droughty
Erosion factor (*K*)	< 0.35	> 0.35	—	Erodes easily
Wind erodibility group	—	—	1, 2	Soil blowing
Texture	—	SCL, CL, SICL	C, SIC, SC	Too clayey
Texture	—	LCOS, LS LFS, LVFS	COS, S, FS, VFS	Too sandy
Weight percent 3 to 10″	< 25	25 to 50	> 50	Too cobbly
Weight percent > 10″	< 5	5 to 15	> 15	Too stony
Layer thickness, in.	> 40	20 to 40	< 20	Thin layer
Organic matter, %	> 1	0.5 to 1.0	< 0.5	Low fertility
Clay activity, CEC/clay	0.24	0.16 to 0.24	< 0.16	Low fertility
Calcium carbonate equivalent, %	< 15	15 to 40	> 40	Excess lime

[1]For definitions, see footnotes for Table 2.10.

Source: Soil Survey Staff, 1992.

land use. The conventional color scheme is green for suitability with none or few limitations not requiring mitigation, yellow for suitability with mitigation of moderate limitations necessary for the land use, and red for unsuitability for the particular land use. In the last case, mitigation is not possible or feasible, economically or ecologically, under present conditions.

An Example of the Interpretation Process

A two-part example is given to provide a better understanding of the interpretation process. The first part illustrates an interpretation based on the various interpretive tables of a county report. The second part deals with an interpretation made "from scratch," using a guideline to develop the interpretation matrix. Interpretation maps for the two examples are not shown because their derivation is similar to what has already been discussed and illustrated.

Procedure. The steps include the following:

(a) Compile a list of the soil mapping units occurring within the area of interest from the appropriate county soil survey report (Column 1, Table 2.14; Fig. 2.7 is the example area of interest).

(b) List the soil characteristics across the top (Row 1, Table 2.14) that influence the soil behavior for the particular interpretation objective if making an independent interpretation, in the example, on-site waste disposal. If using the county soil survey report, obtain the ratings directly from the interpretation table in the report and go to step (d).

(c) List these characteristics for each soil mapping unit (the matrix of Table 2.14). These are obtained from the series descriptions and the engineering tables provided in the soil survey report. For an independent interpretation, other data may have to be obtained from the NRCS.

(d) Apply the ratings from the guidelines (Table 2.9 or from the soil survey report) to the characteristic matrix.

(e) An interpretation matrix is derived and prepared from step (d), as shown in Table 2.15.

(f) An overall rating is assigned to the soil mapping units (column six in Table 2.15). It should be remembered that the NRCS employs the "worst-case scenario." This means that if a soil mapping unit has one characteristic of severe limitation, the unit receives a rating of "severe." This procedure is not always practical, but does create a warning to users of the soil survey reports. A more satisfactory method is to weight the various factors and their ratings. The severe rating would be applied only when the cumulative weighted rating would exceed a predetermined threshold. A modification of this process is now being used by the NRCS. Restrictive features are identified for those soils having

TABLE 2.14 Selected Soil Mapping Unit Characteristics[1] for On to site Waste Disposal

Soil Map Unit	Drainage Class[2]	Bedrock Depth, in.	Flooding	Permeability, in/hr	Slope, %
Angola-Darien (AnB)	Somewhat poorly	30	None	< 0.2	3 to 8
Aurora (AwC)	Moderately well	30	None	0.06 to 0.2	8 to 15
Aurora-Farmington (AXE)	Somewhat excessive	30	None	0.06 to 0.2	25 to 35
Benson (BeB)	Somewhat excessive	18	None	0.6 to 2.0	3 to 8
Benson-Wassaic-Rock crop (BNC)	Excessive	12	None	0.6 to 2.0	8 to 15
Benson-Wassaic-Rock crop (BNF)	Excessive	12	None	0.6 to 2.0	> 35
Cazenovia (CfB)	Well drained	> 40	None	< 0.2	3 to 8
Cazenovia (CfC)	Well drained	> 40	None	< 0.2	8 to 15
Cazenovia (CfD)	Well drained	> 40	None	< 0.2	15 to 25
Herkimer (He)	Well drained	> 40	Rare	0.6 to 2.0	0 to 3
Fluvaquents	Variable	Variable	Frequently	Variable	0 to 8
Honeoye (HnB)	Well drained	> 40	None	< 0.2	3 to 8
Honeoye (HnC)	Well drained	> 40	None	< 0.2	8 to 15
Honeoye, very stony (HSC)	Well drained	> 40	None	< 0.2	8 to 15
Honeoye-Lansing (HoD)	Well drained	> 40	None	< 0.2	15 to 25
Honeoye-Lansing-Ontario (HTE)	Well drained	> 40	None	< 0.2	25 to 35
Honeoye-Lansing-Ontario (HTF)	Well drained	> 40	None	< 0.2	> 35
Howard (HwA)	Well drained	> 48	None	0.6 to 2.0	0 to 3
Howard (HwB)	Well drained	> 48	None	0.6 to 2.0	3 to 8

Howard (HwC)	Well drained	> 48	None	0.6 to 2.0	8 to 15
Howard gravelly (HyA)	Well drained	> 48	None	0.6 to 2.0	0 to 3
Kendaia (KeB)	Somewhat poorly	> 40	Rare	< 0.2	3 to 8
Lakemont (Lk)	Poorly drained	> 40	None	< 0.2	0 to 3
Lima (LtA)	Moderately well	> 40	None	< 0.2	0 to 3
Lima (LtB)	Moderately well	> 40	None	< 0.2	3 to 8
Lyons (Ly)	Poorly drained	> 40	None	< 0.2	0 to 3
Madrid (MdC)	Well drained	> 40	None	0.6 to 2.0	8 to 15
Mohawk (MwB)	Well drained	> 40	None	0.6 to 2.0	3 to 8
Mohawk (MwC)	Well drained	> 40	None	0.6 to 2.0	8 to 15
Mohawk (MwD)	Well drained	> 40	None	0.6 to 2.0	15 to 25
Ovid (OvB)	Somewhat poorly	> 40	None	0.06 to 0.2	3 to 8
Palatine (PaC)	Well drained	30	None	0.6 to 2.0	8 to 15
Palmyra (PgA)	Well drained	> 40	None	0.6 to 6.0	0 to 3
Palmyra (PgB)	Well drained	> 40	None	0.6 to 6.0	3 to 8
Palmyra-Howard (PHD)	Well drained	> 40	None	0.6 to 6.0	15 to 25
Palmyra-Howard (PHE)	Well drained	> 40	None	0.6 to 6.0	25 to 35
Teel (Te)	Moderately well	> 40	Frequently	0.6 to 2.0	0 to 3
Wassaic (WcB)	Well drained	30	None	0.6 to 2.0	3 to 8
Wassaic (WcC)	Well drained	30	None	0.6 to 2.0	8 to 15
Wayland (Wn)	Very poorly	> 40	Frequently	0.06 to 0.2	0 to 3

[1]Guideline characteristics are given in Table 2.9.

[2]Not rated in Table 2.9 but given here as additional information.

TABLE 2.15 The Interpretation Matrix[1]

Soil Map Unit	Bedrock Depth	Flooding	Permeability[2]	Slope	Overall Rating	Restrictions
Angola-Darien (AnB)	Severe	Slight	Severe	Slight	Severe	Shallow, slow perm.
Aurora (AwC)	Severe	Slight	Severe	Moderate	Severe	Shall, slow perm, slope
Aurora-Farmington (AXE)	Severe	Slight	Severe	Severe	Severe	Shall, slow perm, slope
Benson (BeB)	Severe	Slight	Moderate	Slight	Severe	Shallow to bedrock.
Benson (BNC)	Severe	Slight	Moderate	Moderate	Severe	Shallow to bedrock.
Benson (BNF)	Severe	Slight	Moderate	Severe	Severe	Shallow and steep.
Cazenovia (CfB)	Moderate	Slight	Severe	Slight	Severe	Slow perm.
Cazenovia (CfC)	Moderate	Slight	Severe	Moderate	Severe	Slow perm.
Cazenovia (CfD)	Moderate	Slight	Severe	Severe	Severe	Slow perm., slope.
Fluvaquents	Variable	Severe	Variable	Slight	Severe	Frequent flooding.
Herkimer (He)	Moderate	Moderate	Moderate	Slight	Moderate	Can flood, mod. perm.
Honeoye (HnB)	Moderate	Slight	Severe	Slight	Severe	Slow perm.
Honeoye (HnC)	Moderate	Slight	Severe	Moderate	Severe	Slow perm.
Honeoye, very stony (HSC)	Moderate	Slight	Severe	Moderate	Severe	Slow perm.
Honeoye-Lansing (HoD)	Moderate	Slight	Severe	Severe	Severe	Slow perm., slope.
Honeoye-Lansing-Ontario (HTE)	Moderate	Slight	Severe	Severe	Severe	Slow perm., slope.
Honeoye-Lansing-Ontario (HTF)	Moderate	Slight	Severe	Severe	Severe	Slow perm., slope.
Howard (HwA)	Moderate	Slight	Moderate	Slight	Moderate	
Howard (HwB)	Moderate	Slight	Moderate	Slight	Moderate	
Howard (HwC)	Moderate	Slight	Moderate	Moderate	Moderate	
Kendaia (KeB)	Moderate	Moderate	Severe	Slight	Severe	Slow perm.

Lakemont (Lk)	Moderate	Slight	Severe	Slight	Severe	Slow perm.
Lima (LtA)	Moderate	Slight	Severe	Slight	Severe	Slow perm.
Lima (LtB)	Moderate	Slight	Severe	Slight	Severe	Slow perm.
Lyons (Ly)	Moderate	Slight	Severe	Slight	Severe	Slow perm.
Madrid (MdC)	Moderate	Slight	Moderate	Moderate	Moderate	
Mohawk (MwB)	Moderate	Slight	Moderate	Slight	Moderate	
Mohawk (MwC)	Moderate	Slight	Moderate	Moderate	Moderate	
Mohawk (MwD)	Moderate	Slight	Moderate	Severe	Severe	Slope.
Ovid (OvB)	Moderate	Slight	Severe	Slight	Severe	Slow perm.
Palatine (PaC)	Severe	Slight	Moderate	Moderate	Severe	Shallow.
Palmyra (PgA)	Moderate	Slight	Slight	Slight	Moderate	
Palmyra (PgB)	Moderate	Slight	Slight	Slight	Moderate	
Palmyra-Howard (PHD)	Moderate	Slight	Slight	Severe	Severe	Slope.
Palmyra-Howard (PHE)	Moderate	Slight	Slight	Severe	Severe	Slope.
Teel (Te)	Moderate	Severe	Moderate	Slight	Severe	Frequent flooding.
Wassaic (WcB)	Severe	Slight	Moderate	Slight	Severe	Shallow.
Wassaic (WcC)	Severe	Slight	Moderate	Moderate	Severe	Shallow.
Wayland (Wn)	Moderate	Severe	Severe	Slight	Severe	Flooding, slow perm.

[1]The ratings are from Table 2.9.

[2]Following the worst-case scenario, the minimum limit of the range of permeability is used for the rating.

moderate or severe ratings (last column, Table 2.15). The restrictive features give clues to the mitigative action required to make the soil suitable for that use. In some obvious cases, mitigative actions cannot be taken and the proposed use must be discarded and an alternative use selected. In very restrictive cases, it is best to leave the site alone rather than use "*technological blitzkrieg.*" Trying to "fool Mother Nature" is usually destructive and has severe economical as well as environmental consequences.

(g) The tricolor legend is applied to the detailed soil map (three shades in Fig. 2.10) and now becomes the interpretation map.

The soil areas with severe ratings are seen to cover most of the area. In these cases, the map provides a warning to planners and local officials that development plans for the area must be done with extreme care with some land use, and perhaps some zoning alternatives excluded. On-site waste disposal can be evaluated on a lot-by-lot basis by detailed on-site soil and topographic inspection. Favorable conditions for potential filter field sites can be delineated if the lot has sufficient area and there is diversity in soil and topographic conditions. Obviously, the larger the lot, the greater diversity and potential for favorable filter field conditions. The exception to this situation is where soil and topography are both extremely uniform (usually level or uniform slope) or the entire area is shallow to bedrock or watertable. Alternatives such as cluster housing should be considered (Craul, 1992). In reality, the area has

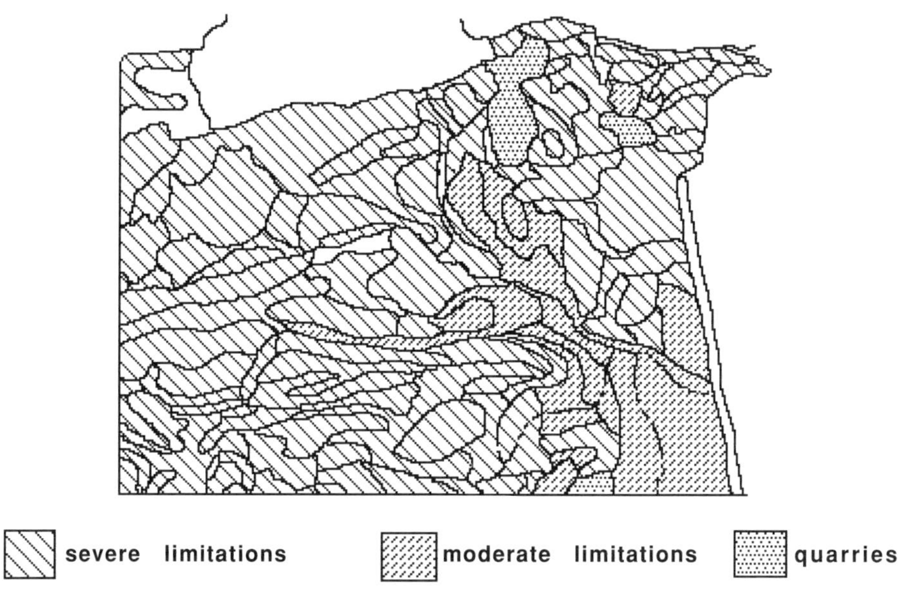

Fig. 2.10 An on-site waste disposal interpretation map for the soil area of Fig. 2.7.

been converted from a dairy farm into 2-acre building lots with an average of 3000-square-foot homes. Sewers and water were installed for most of the area. Outside of the sewer service area, several filter fields have failed within 5 years of construction, one within 6 months.

In summary, site assessment is a necessary process to learn what the site presents in the way of physical opportunities and problems. The process follows one course of action in areas to be developed on natural soils and relatively undisturbed landscapes. The process follows a more intense course of action if on urban or disturbed soils that may include restoration sites with the potential discovery of contamination. The greater the detail and reliability of the information gained from site assessment facilitates the design process and improves the degree of the design success.

3

BIOPHYSICAL ANALYSIS OF SITE

. . . landscaping should be integrated into a good solar design. What is required is a recognition of the diversity and dynamic properties of plant materials. There must be strategies for using the sun as a source of light and heat that recognize the dynamic interaction of the sun, building, and trees.

—Ralph N. Knowles (1981), *Sun Rhythm Form.*

INTRODUCTION

Plants under any environmental situation experience certain naturally induced stresses. These may be from the lack of water, nutrients, or oxygen for the roots, and sunlight. Perhaps, they may be injury-induced stresses due to physical damage from wind, ice, snow, fire and humans (there are some monkeys among us as well as lawnmowers and automobiles!), or the effects of pollution. Stress caused by disease or insect infestation also appear and can be primary causes of stress, or may be secondarily induced by the stresses caused by drought or root suffocation, and so on.

All of these stresses are present in the urban environment. The ones most prevalent are probably due to drought, heat, lack of oxygen, a combination of these, or physical injury. In some cases, the plant is weakened by the effects of one stress and made more susceptible to another form of stress. Usually, when a plant, especially a tree, is examined during an "autopsy," the evidence indicates several factors are involved in its demise, and without details on the life history of the tree, it is difficult to determine the primary cause of death.

Our knowledge of the urban environment clearly indicates that the physical conditions are often not very "plant-friendly." Solar radiation supplemented by radiation from building facades, street surfaces, and heat-generating sources (transportation and industry, etc.), coupled by restricted air flow, combine to create a virtual furnace during the summer and perhaps a freezer in winter for northern cities. Smeltz (1998) cites research by the Department of Meteorology at Pennsylvania State University that shows very significant rise in surface temperatures caused by urban development as detected by satellite imagery. The average increase from 1987 to 1996 over the Chester County, Pennsylvania, area was 36%. In an area enclosing a state park, the increase was 109%. The plants are surrounded by the hot surfaces and many times placed in restricted rooting zones that have an insufficient water and nutrient supply and may even restrict the flow of oxygen to the root systems (Craul, 1992; Thorpe et al., 1978).

In order to obtain a landscape design that is successful in meeting the basic requirements of satisfactory plant growth in urban areas, it is neceesary to assess the physical environment of the proposed project. Here, we are more concerned with the above-ground landscape than with the below-ground landscape of Chapter 2. The latter cannot be totally ignored, and the discussion eventually leads to the design of an adequate soil or an irrigation system to provide needed water for the desired plant palette. Sometimes, it is better judgment to adjust the plant palette to the inherent capabilities of the site rather than attempt to overcome the limitations with an elaborate physical design. There is nothing more unpleasant to the eye (or should be!) of a landscape architect than a wilting or dead plant in a year-old project.

The urban environment is considered a microenvironment being influenced mostly by the temperature, humidity, wind, and radiation that occurs roughly within 1 meter of the ground or the surfaces of plants and structures. These four climatic factors are measured at standard first class weather stations by the National Oceanic and Atmospheric Administration (NOAA, formerly the U.S. Weather Bureau in the Department of Commerce). However, most modern stations are located at airports or other public facilities and do not necessarily characterize the microclimate found within a city or town, and certainly not that of a specific project site. The solution to the problem is the biophysical analysis of the site.

THE SITE HEAT BUDGET

There are several components involved in the energy exchange that takes place at Earth's surface. It is necessary to discuss the components and their contribution to the energy exchange before the process is related to plants and trees. The energy exchange is based on four modes of energy transport. These are *convection, conduction, radiative exchange,* and *latent heat transfer* (Campbell, 1977; Rose, 1966). Convection is the transfer of heat by air movement (wind). Conduction is the transfer of heat within a mass or from one body or mass to another that are in contact with each other. Radiative exchange is the transfer of energy by radiation (longwave or shortwave) with the latter as direct beam radiation from the sun. No intervening molecules are necessary for radiative energy exchange. Latent heat transfer is the loss or gain of energy (heat) by the evaporation or condensation of water.

In a paved, nontranspiring surface, plant water loss depends on the heat load of the site. Most of the loading is due mainly to solar radiation and contributions from other urban heat-generating sources. Since this is a common situation for urban trees, the heat budget method is favored for site analysis and eventual calculation of evapotranspiration (Craul, 1992).

Radiation

Radiation in earth's environment may be divided into short-wave (0.3 to 4 μm) and long-wave (4 to 80 μm) radiation. Most of the solar radiation and that associated with it is in the short-wave range, whereas the radiation (irradiance) emitted by terrestial sources is in the long-wave range.

Solar radiation is the source of most of the energy exchanged at the earth's surface. It determines the climate of an area and the day-to-day weather events. As a result it becomes the most important factor in the heat budget of a site. The other factors are contributors. The radiation reaching a tree is in several forms. These are illustrated in Fig. 3.1,

where S_p = direct solar radiation (surface perpendicular)
S_d = diffuse sky irradiance

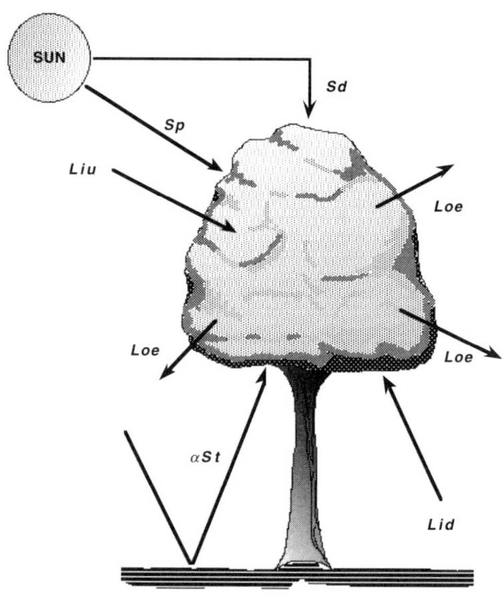

Fig. 3.1 The radiation components of an individual tree. (Adapted from Campbell, 1977.)

S_t = total shortwave irradiance of the surface
L_{oe} = longwave emittance for each side of the object
L_{iu} = longwave irradiance for the up-facing surface
L_{id} = longwave irradiance for the down-facing surface
α = albedo of the surface below the object

The radiation budget may be further explained in Table 3.1. Being primarily concerned with terrestial surfaces having a narrow range of temperatures (compared to the universe), only shortwave and longwave radiation need to be considered for our purposes. The various radiation components will be discussed in order, and later we will return to the topic of the budget itself. The net total radiation is the radiation parameter used later in solving the potential evapotranspiration equation.

Direct Solar Beam Radiation. Direct solar beam energy (S_b) the primary source of radiant (shortwave) energy on the earth's surface; therefore, it's relationship to the surface is discussed first. Because the sun is such a great distance from the earth and much larger than the earth, the sun's rays may be considered as parallel and not focused from a point or to a point as light passing through a camera lens. The radiation received at the outer boundary of the atmosphere and perpendicular to the beam path, S_{po}, is 1.36 kW/m^2. The radiation energy is reduced as the beam passes through the atmosphere so that the radiation received by a perpendicular surface, S_p, becomes

$$S_p = a^m S_{po} \tag{3.1}$$

The coefficient a is an atmospheric transmission coefficient. It is equal to 0.9 on a very clear day, 0.6 on a hazy or smoggy day, and 0.84 on a typical clear day. The exponent m is an optical air-mass number, being the ratio (secant

TABLE 3.1 The Radiation Budget

	Wavelength		
Direction	Shortwave Radiation	Longwave Radiation	Total Radiation
Downward from space: direct solar beam	Direct shortwave[1]	—	—
Downward from the atmosphere	Diffuse shortwave[1]	Downward longwave	Total downward
Upward from the earth's surface[2]	Reflected shortwave	Upward longwave	Total upward
Sum	Net shortwave	Net longwave	Net total

[1]Usually combined in measurement.
[2]Here includes all surfaces, not just the surface below an object. See Fig. 3.8.
Source: Adapted from Reifsynder and Lull, 1965.

of the angle) of the slant-path length to the zenith-path length of the solar beam. The exponent is significant only for sun angles of less than 10°. A correction for terrain altitude may be used for m by substituting $(P/P_o)/(\sin \phi)$, where P is the local atmospheric pressure, and P_o the sea-level pressure.

Incident Radiation. The amount of radiant energy reaching a flat surface is equally distributed over the surface because of the parallel nature of the sun's rays. When the surface is perpendicular to the sun's rays, the radiant energy per unit area is greatest. If the surface is tilted from the perpendicular, the same amount of energy is spread over a greater area, reducing the energy per unit area (Fig. 3.2). Note the spread of the areas over the area. This relationship can be expressed in the formula

$$S_b = S_p \sin \phi \tag{3.2}$$

where S_b is solar incident radiation in kW/m^2, S_p is the solar beam radiation, and ϕ is the angle between the sun's beam and the horizontal surface (defined as the *sun angle*). The angle may be easily found in the field for a moment in time by measuring the length of the shadow of an object with known height and deriving the arctangent of the height over the length. The effect of the sun angle is exhibited on the earth's surface as latitude increases. At the equator, the surface is perpendicular to the sun's rays giving that area of the earth the greatest amount of direct solar incident radiation. As the position changes from the equator toward either pole, increasing the latitude, the surface incident to the sun's rays becomes more acute.

Conversely, the angle between the sun's rays and the line normal to the surface (horizon) increases, and at any given point is equal to the latitude.

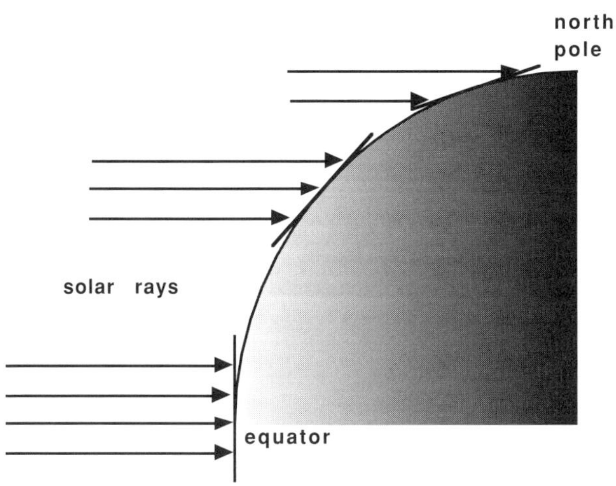

Fig. 3.2 The solar incident radiation on a perpendicular and a tilted surface as distributed over the latitude of the earth.

The incident radiation per unit area is reduced accordingly. The amount of atmosphere that the rays must pass through also increases in the higher latitudes, which further reduces the incident radiation amount. This phenomenon explains one reason why the poles have a cold climate and the equator is hot. The same condition occurs on sloping terrain.

Seasonal Effects of Solar Radiation. The sun angle changes as the earth rotates about the sun due to the tilt of the earth's axis from the vertical. The earth's axis is 23½ degrees from the vertical. Twice a solar year, the earth is vertical with respect to the sun (Fig. 3.3). These positions comprise the vernal (March 20–22) and autumnal (September 20–22) equinoxes (each reversed for the Northern versus the Southern Hemisphere). The two positions are 180° opposite each other. At 90° to these positions are the summer solstice (June 20–22—the longest day) and the winter solstice (December 20–22—the shortest day). At the summer solstice, the earth's axis tilts 23½° toward the sun, and at the winter solstice, the axis tilts 23½° away from the sun. The sun's apparent position on the earth's surface is illustrated in Fig. 3.4. The sun is directly over the equator at the equinoxes and the summer solstice for the Northern Hemisphere is 23½° north which is the Tropic of Cancer. This position is the winter solstice for the Southern Hemisphere. The winter solstice for the Northern Hemisphere is when the position of the sun is 23½° south of the equator which is the Tropic of Capricorn. The date is the summer solstice for the Southern Hemisphere. Also, this is the period of continual darkness above the Arctic Circle in the Northern Hemisphere and midnight sun above the Anarctic Circle. The situation is reversed at the summer solstice for the Northern Hemisphere.

The effects on solar radiation over the earth's surface as the seasons progress through time should be evident based on the earlier discussion of the sun angle effects. The seasonal path traveled by the sun with respect to the earth's surface follows a sine-wave pattern, and is predictable, permitting the calculation of the sun angle at noon for any day of the year. To obtain the sun angle for any time during the day, we must also take into account the rise and fall of the sun from sunrise to sunset.

Daily Cycle. The daily rise and fall of the sun also follows a predictable path. The path from sunrise to sunset is illustrated in Fig. 3.5. Superimposed on this path is also the shift in the azimuth (horizontal direction) of the point of sunrise and sunset. These points vary with the day of the year because of the shift from 23½° north latitude to 23½° south latitude. Thus, the sun rises due east and sets due west only on the dates of the equinoxes. In summer in the Northern Hemisphere, the points are north of east and west. In winter they are south of east and west.

In order to account for all of the conditions affecting sun angles for a given time on a given day, a set of data is required for input to a series of geometric formulas. The data includes the day of the year, the time of day, and the latitude and longitude of the location. The calculations are required to obtain

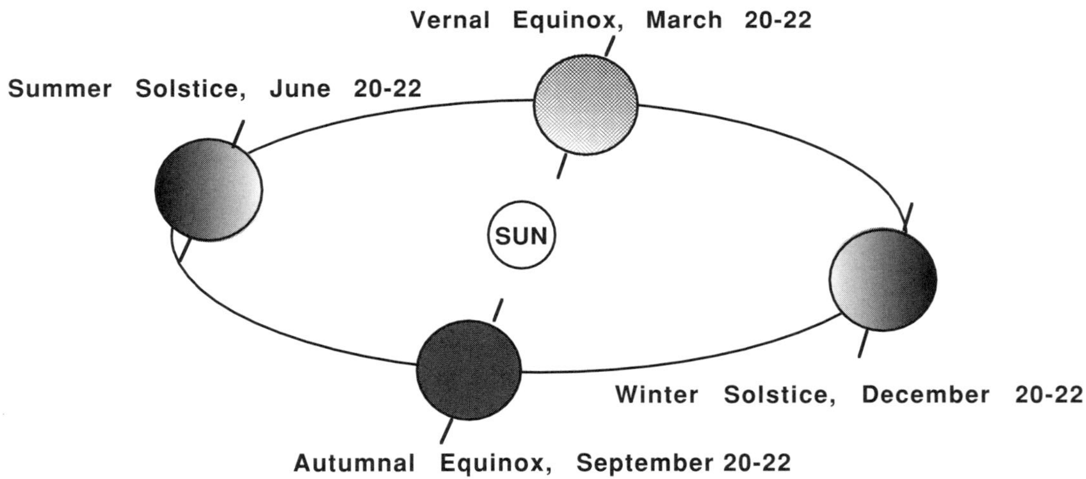

Fig. 3.3 The seasonal position and tilt of the earth.

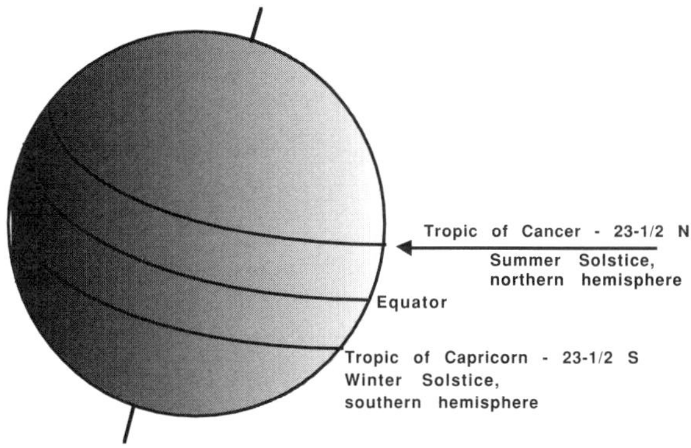

Fig. 3.4 The positions of the sun on the earth's surface at the solstices and equinoxes.

the hourly sun angle for the given location. The following example illustrates the calculations involved.

An Example. The sun angle is desired for a house located at 42° 2′ 19.2″ N (ϕ), 76° 01′ 54.6″ W ($W\lambda$ = west longitude; $E\lambda$ = east longitude) for 10:00 A.M. EST, May 11, 1995. From tables for May 11, the GHA^0 = 180° 54′ 05.3″; May 12 GHA^{24} = 180° 54′ 33.6″ *Decl* δ^0 = 17° 42′ 04.8″; *Decl* δ^{24} = 17° 57′ 32.0″.

All angles must be converted to their decimal equivalents for further calculations. Most scientific calculators or computers have this capability.

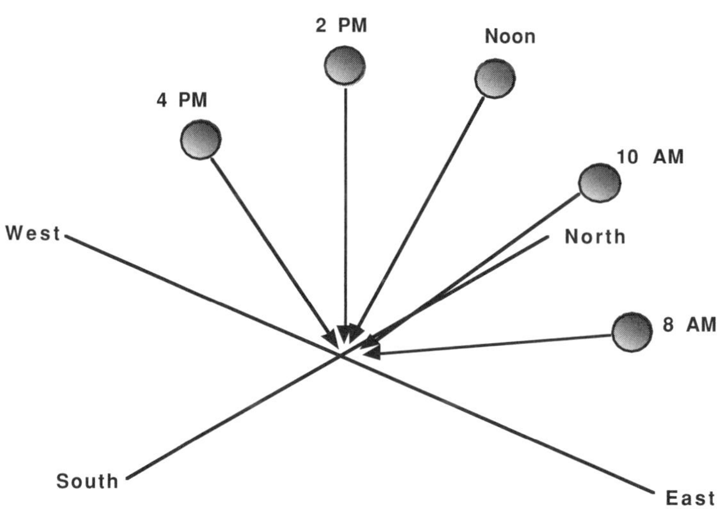

Fig. 3.5 The daily or diurnal path of the sun.

$$
\begin{aligned}
GHA &= GHA^0 + (GHA - GHA^0 + 360)\,(UT1/24) \\
&= 180°\ 54'\ 05.3'' + (180°\ 54'\ 33.6'' - 180°\ 54'\ 05.3'' \\
&\quad + 360)\ (1400/24) \qquad (3.3) \\
&= 180.9014722 + (180.909333 - 180.9014722 + 360)(0.583) \\
&= 30.7973525 = 30°\ 47'\ 50.5''
\end{aligned}
$$

$$
\begin{aligned}
LHA &= GHA - W\lambda \\
&= 30°\ 47'\ 50.5'' - 76°\ 01'\ 54.6'' \\
&= 30.7973525 - 76.0318333 \qquad (3.4) \\
&= -45.23448083 + 360 \\
&= 314.7655192 = 314°\ 45'\ 55.9''
\end{aligned}
$$

$$
\begin{aligned}
Decl\delta &= Decl^0 + (Decl^{24} - Decl^0)(UT1/24) \\
&= 17°\ 42'\ 04.8'' + (17°\ 57'\ 32.0'' - 17°\ 42'\ 04.8'')(1400/24) \qquad (3.5) \\
&= 17.7013333 + (17.9588889 - 17.7013333)(0.583) \\
&= 17.85148821 = 17°\ 51'\ 05.4''
\end{aligned}
$$

$$
\begin{aligned}
AZ &= \tan^{-1} \frac{-\sin LHA}{\cos\phi \tan\delta - \sin\phi \cos LHA} \\
&= \tan^{-1} \frac{-(.7099946)}{(0.74269309 \times 0.3220564) - (0.66963197 \times 0.70420706)} \\
&= \tan^{-1} -3.0554406 \\
&= -71.8775025 + 180^* = 108.1224975 = 108°\ 07'\ 21''^* \qquad (3.6)
\end{aligned}
$$

$$
\begin{aligned}
h &= \sin^{-1}(\sin\phi \sin\delta + \cos\phi \cos\delta \cos LHA) \\
&= \sin^{-1}\,[(\sin 42.03866667 \times \sin 17.85148821) \\
&\quad + (\cos 42.03866667 \times \cos 17.85148821 \times \cos 314.7655192)] \qquad (3.7) \\
&= \sin^{-1}\,(0.2052762 + 0.4978291) = 0.7031053 \\
&= 44.6766753 = 44°\ 40'\ 36.03''
\end{aligned}
$$

*The addition of 180° is a necessary correction from Table 3.2.

TABLE 3.2 Azimuth Angle Correction

When LHA is	Correction: If AZ is Positive	Correction: If AZ is Negative
0° to 180°	180°	360°
180° to 360°	0°	180°

Thus, on May 11, 1995, at 10:00 A.M., the sun is at an azimuth of 180° 07′ 21″ (north being 0°) and at a height above the horizon of 44° 40′ 36.03″ for the given geographical location (Fig. 3.6).

Fortunately, there are various computer software programs available (e.g., Specular *Infini-D* and *3D-Studio-Vis*) and others on the web that will calculate sun angles and plot sun-shade patterns.

Slope and Aspect Effects on Solar Radiation Received by a Surface. As inferred by the radiation components illustrated in Figs. 3.1 and 3.2, the amount of radiation received by a surface not perpendicular to the solar beam will be less than that for the perpendicular surface. Those surfaces facing the sun beam will receive greater radiation than those facing away from the solar beam. The ground sun angle is negative on a north-facing slope for the Northern Hemisphere (Fig. 3.7). Those surfaces sloping at right angles, such as west- or east-facing slopes at noon, will receive an intermediate amount of radiation. The east-facing slope will receive more direct radiation before noon and the west-facing slope will receive direct radiation after noon. But the amount received in these two cases is less than at noon because the solar

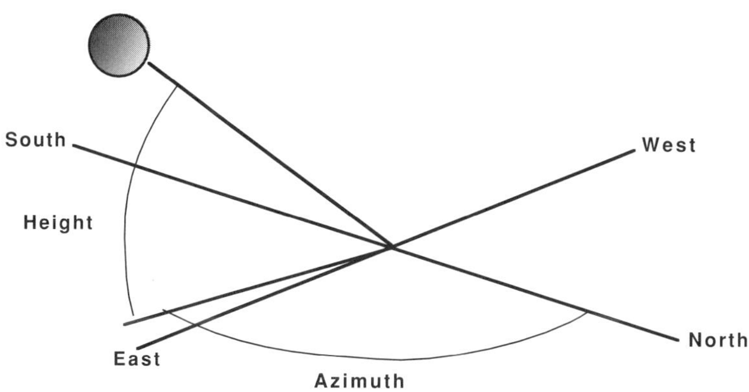

Fig. 3.6 The sun angle for 10 A.M., May 11, 1995 at 42° 2′ 19.2″ N, 76° 01′ 54.6″ W. The azimuth is 108° 07′ 21″ and the height is 44° 40′ 36.03″.

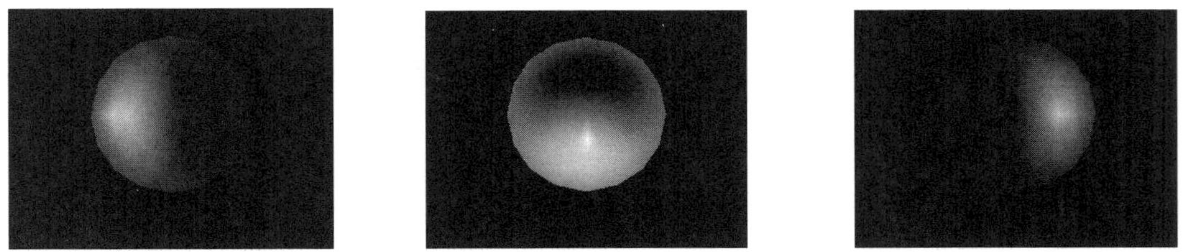

a. hillside sunset b. hillside noon c. hill side sunrise

Fig. 3.7 Several ground sun angle situations related to slope and aspect. North for a, b, and c is at the top.

beam must travel through a thicker layer of the earth's atmosphere. If the solar beam angle is low or the slopes are steep, they may not receive direct beam radiation at noon. The north-facing slope will receive very little direct beam radiation and that will usually occur during mid-summer in mid- and low-latitudes; during winter, the slope may receive no direct beam radiation. Vertical walls of structures will have similar effects on direct solar beam radiation. More hillside surface is illuminated at noon than at sunrise and sunset because of the higher angle of the sun. Also, the radiation intensity (flux) is greater at noon.

Diffuse Sky Radiation. The solar radiation that passes through the atmosphere is partially diffused by the various molecules of gasses and water vapor. The diffused radiation (S_d of Fig. 3.1) eventually reaches the earth's surface. The light reaching the surface on an overcast day is totally diffused radiation. The light received in a courtyard or a mountain valley without direct solar beam radiation early or late in the day is primarily diffused radiation. Obviously, its energy level is far less than that of direct solar beam radiation. It has more significance to photosynthesis than it does from the radiant heating effect but still must be considered in the radiation balance. There is also a shift to the red portion of the spectrum represented by this light due to atmospheric refraction effects; hence, red sunrises and sunsets and the "alpenglo" of snow-capped mountains. Diffuse sky radiation may be calculated from

$$S_d = 0.5S_{po}(1 - a^m)\sin \phi \tag{3.8}$$

where S_d is diffuse sky radiation in kW/m^2, S_{po} is direct beam radiation in kW/m^2, a is the atmospheric transmission coefficient, and m is the optical air-mass number. The diffuse radiation on a cloudy day will be about 146 to 292 kW/m^2 or about one-quarter to one-half of noon intensity of a clear day. When clouds obscure the sun, $S_t = S_d$. Table 3.3 gives the diffuse radiation for various sun angles.

Total Shortwave Radiation of the Surface. The radiation received by the surface (S_t of Fig. 3.1) includes the direct beam radiation and diffuse sky irradiance, or $S_t = S_b + S_d$. Some of this energy is absorbed and some is reflected. The proportion of each is related to the sun angle and the albedo of the surface. The amount of energy absorbed by the surface may be calculated from the formula

$$S_a = S_b(1 - a)\sin \phi \tag{3.9}$$

where S_a is the radiation absorbed by the surface or absorptivity, S_b is the incident radiation, a is the albedo of the surface, and ϕ is the sun angle. The equation provides the amount of heating experienced by the surface and re-

TABLE 3.3 The Diffuse Sky Radiation on a Horizontal Surface by Sun Angle

Sun Angle, Degrees	Diffuse Radiation, kW/m^2
5	20.4
10	40.8
20	68.0
30	88.4
40	95.2
50	102.0
60	102.0
70	108.8
80	108.8
90	108.8

sults in surface-temperature changes. Ignored here is the conduction of heat within the surface material.

Albedo (Reflectance) of the Surface. The incident radiation received by a surface is both absorbed and reflected. The fraction of the radiation reflected is called the *albedo* (or reflectivity). The term albedo usually refers to the reflected portion of a specified spectral band, such as the entire solar spectrum or the visible portion, whereas the term *reflectivity* is usually restricted to specified monochromatic reflection (Reifsynder and Lull, 1965). The albedo of various materials vary depending on the surface color and roughness. The simplest contrast is that between white and black surfaces having the same roughness. White has a greater albedo than black. A smooth surface has a greater albedo than a rough surface. However, color seems to be the dominant factor. Vegetation has a relatively low albedo as compared to light-colored paving. Table 3.4 lists the albedos for some common materials.

Absorptance and Emittance. Absorptance (or *absorptivity* as used by some authors; Campbell, 1977) is defined as the fraction of incident radiation at a given wavelength that is absorbed by a material. A black body absorbs all radiation applied to it and has an absorptivity of 1, whereas a mirror reflects all radiation applied to it and has an absorptivity of 0. *Emittance* (or *emissivity* to be more specific) is the fraction of a black-body emission at a given wavelength emitted by a surface. These terms are applied to surfaces for evaluation of surface heating. The absorption and emission of radiation follow the same process of energy status change of atoms or molecules within the material. Absorption of radiation always results in the transformation of the energy to radiation of another wavelength or to some other form of energy such as heat. Most of the energy in solar and terrestial wavelengths absorbed by material

TABLE 3.4 Albedo of Various Materials

Material	Albedo
Soil, dry	0.08 to 0.35[1]
Sand, dry	0.35 to 0.75
Sand, wet	0.20 to 0.30
Brick	0.30 to 0.65 (glazed)
Granite	0.45
Sandstone	0.27 to 0.46
Limestone	0.50 to 0.65
Concrete, dry	0.17 to 0.27
Asphalt	0.05 to 0.10
Turfgrass	0.21 to 0.24
Conifer forest	0.16 (0.10 to 0.20)
Hardwood forest	0.18 (0.05 to 0.15)
Swamp forest	0.12
Tundra	0.15 to 0.30
Open water	0.05 to 0.09
Fresh snow	0.75 to 0.95
Old snow	0.40 to 0.70

[1]Varies by soil color: dark topsoil is 0.05 to 0.15 and light-colored dry soil is 0.32.

Sources: Compiled from Campbell, 1977; Marsh, 1991; Mazria, 1979; Reifsnyder and Lull, 1965; Rose, 1966.

is converted into thermal energy that manifests itself in the temperature of the material. Thus, the radiation would raise the temperature of the material. For most opaque solids at terrestial temperatures, the absorptivity is very weakly dependent on the temperature of the absorbing material.

Longwave Radiation from the Atmosphere. The atmosphere produces longwave radiation that potentially reaches terrestial objects and surfaces. Atmospheric emissivity may be evaluated by

$$\varepsilon_A = 0.72 + 0.005T_a \tag{3.10}$$

where ε_A is the atmospheric emissivity, and T_a is the absolute air temperature, Kelvin. Clouds have an emissivity of 1, so when clouds are present, the atmospheric emissivity is higher than for a clear sky. Thus,

$$\varepsilon_{AC} = \varepsilon_A + C(1 - \varepsilon_A - 4\Delta T/T_a) \tag{3.11}$$

where ε_{AC} is the emissivity of the cloudy sky, ε_A is the emissivity of the clear sky, C is the fraction of the sky covered by clouds, and ΔT is the difference between T_a and the cloud-base temperature. The Stefan-Boltzmann equation is now applied to obtain the longwave radiation from the atmosphere.

Longwave Radiation from the Object. All objects have their own temperatures. Inanimate materials have very low internal temperatures and the apparent temperature is due more to external influences such as the absorption of radiant energy. The absorptivity of a material depends on the roughness of its surface. Emissivity likewise is dependent on surface roughness in addition to its temperature. Living systems exhibit higher internal temperatures due to their biochemical processes, and reflect radiant energy as well. In both cases, the radiation emitted is primarily long-wave as short-wave emission is negligible.

In the case of the plant crown (more specifically, the leaf or needle) radiation from the object is subdivided into the sides, the upper and the lower surfaces. The latter two are the sunlit and shade surfaces and the side surfaces may have sunlight or shade to varying degrees as the sun follows its daily and seasonal paths. If the crown model is assumed as a cone, the side effects are reduced. A cube model or inverted cone would exhibit greater side effects.

Longwave radiation from any material follows the Stefan-Boltzmann law as

$$\Phi = \varepsilon \sigma T^4 \tag{3.12}$$

where Φ is the long-wave radiation emitted by the body in W/m^2, ε is the emissivity and obtained from tables, σ is the Stefan-Boltzmann constant (5.6697×10^{-8} W/m^2), and T is the absolute temperature (Kelvin) of the body. Φ is always less than 1 for "nonperfect" black bodies.

A surface radiates outward at a rate that is dependent only on its own temperature. The radiation balance on the surface is determined largely by the relative amounts of radiation received from the sky and other surfaces. The transmission of radiation from one body to another is given by (McAdams, 1954)

$$dW_{1\to 2} = \frac{I_1\, dA_1 \cos \beta_1\, dA_2 \cos \beta_2}{r^2} \tag{3.13}$$

where I_1 is the radiation flux density (W/m^2) from dA_1 normal to the surface, cos β_1 is the angle of radiation between the two surfaces (from 1 to 2), cos β_2 is the angle from surface dA_2 to dA_1, and r is the distance between the two surfaces, m. The total radiation leaving dA_1 can be found by integrating the equation over the entire hemisphere above dA_1. Similarly, the radiation received by any portion of the hemisphere can be found by a suitable integration. The relation is illustrated in Fig. 3.8.

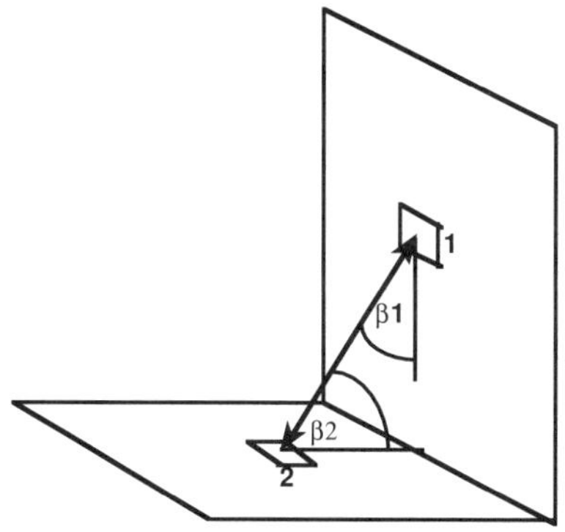

Fig. 3.8 The cosine relationship between two radiating surfaces.

Radiation in Enclosures

The amount of radiation emitted by a vertical or horizontal surface toward and received by an object, a tree, involves the analysis of angles between the boundaries of the surface and the location of the tree. The concept is simply the application of the cosine law to finite surfaces such as building facades and walls. The topic has been exhaustively elucidated by Fanger (1970). His application is for the calculation of mean radiant temperature for persons sitting and standing in a heated or cooled room. The analysis as modified here permits the calculation of the mean radiant temperature of the tree(s) from heat emitted from all surfaces reaching the tree.

Mean Radiant Temperature. To understand this approach to angle factors, it is necessary to begin with the concept of mean radiant temperature. It is simply a different approach to the heat transmission method of McAdams (1954), given earlier, and is more comprehensive in application. Mean radiant temperature in relation to an object at a given point in an enclosure is defined as that uniform temperature of black surroundings that will give the same radiant heat loss from the object. The object's location in the space must be known, since the mean radiant temperature will often vary from point to point. Keep in mind that the approach deals with longwave radiation.

For normal longwave radiation, the absorbance is equal to the emittance, and the emittance of the object has no influence on the mean radiant temperature. For spaces in which all surfaces have a uniform temperature, as on a cloudy day or in full shade, the emittance has no influence on the mean radiant temperature. If there are significant temperature differences among the surfaces as with sunlit and shaded walls, the emittances will affect the relative influence on the mean radiant temperature of the various surfaces in the enclosure.

The analysis process makes several assumptions:

1. The tree is placed in an enclosure consisting of N surfaces, each of which is isothermal and have temperatures $T_1, T_2, \ldots, T_N$ (K).
2. The corresponding emittances are $\varepsilon_1, \varepsilon_2, \ldots, \varepsilon_N$.
3. The angle factors between the tree and the surfaces are $F_{T-1}, F_{T-2}, \ldots, F_{T-N}$.
4. If a surface is found not to be isothermal, it is subdivided into smaller isothermal surfaces.
5. Each surface is assumed to be gray, so that emittance equals absorptance.
6. Radiation emitted and reflected from any surface is diffusely distributed following Lambert's Cosine Law.

Emitted and reflected radiation for nonmetallic surfaces are directionally indistinguishable, and it is sufficient to analyze only their sum, the radiosity, B, which is defined as

$$B = \varepsilon\sigma T^4 + \rho H \tag{3.14}$$

where H = incident radiant energy received by the surface per unit time and unit area
ρ = reflectance, which for a gray body is equal to $1 - \varepsilon$
ε = emittance of the surface
σ = Stefan-Boltzmann constant
T = absolute temperature of the surface

The radiosity is the rate at which radiant energy leaves a surface: The first term on the right in Eq. (3.14) is the radiation emitted by the surface and the second term is the reflected radiation.

By taking into consideration all that has been discussed in assumptions 1 to 6 and the radiosity concept, the following equation can now be derived:

$$\varepsilon\sigma T^4_{mrt} - T^4_T = \varepsilon_t[B_1F_{T-1} + B_2F_{T-2} + \cdots + B_NF_{T-N}] - \varepsilon_T\sigma T^4_T \tag{3.15}$$

where ε_T is the emittance of the tree crown, and T_T is the outer temperature of the tree crown. As Fanger points out, T_T has no influence on T_{mrgt}, then

$$T^4_{mrt} = \frac{1}{\sigma}(B_1F_{T-1} + B_2F_{T-2} + \cdots + B_NF_{T-N}) \tag{3.16}$$

Radiosities for all surfaces in an enclosure can be calculated when the emittance and the temperature for each surface is known. Most building materials have a high emittance that dominates the reflected energy, so the latter can be neglected. Equation (3.16) can then be simplified to

$$T_{mrt}^4 = T_1^4 F_{T-1} + T_2^4 F_{T-2} + \cdots + T_N F_{T-N} \tag{3.17}$$

If there are relatively small temperature differences between the enclosure surfaces, Eq. (3.17) can be linearized as

$$T_{mrt} = T_1 F_{T-1} + T_2 F_{T-2} + \cdots + T_N F_{T-N} \tag{3.18}$$

Equation (3.18) is used on cloudy days or if the enclosure is in full shade, whereas Eq. (3.17) must be used on sunny days or when there are large differences in enclosure surface temperatures.

Because of the high solar radiant temperature, the radiant exchange with the surroundings is more or less independent of the temperature of the surroundings. But the mean radiant temperature of the unirradiated tree must be determined. Fortunately, one can assume T_{umrt} is equal to the air temperature. Obviously, the tree and its surroundings are affected by solar shortwave radiation and it must be included in the total thermal analysis. Equation (3.19) is then used to calculate the mean radiant temperature of the irradiated tree:

$$\varepsilon_T \sigma A_{\text{eff}}(T_T^4 - T_{mrt}^4) = \varepsilon_T \sigma A_{\text{eff}}(T_T^4 - T_{umrt}^4) - A_T \alpha_{\text{ir}} q_{\text{ir}} \tag{3.19}$$

where A_{eff} = effective radiation area of the tree crown
α_{ir} = absorptance of the tree crown
q_{ir} = irradiance flux density (solar), kcal/m^2/hr
A_T = area of the tree crown projected on a plane perpendicular to the Sun, m^2

Projected Area Factor. The effective radiation area of the tree crown, A_{eff}, is the total area of the tree crown as viewed horizontally (Fig. 3.9a). The area of the tree crown irradiated, *AT*, is at the solar angle for a point in time (Fig. 3.9b). Obviously, the latter will change seasonally and daily as the solar angles change.

The ratio $f_T = A_T/A_{\text{eff}}$ expresses the effect of the solar angle on the irradiated surface versus the total radiation area of the crown. At small solar height angles, approximately 0° to 45°, f_T is nearly 1; for greater solar heights, f_T decreases.

The shape of the tree crown must be considered when calculating the total radiation area. Some common crown shapes are illustrated in Fig. 3.10. When the solar angle is directly overhead, the total crown radiation area is a circle for most cases. The spherical crown will have a spherical irradiated surface for all solar heights. The ellipsoidal shape and paraboloidal shape will be maintained at low solar heights and will become shortened and transposed into a more circular shape as the height increases to overhead. The transformation occurs when the solar angle equals the inverse slope of the crown surface. For conical crowns, the irradiated shape is a triangle for lower solar

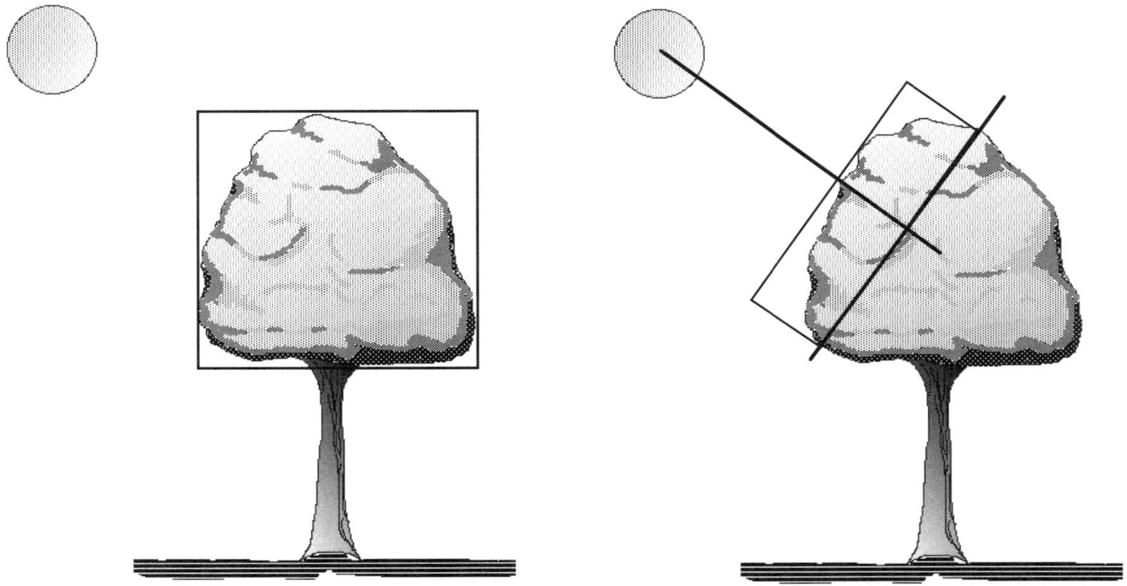

Fig. 3.9 The (a) effective and (b) irradiated areas of the tree crown.

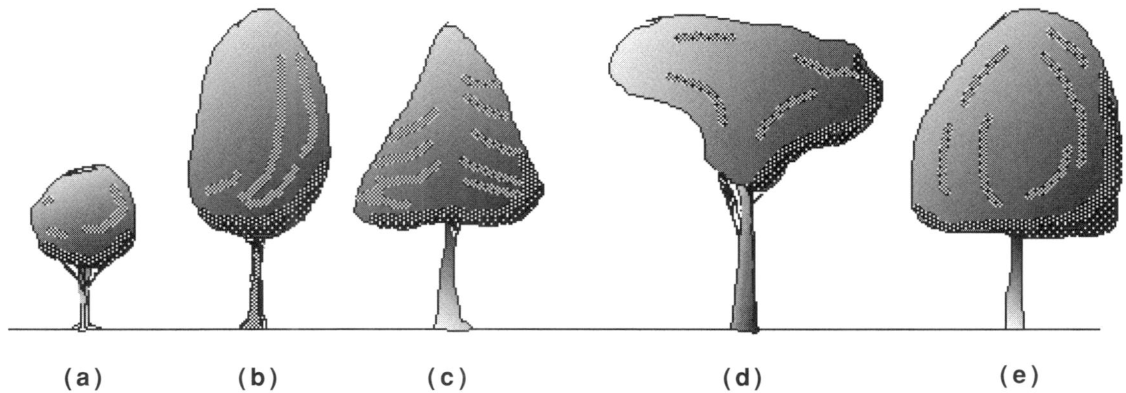

Fig. 3.10 Several tree crown shapes: (a) spherical; (b) ellipsoidal; (c) conical or neoloid; (d) inverted neoloid; and (e) paraboloidal.

heights. The irradiated triangular area becomes smaller and also transposes into a circle as the height of the sun increases to overhead. The radiation area for the inverted neoloid and cone is the inverted neoloid and inverted cone. When the sun is overhead, the area becomes a circle as in the other cases. The area formulas for these shapes are listed in Table 3.5.

Angle Factors. Enclosure surfaces are normally rectangular, and the angle factors for radiant exchange are between the tree and vertical and horizontal rectangles. The latter are the sky and the ground surface. The appropriate rectangles for a tree at a known location are shown in Fig. 3.11. The angle factor between the tree and the entire surface is

$$\overline{F}_{T-A} = \frac{1}{2\pi^2} \int_{x/y=0}^{x/y=a/c} \int_{z/y=0}^{z/y=b/c} \int_{\alpha=0}^{\alpha=2\pi} \frac{f_p}{\left[1 + \left(\frac{x}{y}\right)^2 + \left(\frac{z}{y}\right)^2\right]^{3/2}} d\left(\frac{x}{y}\right) d\left(\frac{z}{y}\right) d\alpha \tag{3.20}$$

F_{T-A} represents the mean value of the angle factor between the tree and a rectangle when the tree is rotated about a vertical axis (Fanger, 1970). Equation (3.20) has been reduced to two graphs for an array of a/c and b/c ratios (Fig. 3.12). A calculation of the angle factors for Fig. 3.11 is given in Table 3.6.

Incident Radiation. Because the incident longwave radiation H_i consists of emitted radiation as well as rereflected radiation from other surfaces, radiosity may be reintroduced. Thus,

$$B_i = \varepsilon\sigma T_i^4 + \rho_i H_i \tag{3.21}$$

where ρ_i is the reflectance of surface A_i. If A_j is any other surface in the

TABLE 3.5 Formulas for Area Determination of Various Tree Crown Shapes

Crown Shape	Area Formula
Spherical	$A = \pi r^2$
Ellipsoidal	$A = \pi ab$
Conical (triangle)	$A = \frac{bh}{2}$
Neoloidal	$A = \frac{h(a + 2b)}{4}$
Paraboloidal	$A = \frac{4}{3} ab$ or $\frac{4\ ab}{3}$

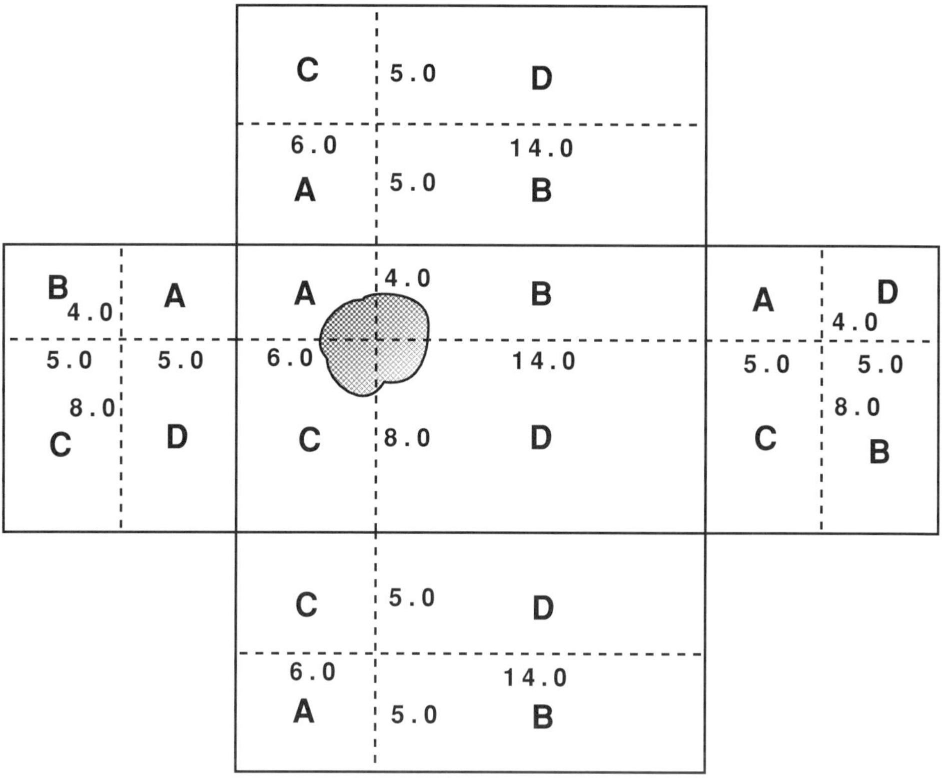

Fig. 3.11 The diagram of angle factors for a tree of known location within an enclosure.

enclosure, contributions arriving at A_j from all surfaces are expressed as a sum:

$$H_i = \sum_{j=1}^{N} B_j F_{A_i - A_j} \tag{3.22}$$

and Eq. (3.21) becomes

$$B_i = \varepsilon_i \sigma T_i^4 + (1 - \varepsilon_i) \sum_{j=1}^{N} B_j F_{A_i - A_j} \tag{3.23}$$

Equation (3.23) needs to be written for each of the N surfaces in the enclosure, creating an equation series. For A_i, the total radiant heat exchange can be determined by

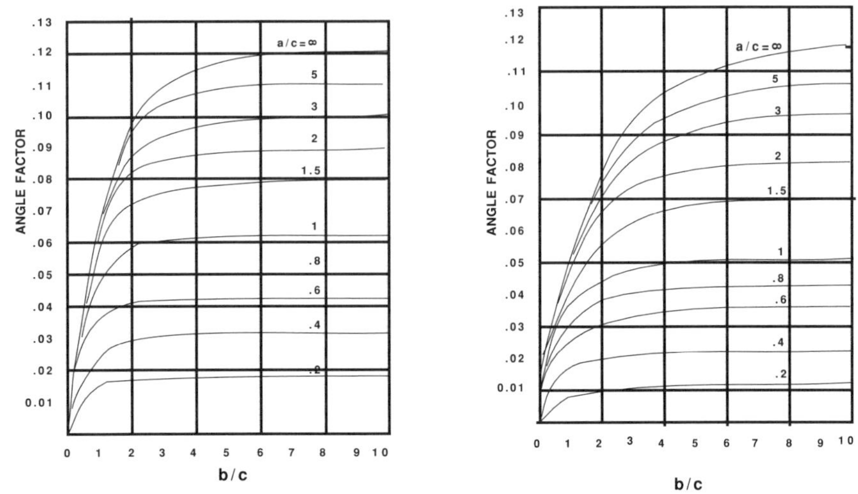

(a) (b)

Fig. 3.12 The graphs of mean value of the angle factor between a tree and (a) a vertical rectangle and (b) a horizontal rectangle. (Redrawn from Fanger, 1970.)

$$q_i = \frac{\varepsilon_i}{1 - \varepsilon_i} (\sigma T_i^4 - B_i) \tag{3.24}$$

leading to a second series of equations for the enclosure surfaces. Since emittances are high, the reflections can usually be disregarded. After this omission and linearization of the equations, the basic equation for the series becomes

$$q_i = \varepsilon_i \sigma \sum_{j=1}^{N} \theta_{ij}(T_i - T_j) F_{A_i - A_j} \tag{3.25}$$

where

$$\theta_{ij} = \frac{T_i^4 - T_j^4}{T_i - T_j} \tag{3.26}$$

If there are only small temperature differences among the surfaces, as on a cloudy day or complete shade, θ_{ij} can be set to one fixed value. If there are large differences, θ_{ij} must be individually calculated. Table 3.7 provides some θ_{ij} values.

Radiosity may also be introduced to shortwave radiation. The equation for shortwave radiosity, B^*, is written as

TABLE 3.6 Calculation of Angle Factors for Fig. 3.10

North Wall	$F_{T-ABCD} + F_{T-A} + F_{T-B} + F_{T-C} + F_{T-D}$		
	F_{T-A}	$(b/c, a/c) = (1.25, 1.50)$:	0.063
	F_{T-B}	$(1.25, 3.50)$:	0.081
	F_{T-C}	$(1.25, 1.50)$:	0.063
	F_{T-D}	$(1.25, 3.5)$:	0.081
			0.288
East Wall	$F_{T-ABCD} + F_{T-A} + F_{T-B} + F_{T-C} + F_{T-D}$		
	F_{T-A}	$(b/c, a/c) = (0.36, 0.29)$:	0.006
	F_{T-B}	$(0.36, 0.57)$:	0.011
	F_{T-C}	$(0.36, 0.57)$:	0.011
	F_{T-D}	$(0.36, 0.29)$:	0.006
			0.034
South Wall	$F_{T-ABCD} + F_{T-A} + F_{T-B} + F_{T-C} + F_{T-D}$		
	F_{T-A}	$(b/c, a/c) = (0.63, 0.75)$:	0.034
	F_{T-B}	$(0.63, 1.75)$:	0.048
	F_{T-C}	$(0.63, 0.75)$:	0.034
	F_{T-D}	$(0.63, 1.75)$:	0.048
			0.164
West Wall	$F_{T-ABCD} + F_{T-A} + F_{T-B} + F_{T-C} + F_{T-D}$		
	F_{T-A}	$(b/c, a/c) = (0.83, 0.67)$:	0.034
	F_{T-B}	$(0.83, 0.67)$:	0.034
	F_{T-C}	$(0.83, 1.33)$:	0.048
	F_{T-D}	$(0.83, 1.33)$:	0.048
			0.164
Ground	$F_{T-ABCD} + F_{T-A} + F_{T-B} + F_{T-C} + F_{T-D}$		
	F_{T-A}	$(b/c, a/c) = (0.80, 1.2)$:	0.027
	F_{T-B}	$(0.80, 2.8)$:	0.040
	F_{T-C}	$(1.2, 1.6)$:	0.044
	F_{T-D}	$(1.6, 2.8)$:	0.062
			0.173
Sky	$F = \sin^2[(\arctan d/2h)] = \sin^2\left(\arctan \dfrac{16\text{ m}}{2 \times 10\text{ m}}\right) = 0.390$		

TABLE 3.7 Some Values for θ_{ij}

	t_i (°C)					
t_j (°C)	0	10	20	30	50	100
0	0.81	0.86	0.91	0.96	1.07	1.38
10	0.86	0.91	0.96	1.01	1.12	1.44
20	0.91	0.96	1.00	1.06	1.17	1.50
30	0.96	1.01	1.06	1.11	1.23	1.56
50	1.07	1.12	1.117	1.23	1.35	1.70
100	1.38	1.44	1.50	1.56	1.70	2.08

Source: Fanger, 1970.

$$B_i^* = (1 - \alpha_i^*) \sum_{j=1}^{N} B_j F_{A_i - A_j} \tag{3.27}$$

producing a series for the enclosure surfaces. If one or more of the surfaces is an "emitter," the appropriate equation in the series is expanded by an extra term. After determination of the radiosities from the series in Eq. (3.27), the absorbed shortwave radiation for each surface is calculated as

$$q_i^* = \alpha_i^* \sum_{j=1}^{N} B_j^* F_{A_i - A_j} \tag{3.28}$$

Now the shortwave and longwave radiations may be combined into

$$q_{ri} = q_i + q_i^* \tag{3.29}$$

for each surface, providing another series of equations.

The convective heat transfer must now be considered. The convective heat loss from the surface per unit time per unit area is

$$q_{ci} = h_{ci}(t_i - t_a) \tag{3.30}$$

where h_{ci} is the convective heat transfer coefficient, and t_a is the air temperature. When air velocities are low and free convection occurs, the convective heat transfer coefficient can be calculated from the usual formula:

$$\mathrm{Nu} = c(\mathrm{Gr} \times \mathrm{Pr})^n \tag{3.31}$$

where Nu is the Nusselt number, Gr the Grashof number, Pr the Prantl number, and c and n are constants. The constants must be derived for the outside conditions. For laminar-free convection, $c = 0.54$ and $n = 1/4$. A different approach is required when transient conditions prevail.

Heat transfer at the surface also occurs by conduction to the surface. In the situations dealt with here, the longwave radiation is more the result of daytime surface temperatures caused by external radiation reaching the surface than by conduction within the wall. Daytime heat is conducted away from the surface when the air temperature is greater than the surface temperature. Surface temperature at night is very much determined by heat conduction to the surface when the air temperature is less than the surface temperature. It is based on the amount of heat stored during the daytime and related to the heat capacity and volume of the wall material. For most purposes, the Stefan-Boltzmann equation, Eq. (3.12), is applied here.

The Sky View Factor. The ratio of radiation received by an object to that radiated to the entire hemisphere above it is the view factor, F (Fig. 3.13). It receives radiation partly from the sky and partly from the objects around it such as building facades. The proportion between the two sources is related to the diameter d of the opening and the height h of the surrounding objects or walls. The fraction of the radiation received by the sky hemisphere from the tree is

$$F = \sin^2[\arctan(d/2h)] \tag{3.32}$$

The reciprocal of F is the view factor of the walls or surrounding objects. If the effective sky temperature and wall temperatures are known, the incoming radiation to the tree can be calculated by the Stefan-Boltzmann law as previously given. This factor is applied to the longwave radiation analysis of all objects in the opening. The view factor is constant throughout the courtyard; it is not position-dependent because all the vertical angles of the upper hemisphere between the primary object and its neighbors are complementary and the total remains the same. The case for a free-standing tree in an incomplete forest canopy is treated by Waggoner and Reifsynder (1961).

The Radiant Energy Budget (The Radiation Balance). Radiant energy at a given location is both gained and lost, resulting in a radiation budget or balance. For example, a tree is surrounded by an enclosure (building facades or in a courtyard) of uniform temperature. The tree is in thermal equilibrium with the enclosure. The tree will gain heat through its absorptivity and incident radiation. It will lose heat by its emissivity and temperature. If the absorptivity were larger or smaller than the emissivity, it would be expected that the tree would exhibit a net gain or loss of heat. Since absorptivity and emissivity are the same (second law of thermodynamics), the tree would re-

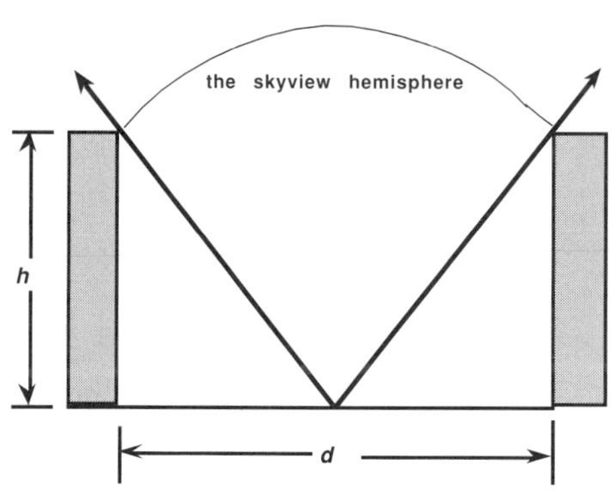

Fig. 3.13 The skyview factor of a courtyard.

main in equilibrium with its environment. As the environment heat changes, so would the tree heat change. It can then be concluded that the tree heat loading is that of the environment in which it exists. The heat loading is roughly equivalent to the net radiation as

$$R_n = 2L_{oe} - a_S S_t(1 + \alpha) - a_L(L_{iu} + L_{id}) \qquad (3.33)$$

where R_n is the net radiation in kW/m^2, a_S is the shortwave absorptivity, S_t is the total shortwave radiation, L_{oe} is the longwave radiation of each side of the object, L_{iu} is the longwave radiation of the upper side of the object, L_{id} is the lower-side longwave radiation, a_L is the longwave absorptivity, and α is the albedo of the surface below the object.

The application to an individual tree of the mean radiant temperature requires the subdivision of the crown into components, as suggested by Eq. (3.33). The tree crown is divided into irradiated and unirradiated segments, shade segments on the bottom and the sides, creating five or six segments. The resulting equation, similar to Eq. (3.33), would have additional terms.

To summarize, the direct solar beam is complex in its movement over the surface of the earth and how it is influenced by atmospheric conditions and obstructions (sunlight-shade). Diffuse sky radiation originates from the entire sky hemisphere unless decreased by the view factor. Longwave sky radiation depends on water droplets, carbon dioxide, and other gases. The amount of shortwave radiation from the ground surfaces depends on the albedos of the materials. The longwave radiation from the surfaces depends on their individual temperatures, emissivities, and reflectivities to incoming longwave radiation, but they combine to produce a collective energy level.

The energy exchange that takes place in a plant canopy among the mass of leaves, individually and collectively, and their various layers is far too complex to discuss within the scope of this text. Campbell (1977) provides a detailed explanation. However, the other exchange surfaces involved in the budget are less complex and can be evaluated with less difficulty and are discussed in the following section.

HEAT STRESS OF URBAN TREES

A close scrutiny of the heat stress of the tree or a group of trees may be an approach to the evaluation of evapotranspiration on a particular site. The approach leads to a clearer picture of the type of soil and volume required to maintain plants from one precipitation event to another. This is an indirect approach since there is real difficulty in determining the actual water utilization of individual trees under urban conditions. A suggested initial step is to recognize the various urban heat-stress conditions, as illustrated in Figs. 3.14 and 3.15.

The single, open tree (Fig. 3.14) is exposed to solar radiation and also to a flow of air that tends to cool crown temperatures and reflected heat from

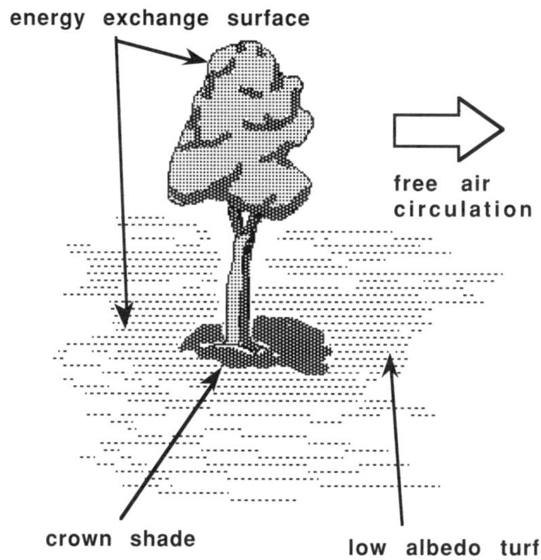

Fig. 3.14 The heat situation of the open-grown tree on turf. (Craul, 1994.)

the low-albedo grass surface. The air flow removes moisture from the crown, reducing vapor pressure of the air and increasing evapotranspiration. This situation is relatively cool in the urban environment.

The single, enclosed tree (Fig. 3.15) is exposed to direct solar radiation in addition to reflected radiation from the pavement below the crown and the building facade. Heat is probably transported by turbulent air (convection) from the street, which, more than likely, has higher summer temperatures than the enclosure. Crown temperatures are quite high on a sunny summer day.

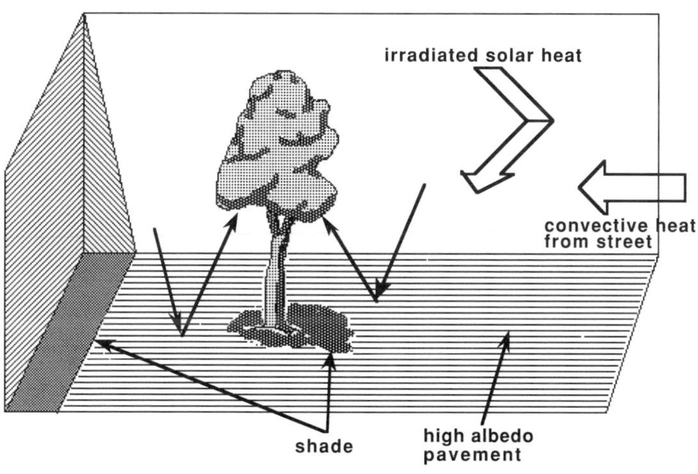

Fig. 3.15 The enclosed isolated tree. (Craul, 1994.)

Even soil temperatures beneath the pavement could attain high levels here. Reduction of temperatures by meso-scale windflow is probably minimal. As Waggoner and Reifsnyder (1961) point out, solar energy is used to heat the soil (surface) in openings between the trees. Upward longwave radiation may be so high that the net radiation is smaller than that of a closed tree stand. However, evapotranspiration per unit of leaf surface in the open tree may be greater as the leaf intercepts reflected energy from the sunlit surface. The copse (Fig. 3.16) has a closed canopy (essentially so in the diagram; in a closed canopy, the crowns would touch or be intermingled with each other), shading the surface below the crowns. The exchange (the boundary layer) occurs in the upper crowns rather than around the entire crown, as in the single-tree situation. Crown temperatures are reduced accordingly. If in a natural setting, the soil surface litter acts as a mulch, cooling the soil. The trunk space has lower air flow than above the canopy, but the temperatures are moderated by the shading of the crown and is the coolest portion of the copse. The stress on each tree is quite different for each situation, therefore, the underground landscape needs to be different for each. The details for each situation require elucidation.

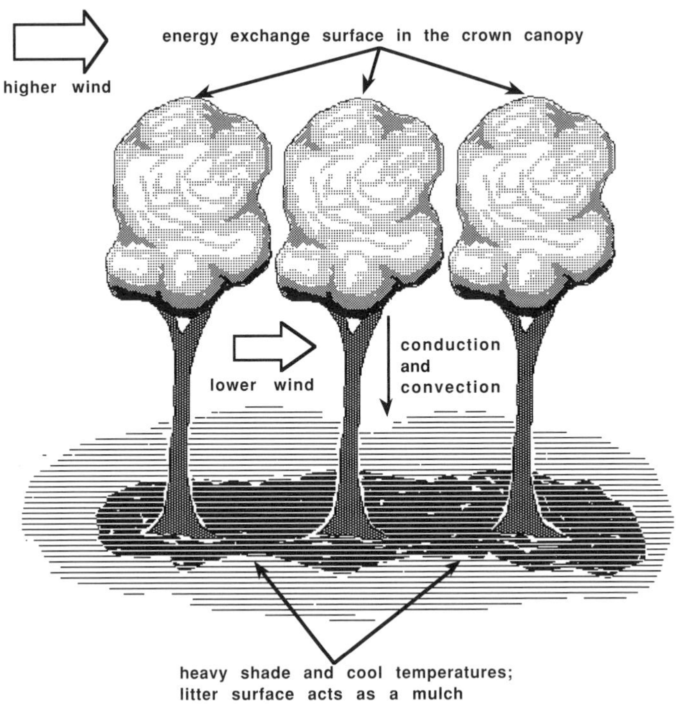

Fig. 3.16 The tree copse. (Craul, 1994.)

Detection of Plant Stress

Plant stress may be detected by various methods. One indirect method is by color infrared photography. In these images, healthy tree foliage appears as a bright magenta, whereas stressed trees are gray. Those trees moderately stressed are partially gray (Estes, 1983). Unfortunately, color infrared film is expensive, unstable (needs to be frozen in storage), very light-sensitive, and very expensive to develop. The author is aware of only one laboratory in the entire United States that develops 70- or 35-mm infrared film.

Moisture stress may be determined by xylem sap-pressure measurements (Bassuk and Whitlow, 1985; Waring and Cleary, 1967). The water column within the tree from the roots to the leaves is under tension (negative water potential, Ψ) most of the time. The tension in the leaves is usually in equilibrium (but still greater in magnitude) with the soil tension at dawn or when the soil supplies adequate moisture at a sufficent rate to meet transpirational needs, or the stomates are closed (Hillel, 1980). The greatest tension is detected in the leaves or needles. Variation in tension is exhibited throughout the column as the soil moisture and leaf transpirational stress vary. Field measurement requires special equipment and multiple observations on *in situ* leaves; thus, is not generally practical under ordinary urban design situations.

For these and many other reasons, the estimation of potential evapotranspiration is preferred in most cases, even with the inherent limitations of the technique.

EVAPOTRANSPIRATION

Evapotranspiration (ET) is defined as the water loss by evaporation from the soil surface and by the transpiration of water vapor from the leaves of plants. The origin of the transpired water is also the soil. It is difficult to measure each of the two losses. The climatic factors affecting evapotranspiration, including radiation, air temperature, humidity, and wind, are easier to measure. Thus, equations expressing the relationships must be employed.

The precise estimation of evapotranspiration is difficult at best and not without problems (Federer, 1970). However, general estimates are possible if the appropriate factors are measured or calculated with a careful regard for a reasonably accurate representation of the climatic elements for critical periods of the growing year.

Potential Evapotranspiration (ET_o)

An international standard for potential evapotranspiration has been adopted by the UN Food and Agricultural Organization. It is the amount of water lost from a continuously watered soil in a pan and planted to fescue clipped to

12 cm (approximately 5 in.). The fescue was chosen because it represents a broader range of crop plants than would a tree or woody shrub.

Wind is less important to tree ET than it is to open turf. The difference has to do with the two boundary layers. The boundary layer over turf is a more continuous, simple phenomenon, whereas the boundary layer around or over a tree crown is more complex and variable. Surface evaporation does not apply to trees. Stomatal control is more important in trees and woody shrubs. The stomates may close at low humidity or high crown temperatures. Stomatal activity (sensitivity) moderates tree water loss at high evapotranspiration levels in most cases. Tree transpiration is highly variable at high evapotranspiration levels.

Three methods exist for the calculation of potential evapotranspiration. These are the water conservation equation, the flux equation, and the heat budget equation. It is the last equation that is most useful for our needs. The equation is

$$\lambda E_p = \frac{s}{s + \gamma}(R_n + G) + \frac{\gamma}{s + \gamma} f(u)(\rho'_{va} - \rho_{va}) \tag{3.34}$$

where λ = latent heat of vaporization
E_p = potential evapotranspiration
γ = thermodynamic psychrometer constant ($\rho c_p/\lambda$)
s = slope of the saturation-vapor-density curve
R_n = net radiant flux density
G = heat storage rate, measured or estimated
$f(u)$ = λ/r_H, empirical wind function from the Penman equation, r_H = sensible heat-transport resistance
ρ'_{va} = saturation-water-vapor density, air
ρ_{va} = water-vapor density, air

Tabular data provide s, $s/(s + \gamma)$, $\lambda/(s + \gamma)$, and ρ'_{va} as functions of temperature. The required parameters are R_n, G, $f(u)$, and ρ_{va}. The daily average of G may be assumed to be 0; for crops, the daytime value may be assumed as $0.1R_n$. If the daily average of $f(u)$ is in m/s at 2 m above the surface, then, $f(u) = 5.3(1 + u)$.

Actual (Nonpotential) Evapotranspiration (ET_a)

Only open-water surfaces will evaporate continuously at potential rates. Constraints on water loss from a surface and plants prevent potential levels from occurring except for a fraction of the time. The constrained level of evapotranspiration is considered actual or nonpotential evapotranspiration. The actual evapotranspiration is more important to the water budgets of many surfaces than potential evapotranspiration estimates.

Actual transpiration is less than potential because the plant stomates may close under excess stress. This physiological reaction is a protection by the plant against severe dessication and an attempt to maintain water balance within the plant. Severe reduction in soil moisture, high evaporative demand from low humidity, intense solar radiation, or high winds will cause the stomates to close. Campbell (1977) suggests that the best approach to the estimation of actual evapotranspiration is to consider the rate of water supply rather than the rate of water loss. Successful computer simulations of soil drying and water extraction by plants have been attained. Other attempts have been made to express actual evapotranspiration as a fraction of potential evapotranspiration. These attempts are not fruitful if the soil or the soil-plant system is controlling water loss. This is the usual case in urban situations. The successful simulations have been applied to crops that cover the soil surface and not isolated plants with the soil covered by impermeable surfaces. In the latter case, it is more appropriate to estimate the potential evapotranspiration and design the planting accordingly, as is done here. Obviously, it is a conservative approach.

Evapotranspiration Estimation from Climatic Data

It is possible to estimate potential evapotranspiration from average climatic data as collected and compiled by the National Oceanic and Atmospheric Administration (National Weather Bureau). For most reporting stations, the data include daily and monthly temperature averages and extremes, dewpoint, relative humidity, wind direction and speed, precipitation, cloudiness, average barometric pressure, average vapor-pressure deficit, and others. Some first-class stations may report sunshine duration, average direct solar radiation, net radiation, and Type A pan potential evapotranspiration. The best and most comprehensive measurements are measured at universities and research stations. However, the compiled data are not always in convenient form or readily available to the public.

There are two major limitations to the use of this method, though lots of data are available. One is that the derived average values may not reflect the true biophysical situation for a region even after working through a large mass of data; are annual or only growing season data used? The second reason, and perhaps the more important one, is that the reporting station location may be far removed from the site to be characterized and significantly different climatically from the observation point.

The best approach involves the measurement of climatic factors on the site for several intervals during the year, as suggested in an earlier section of this chapter.

SUN-SHADE RELATIONSHIPS

In order to determine radiation levels of a site that contain facades giving sun and shadow relationships, it is necessary to calculate or measure daily sun angles on the site for the growing season and seasonal variations outside the growing season (Fig. 3.17).

Sun and Shade Areas An object is created in the landscape plan such as a tree, wall structure, and so on. The program asks for a latitude-longitude point and date. The program then runs the sun angle for the day illustrating the sunlight-shadow relationship for the sunlit hours. Cloudiness and diffused radiation are not taken into account.

The sun angles are necessary for plotting the sunlight versus shade areas over time. The information is needed to calculate the amount of radiation received by the plant palette. The area, location, and duration of the sunlight-

Fig. 3.17 The solar viewgraph for determining the view factor (sunlight-shade) for a given point.

shade areas vary not only with the time of day, but also the season of the year (Figs. 3.18 and 3.19). Generally, in the past, only several critical days were chosen for sunlight-shade analysis because of the large amount of computation involved. The desktop computer now enables hourly analysis for every day of the year if necessary.

CALCULATION OF POTENTIAL EVAPOTRANSPIRATION OF A SITE

Create a Computer-Site Model

The first step is to create a scale model of the project site in the computer. The model should show structure arrangement and size, including the heights of the structures. Proper azimuthal orientation of the model is important (Fig. 3.20). Proposed vegetation (the planting plan and its plant palette) should be accurately located on the site and drawn to scale.

Notation should be made of the materials comprising the structure facades (objects) and those covering the ground surface, the latter being either paved, bare soil, or vegetated (type—turf, woody ground cover, mulch, etc.). The albedos for the various surface materials should be assigned at this time.

The Calculation Procedure

Calculate Hourly Sun Angles and Creation of Sun-Shade Areas. Data input for these calculations include the latitude and longitude of the site location and day of the year numbered from 1 to 365. The program will then run through the daylight hours by the hour, obtaining the sunlight-shade image for each hour and storing it individually for later compilation with the remaining images to form a daily composite. The images must be translucent

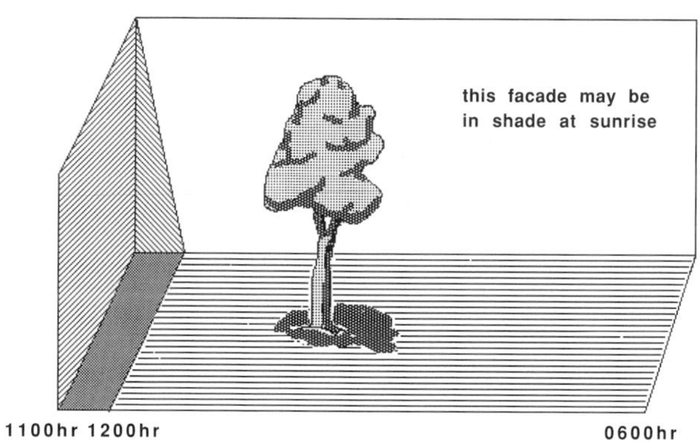

Fig. 3.18 The accumulated solar radiation from 6 A.M. to noon, simplified.

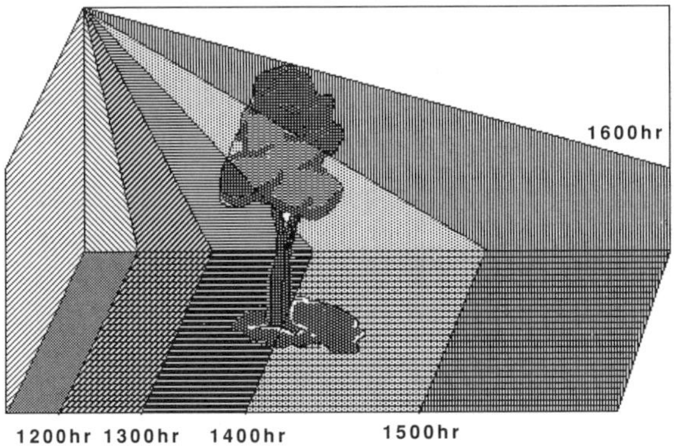

Fig. 3.19 The accumulated afternoon shade from noon until 4 P.M., simplified.

so that boundaries between the progressing sunlight-shade areas may be identified for later use; the differential area for each hour is stored and used later to calculate the net radiation for that hour. The sunlight-shade areas are applied to the vertical surfaces as well, thus requiring a three-dimensional plot

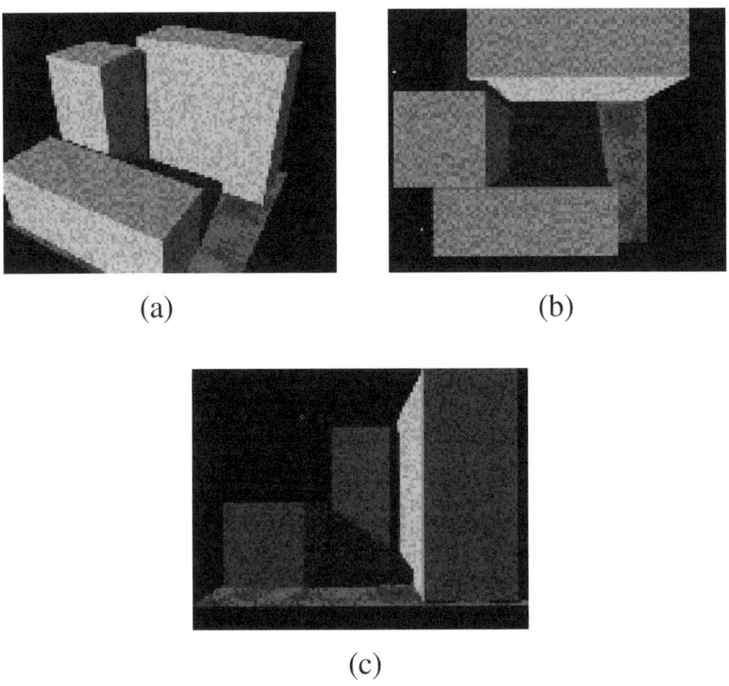

(a) (b) (c)

Fig. 3.20 Three views of a shaded courtyard at approximately 42° north latitude at 1000 hr, vernal equinox: (a) 3-D view, (b) top view, and (c) east view.

on a flat surface. If radiation is required for additional days, an additional run of the program for each day is necessary.

Determine Solar Heating for Sun-Shade Areas. The solar heating of the sun areas, including all surfaces, is derived from the direct beam radiation calculation. A calculation is made for each hourly sun segment as the sunlight progresses during the day. Thus, a series of sun-lit patches with varying levels of heating and cooling is developed (Fig. 3.19). Each of these patches may contain absorbing surfaces of different materials (different albedos) and at different ground angles. Calculations must be made for each surface area for each hour and the heat loading accumulated for each hour. Heat loss from shaded surfaces is calculated for the shade periods.

The areas remaining in shade throughout the day must be treated as heated by diffuse radiation and by convection. It is unlikely that heat conduction through shaded building facades contribute significant amounts of heat to the plant environment on the project.

The calculations are as follows:

A. Direct perpendicular beam shortwave radiation:

$$S_p = a^m S_{po}$$

where $S_{po} = 1.36\ \text{kW/m}^2$. Data input for this formula are $a = 0.84$ for a typical clear day, 0.90 for a very clear day, 0.60 for a smoggy day; $m = 57.30$ for 1°, 11.4 for 5°, 5.76 for 10°, and becomes insignificant for angles greater than 10° and may be neglected.

B. Direct beam incident shortwave radiation:

(1) horizontal surface:

$$S_b = S_p \sin \phi$$

where ϕ, the *sun angle,* $= h$ as calculated from the solar angle and applied to a horizontal surface.

(2) vertical or nonhorizontal surface: a more complex equation must be applied because two angles are involved: ϕ, the Sun angle, and φ, the difference angle between the Sun's azimuth and the azimuth of the object surface, as illustrated in Fig. 3.21:

$$S_b = S_p \sin \phi \sin \varphi$$

The analysis of all surfaces presents a very complex computational array. The program requires a large amount of input data to complete the computations for the project. The data input is derived mainly from the computer construction of the project design.

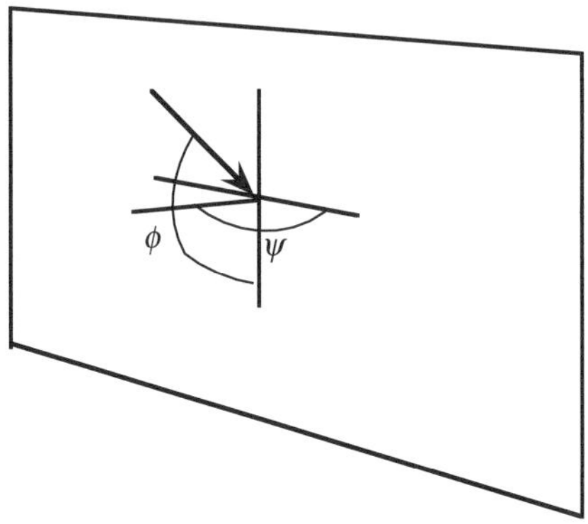

Fig. 3.21 The complex angle of direct beam radiation on a vertical wall.

C. Diffuse sky shortwave radiation:

$$S_d = 0.5S_{po}(1 - a^m)\sin\phi$$

D. Sky longwave radiation:
(1) for clear sky:

$$\varepsilon_A = 0.72 + 0.005T_a$$

(2) or for cloudy sky:

$$\varepsilon_{AC} = \varepsilon_A + C(1 - \varepsilon_A - 4\Delta T/T_a)$$

then,

$$\Phi = \varepsilon_A\sigma T^4 \qquad \text{or} \qquad \Phi = \varepsilon_{AC}\sigma T^4$$

E. Terrestrial longwave radiation:
(1) from the ground surface:

$$\Phi = \varepsilon\sigma T^4$$

(2) from the object surface(s): The mean radiant temperature of Fanger (1970) is applied in this step, which includes the ground surface and sky view.

All steps are required for the sunlight segments of the project surfaces, whereas only Steps C through E are required for the shade segments.

Obtain Daylight Wind Speed, Air Temperature, and Vapor Pressure. Hourly observations of wind speed, air temperature and vapor pressure are parameters to be entered into the ET_o equation for each segment of the patchwork to obtain hourly ET_o.

Since micrometeorological data are hardly ever available for locations other than specific research sites, it is necessary to deliberately obtain the data. The cost of instrumentation with acceptable accuracy has been reduced in recent years. Therefore, it is possible to install a fairly complex instrumentation suite for intermittent intervals on the site. Continuous observations are the best source of data, but are costly and cannot be justified for many projects. Observations made for the critical-season periods are recommended. These would be the solstices and the equinoxes. Intermediate values can be usually safely extrapolated from the sine-wave curves characteristic of the time progression of air temperature and solar radiation. Wind speed and vapor-pressure deficit are more variable and less predictive. Unfortunately, this method misses the periods of the occurrence of temperature extremes such as July or August and January or February in the Northern Hemisphere. A simple solution is to supplement the four seaonal day observations with air temperature, wind speed, and vapor-pressure observations during the expected extreme periods as predicted by climatological data for the region.

The length of observation, if not continuous for the year, can usually be 7 days centered on the four annual seasonal days and other critical periods. The interval covers the two weather cycles that have the most significant effect on climatic factors within a season, the 3-day cycle within the 7-day cycle.

Calculate Potential Evapotranspiration for the Area. The ET_o is then averaged to obtain the daily ET_o. The net radiation is obtained for each sunlight-shade segment for each location of concern (e.g., each tree location) and averaged for the day. The average value together with the average climatic data obtained from local measurements or estimations from published climatic data are inserted into the potential evapotranspiration equation. The derived ET_o is applied to each location and used to design the soil and the irrigation system, if used, for that location. The formula is repeated here as

$$\lambda E_p = \frac{s}{s + \gamma}(R_n + G) + \frac{\gamma}{s + \gamma} f(u)(\rho'_{va} - \rho_{va})$$

with the components previously defined.

DYNAMIC WATER RELATIONS OF DESIGNED SOIL

We still do not have firm ideas about the retention and flow of water in disturbed or designed urban soil profiles. Their hydraulic properties do not permit extrapolation of water relations from known soils. Thus, designed soils themselves must be studied. For the present, valid application of general soil physics knowledge must be applied to the urban situation with the understanding that not all conditions can be described, resolved, or anticipated. Some proposed questions include: How rapidly is a soil in confined space dried out by tree roots; or what is the hydraulic conductivity of the several horizons in the designed profile; or how does the wetting front, if one exists, move down through the profile when irrigation or precipitation occurs? These and other questions are partially answered in Chapters 4, 7, and 8.

The potential evapotranspiration obtained through the preceding is inserted into the process of irrigation design in Chapter 8.

THE USE OF SOLAR RHYTHM IN LANDSCAPE DESIGN

Knowles (1981) presents some appropriate and relevant ideas concerning the *solar envelope* of a project and the consideration of it in building and landscape design. He defines the solar envelope as the space within which solar radiation is available to the project and does not impinge on the abutting parcels. It is "a container to regulate development within limits derived from the sun's relative motion." Development within the envelope will not shadow its surroundings during critical periods of the day, being defined by the passage of time as well as the constraints of property. His major idea is that the building should take advantage of the sun's rhythm and that the landscaping outside the building should be designed with the same attitude. Most of our current architectural approaches do not make use of solar rhythm; the buildings are "static" and the landscape is in shadow or sun at critical times during the day, either when it is too hot or too shaded for the plant palette.

Shadow and Sun in Street Grids

A contrast between the Jeffersonian and Spanish street grids (Fig. 3.22) provides an interesting example of differences in solar patterns in the urban environment. The contrast has significant implications to the expected growth of trees in the streets. It provides a strong reason why uniform growth of the same species is not obtained within a city block. There are other reasons as well, such as soil differences, but differences in solar pattern is the overriding one.

To summarize, Knowles's comparison in the Jeffersonian grid east-west streets will be shadowed the entire winter day in northern latitudes. North-south streets are lighted and warmed during midday (Fig. 3.23). In summer, the Sun's rays are more directly east in the morning and west in the afternoon, and nearly overhead at midday. East-west streets will have little shadow at midday and much less in the morning and afternoon. North-south streets will

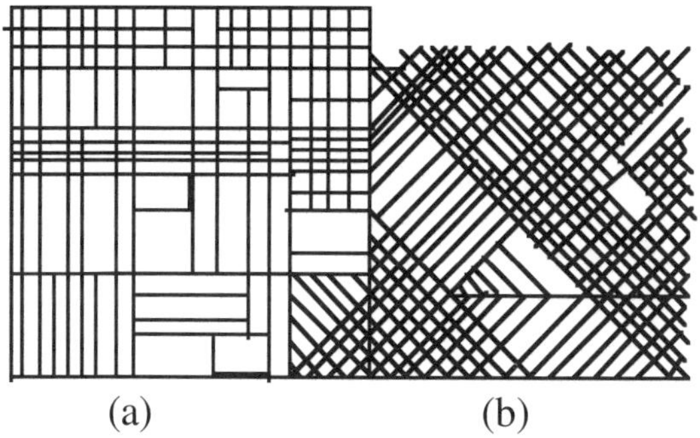

Fig. 3.22 The contrast between the (a) Jeffersonian and (b) Spanish grids. (Adapted from Knowles, 1981.)

have shadow in the morning and afternoon but have full sun at midday. In the Spanish grid, every street receives some direct winter sunlight and heat sometime between 9 A.M. and 3 P.M. All streets have some shadow, but it is narrow, giving more sunlight than east-west streets. In summer, the Spanish grid provides some shadow during most of the day except for a short period from late morning to early afternoon. The colonade or veranda architecture of Spanish building facades enhances the effect. As Knowles points out, street orientation is almost never considered in land-use and planning decisions.

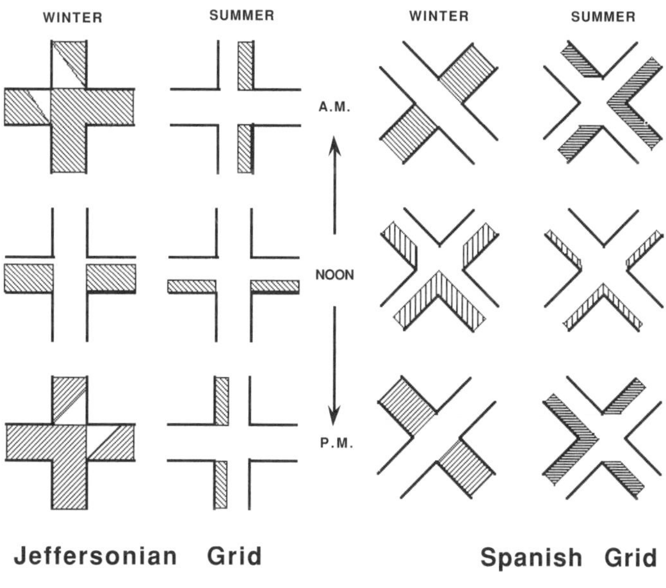

Fig. 3.23 The daily and seasonal sunlight-shade relationships of the Jeffersonian and Spanish grid patterns. (Adapted from Knowles, 1981.)

Implications to Planting

Street Planting. There are enough research results to tell us that trees vary in their growth from one side of the street to the other, depending on street orientation and heights of abutting structures (Bassuk and Whitlow, 1985). Trees with heights shorter than structure heights may not receive any sunlight or only a few hours at midday, during the winter months in more northerly latitiudes on the south side of east-west streets, whereas the north-side trees may receive partial or total sun depending on the height of the south-side structures and the width of the street. The general ratios are illustrated in Fig. 3.24. Obviously, the dormant season sunlight is not as critical as growing season sunlight, but some deciduous trees carry out photosynthesis through their bark in winter.

The figure shows (not illustrated directly) that a 15-foot-tall tree within 10 feet of a 50-foot-high or higher building would be in shade at summer solstice for all four latitudes. Though not illustrated, summer solstice sun angle data show that a 100-foot-high building at 30° latitude has the same shadow as a 50-foot-high building at the equinoxes.

The sunlight situation changes in the summer. The trees on the northern side of the east-west street may receive sunlight all day and the southside trees may receive sunlight most of the day.

The Spanish grid provides a different ratio situation on different orientations of the streets. These differences may be observed in Fig. 3.26, with southwest-northeast streets.

On-Site Planting. The shade cast by structures or other trees should be considered when designing a project layout and plant palette. Because on-site

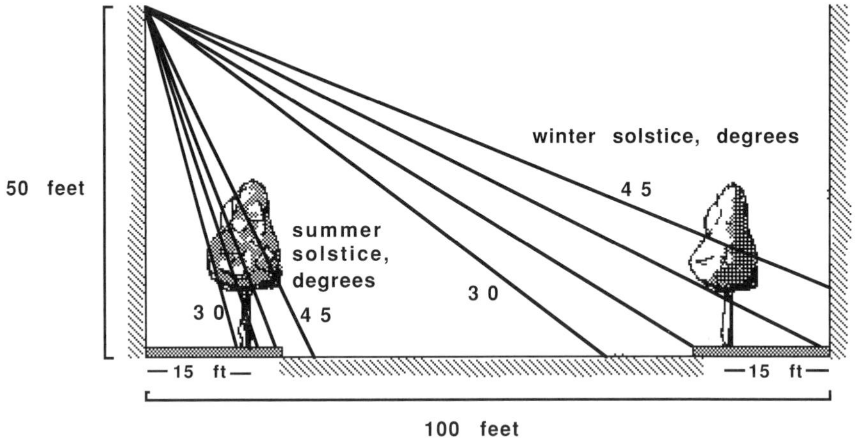

Fig. 3.24 Distance/height sunlight-shade ratios for 30, 35, 40, 45 degrees latitude at 76 degrees longitude, summer and winter solstices.

plantings usually require the plants to be located close to structure facades, the shading area of the structure is the most important factor. If the structure area is a very small proportion of the project area and not of great height, a large proportion of the planting area may receive a large amount of sunlight. Some distance/height ratios for four structure walls are given in Figs. 3.24, 3.25, and 3.26. Obviously, the orientation of the structure plays an important part in the determination of the sunlight-shade situation within the project area. Knowles (1981, pp. 133–135) discusses the interrelationship between the structure and on-site planting within the solar envelope. The reader should refer to his text for more details.

For the creation of shading by trees, which is one of the primary objectives of including trees in the plant palette, the crown shape and tree distribution have an effect on the spatial distribution of shade (Kuuluvainen and Pukkala, 1987). The greatest amount of spatial shade is cast by the umbrellalike, horizontally extended crowns much like that of the mature American elm (*Ulmus americana,* L.). At low density or for trees with narrow crowns, such as the columnar English oak (*Quercus robur,* L.) or the various columnar cedars, the

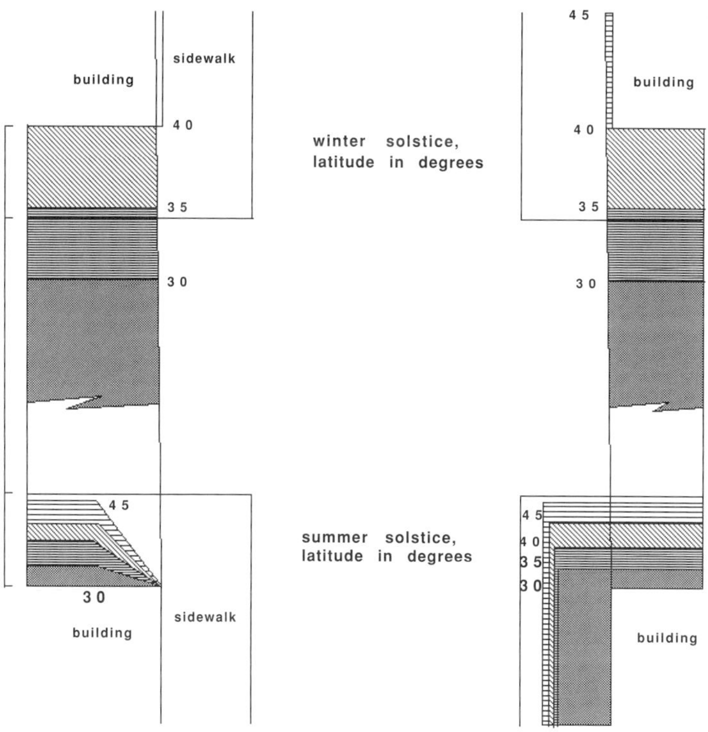

Fig. 3.25 The ratios of tree to structure heights in east-west streets (Jefffersonian grid) within summer and winter sun angle limits for several latitudes. Same relative scale as Fig. 3.24.

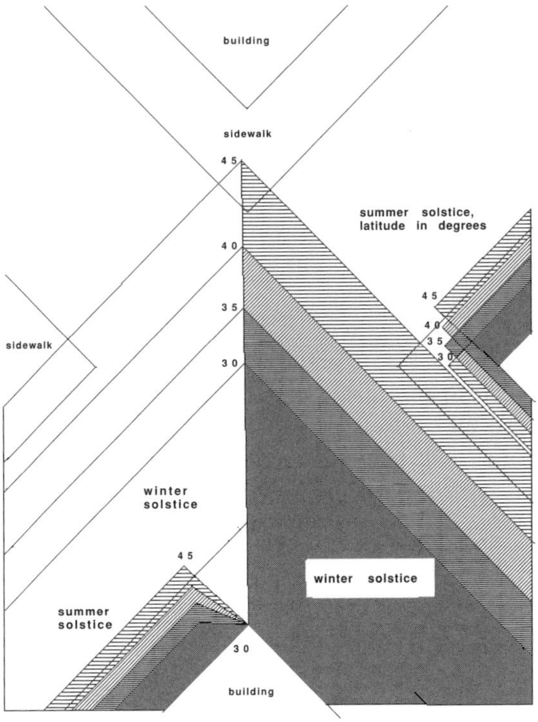

Fig. 3.26 The ratios of tree to structure heights in southwest-northest streets (Spanish grid) within summer and winter Sun angles for several latitudes. Relative scale is the same as Fig. 3.24. Buildings are 50 feet tall, sidewalks 20 feet wide, and the street is 60 feet wide.

spatial distribution of trees has only a small effect on the amount of shading. Further, the total amount of shading in a random distribution of columnar trees is almost the same as in a systematic distribution. In a stand where the total horizontal projection area of crown cones is great, the spatial distribution of trees should be systematic for effective light interception and shade projection.

EVAPOTRANSPIRATION EXAMPLES

Isolated Tree

An isolated tree existing in the unshaded open is subjected to heating by direct solar radiation throughout the day. Shortwave and longwave sky radiation from the entire hemisphere above the tree and the longwave radiation from the ground surface are the other radiational factors involved. Thus, the longwave radiation calculation is simplified, being only from the sky and the ground surface. The meteorological effects of wind, air temperature, and vapor pressure have a more significant direct effect than when structures or building facades influence the radiation situation (Fig. 3.14).

A Copse

The tree copse becomes somewhat more complex because of the shading effects of tree crowns on each other, the creation of a crown canopy acting as a boundary surface, and the trunk space below (Fig. 3.16). The effective radiation area becomes the crown canopy of the copse. It is subdivided in much the same way as an individual tree crown but obviously has greater dimensions. The ground radiation becomes more complex since the ground surface material is not uniform about the copse. Beneath the copse may be an understory or at least the rudiments of a forest floor if it is not turf or covered by paving.

An Avenue

The trees on an avenue must be handled as a complex array of individual trees. The direct shortwave incident radiation varies from one tree to another (Fig. 3.27) being dependent on shadows cast by buildings and other obstructions. Secondly, the amount of longwave radiation received from the building facades varies with position according to their angle factors. Wind, humidity, and air temperature differences are to be expected. Obviously, the orientation of the avenue will have the most effect on the physical environment.

CONCLUSIONS

To provide the proper conditions for tree growth in urban areas, more attention must be given to the environment in which trees or other plants are to be grown. The effects of that environment must be measured. Then provision

Fig. 3.27 A treed east-west avenue.

can be made for adequate soil rooting volume and irrigation, if needed, so that uniform growth is achieved. Each tree must be treated as an individual rather than treating all trees in a planting project exactly the same. Exception to this approach occurs when diversity in growth and development is a desired result. However, aesthetic or palette diversity still requires the landscape architect to make certain that conditions are present for satisfactory or even optimum growth of all palette components. The material discussed in this chapter provides a guide to the landscape architect and urban forester to attain the desired result.

4

DESIGN AND CONSTRUCTION OF THE URBAN SUSTAINABLE SOIL PROFILE

INTRODUCTION

Soil morphology consists of a series of horizons with varying physical, chemical, and biological characteristics. These provide attributes that are beneficial, in most cases, to plant growth. Further, there are important relationships and interactions between the horizons that give an overall set of characteristics to each profile, such as waterholding capacity, fertility, drainage and aeration, compaction resistance and soil strength, and so on. Making use of the knowledge of the kind of soil profile that is created under a given set of environmental conditions, leads to the conclusion that a soil can be designed to meet the conditions of a project and its designed plant palette. Because the soil is designed specifically, the potential for a successful and longlasting landscape design is greatly enhanced.

As Spirn (1992) rightly points out, many, if not most, landscape projects of the future will be on sites formerly occupied by urban structures and facilities. Because these sites normally will contain urban soils, it is imperative to have knowledge of the potential solutions to the problems presented by the less than favorable conditions of the urban sites. A soil must be designed for the site taking into consideration the materials available to the designer. Three alternatives are generally available:

1. Import topsoil from another location.
2. Modify existing soil material with amendments.
3. Design a topsoil or an entire soil profile.

Each of these alternatives has advantages and some distinctive disadvantages and will be discussed in detail in this chapter. However, the first alternative is not considered unless the topsoil is to come from an area to be developed and the soil eventually covered over. Then, it is appropriate to "save the topsoil." It must be analyzed to make certain it meets the requirements of the plant palette and the site. The deliberate importation of topsoil from a natural landscape is to be discouraged.

The discussion eventually leads into the Soil Design Decision Process of Chapter 12 after covering some general principles in soil design followed by construction and installation details.

THE SOIL MODEL

Ecosystems increase in order, or have decreased *entropy,* as they mature toward a climax or steady state. This progression is reverse of that in open systems existing in the universe. The entropy, or order, tends to decrease, leading to more disorder.

When designing an ecosystem, it is appropriate for landscape designers to strive for a *climax* or *steady-state* system, one that is relatively stable over time. Maintenance and restoration of the design are kept to a minimum, thus reducing costs and assuring a more lasting design. A plant design containing a *pioneer* palette on a *mature* soil for a given landscape environment will tend to change over time and will be more difficult to maintain. The same principles of ecosystem entropy and steady state should be applied to the design of a soil, otherwise problems may arise over time.

Soil components that are considered in designed soil are as follows:

1. The fabric: the mineral skeleton or matrix of the soil; a sand, an existing soil material, or glass (see the section on "Sustainable Soil Design").
2. Organic component: the structural enhancer and energy source for the soil organisms and the protective surface layer (mulch); composted organic materials.
3. Pore-space (structural) amendments: needed to preserve or enhance the structure of the soil and for rooftop applications, a lightening material; inorganic or manufactured synthetic materials such as sintered and crushed diatomaceous earth.

The type and proportion of these components are determined by the conditions present in the project site and the plant palette together with the steady-state principle.

The greatest problem involved in designing soil is to create the physical condition of favorable structure to the soil. Natural soil forms structure in the pedogenic process (Buol, Hole, and McCracken, 1989). Materials used in creating designed urban soils do not provide natural structural characteristics. Even natural topsoil used as a component has lost much of its structure due to the excavation, hauling, and respreading operations (Craul, 1992). The process of structure formation is difficult to duplicate in designed or built soils although many attempts have been made to find substitutes to create structural attributes. Soil amendments to provide structural characteristics include various synthetic polymers (e.g., VAPAM), diatomaceous earth products, sintered flyash, expanded slates and shales, vermiculite, and Perlite. None is a totally satisfactory substitute, but all are useful for certain applications. Therefore, the texture of designed soil tends to be best within the sandy classes to provide the proper pore space, drainage, and compaction-resistance requirements to compensate for the lack of relative structure.

The use of any of these components is strongly dependent on availability and suitability. Availability is determined by marketing surveys. Fortunately with the increase in interest and need for recycling materials, component supplies should increase with time. Transportation will remain a major cost in most localities, but may be offset by the lack of suitable alternatives. Suitability of the materials should improve as manufacturing processes are refined and standard specifications are developed (Craul and Switzenbaum, 1996). Finally, intelligence must be used in the design and actual construction of the soil profile within the space (volume) constraints of the project.

The Natural Soil Profiles

In order to understand the design and construction of an urban soil, it is well to know the natural soil profiles and their horizons. Appropriate natural soil profiles may act as models for urban soil design for various situations.

Introduction to Horizons. There has been a tendency in the past for landscape architects to specify a homogenous soil no matter what the depth. Below a certain minimum soil thickness, about 18 inches (45 cm), homogenous soil is probably satisfactory. However, most applications permit a greater soil depth and there is no reason to specify topsoil-quality material for the entire depth. It is both expensive and unnecessary. The better approach is to subdivide the soil depth (profile) into horizons just as with natural soils.

The Idealized Profile. The vegetated soil profile is basically comprised of organic horizons (*O horizons*) lying on the surface composed of organic litter in various stages of decomposition (Fig. 4.1). The upper or surface mineral horizon, composed mainly of mineral particles with incorporated humus, is the topsoil, or *A horizon.* Its thickness will vary greatly from one soil to another and may be totally absent in some. Below this horizon may occur a leached *E horizon.* Several transitional horizons may follow to the *B horizon(s)*, or subsoil. The *C horizon,* or parent material, is the relatively unweathered material in which the soil profile formed. Bedrock may or may not lie at the bottom of the profile. Some soils may have different parent materials for the several horizons forming lithologic discontinuities. This condition is common in alluvial and glaciated soils.

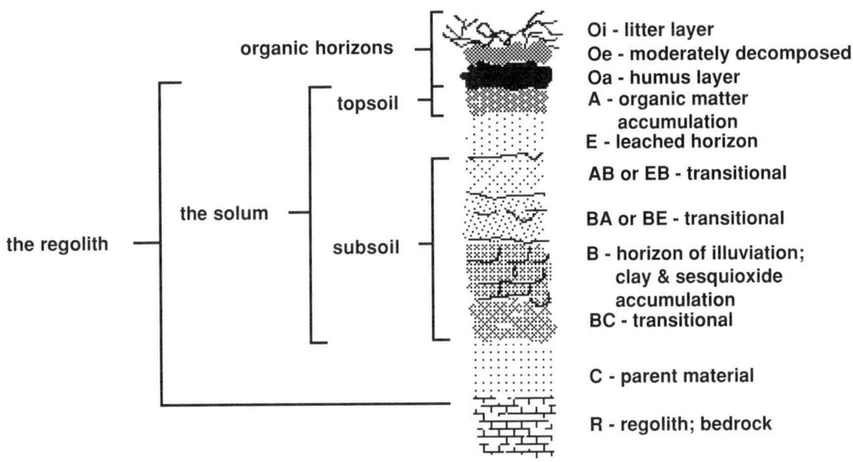

Fig. 4.1 The idealized soil profile.

In the design of urban soils, the primary concern is with only several facets of the ideal profile (Chapter 11):

(a) mulch emulates the organic horizons in an open-soil application
(b) a surface horizon designed much like the A horizon (topsoil) as the primary rooting medium for nearly all applications
(c) a subsoil that provides added rooting volume for plant stability, and nutrient and water storage to supplement the topsoil
(d) a drainage layer in deeper urban profiles that serves in place of the parent material

The horizonal design sequence would be O, A, B, and C in the more complex applications. The simplest would be a thin, homogeneous topsoil or all A horizon material. The omitted horizons of the ideal profile are generally not necessary in urban soil design, and depth limitations may prohibit their inclusion. Omitting the transitional horizons may present problems with interface creation between the horizons, but this disadvantage can be overcome by appropriate design and installation techniques discussed later in the chapter.

The Soil-Forming Processes. How soils form on the landscape is a valid approach to understanding the soil profile. There are five soil-forming processes that have the potential to occur anywhere on earth's surface. How they vary from one locality to another will be covered in the next section.

Table 4.1 briefly describes the five soil-forming processes that occur within the soil profile. The *leaching* of water-soluble and dilute-acid-soluble compounds occurs as water percolates through the soil in response to gravity and

TABLE 4.1 The Five Profile-Forming Processes

Process	Effect in the Profile
Leaching	Removal of water- and acid-soluble compounds; may precipitate below
Translocation of materials (Lessivage)	Physical movement and deposition of materials and substances
Synthesis of new compounds	Formation of substances from pedogenesis
Cycling of organic matter	The synthesis and decomposition of carbon compounds by organisms
Formation of structure	Aggregation of primary soil particles into the soil fabric

capillary forces. The compounds are removed from the upper portion of the profile and may precipitate out in a lower portion. *Translocation* is the physical movement of particles and substances, without solution, from one portion to another within the profile. Translocation may occur in any direction, but is dominantly in the vertical direction. The process is similar to "piping" in soil engineering but less drastic in magnitude. *Synthesis of new compounds* is the result of biochemical processes by soil-inhabiting organisms and the processes of disintegration and decomposition of minerals into new compounds. The *cycling of organic matter* from plant litter through the carbon cycle with the release of nutrients, heat energy, and water, leaving the residue of humus, is one of the more important cycles in nature. *Formation of structure* is the aggregation of soil primary particles of sand, silt, and clay into aggregates or peds, forming the structural form so necessary to the tilth of the soil and the formation of pore space.

The Factors of Soil Formation. Soils vary in their profile characteristics from one locality to another. To understand the spatial differentiation, five factors must be examined, including the climate of the locality, organisms (vegetation) found on the site and involved over time in the formation of the soil, the parent material from which the soil is derived, the topography on which the soil site occurs, and the time that the processes have been effective at the site. Table 4.2 indicates the active and passive factors of formation. Thus,

TABLE 4.2 The Factors of Soil Formation

Factor	Parameter	Relative Action
Climate	Rainfall, temperature	Active
Vegetation	Organic matter, organisms	Active
Parent material	Fabric, nutrients	Passive
Topography	Exposure, drainage	Modifier
Time	Duration, age	Integrator

climate as rainfall and temperature provide (or cause lack of) water and energy for physical and chemical weathering and biochemical processes by organisms acting on the passive parent material of the earth's mantle. The parent material situation is modified by the influence of topography on aspect, exposure, drainage, and slope position. Time is the integrator of all the others; the longer the time the processes have occurred, the stronger the influence of the climate and organisms.

Even though the discussion may appear to be more applicable to students in soil science, it is important for the landscape architect to recognize that a soil designed for the northeastern United States is not applicable to the southwestern United States unless a northeastern plant palette *and* soil is used with copious amounts of irrigation. It is strongly suggested that the plant palette *and* the soil be designed for the given climatic region or zone. Thus, it should be obvious that a xeric plant palette should have a xeric soil with little or no irrigation.

Examples of Natural Profiles. Brief descriptions of several major natural profiles provide models for the design of urban soils. If the soil profile has formed in nature under the factors of soil formation for a given region or locality, then why not attempt to emulate that profile for a specified palette? Obviously, some modifications may be required to "fit" the soil to the design as all natural soils are not necessarily satisfactory for models of designed soil. The profiles described are illustrated in Fig. 4.2.

Podzolization (Spodosols)

1. Occurs under cool, humid climate with forest vegetation. Leaching predominates with translocation of iron and humates the major profile processes. The leached horizon (E) is gray to almost white in color. A rusty reddish-brown color dominates in the accumulation zone (Bs). There is little structure development.
2. Organic matter tends to accumulate on the soil surface with decomposition slow and little, if any, incorporation into the surface mineral horizon. Organic horizons, especially the Oa is evident.
3. Mostly fungi-dominating, few actinomycetes and earthworms; ants, mites, centipedes and millipedes, voles and mice present. Ectomycorrhiza is usually abundant on conifers. Nitrogen-fixing nodules are usually lacking.
4. Energy absorption is relatively high but stored in an accumulated biomass with little dissipated by decomposition.
5. Soil tends to be acid, and soluble nutrients are in limited supply, especially calcium, phosphorus, and some micronutrients such as boron; nitrification is limited, with ammonium-nitrogen prevalent. The pH usually decreases with depth in the soil depending on the origin of the parent material. The dominant mineral residue is silica.

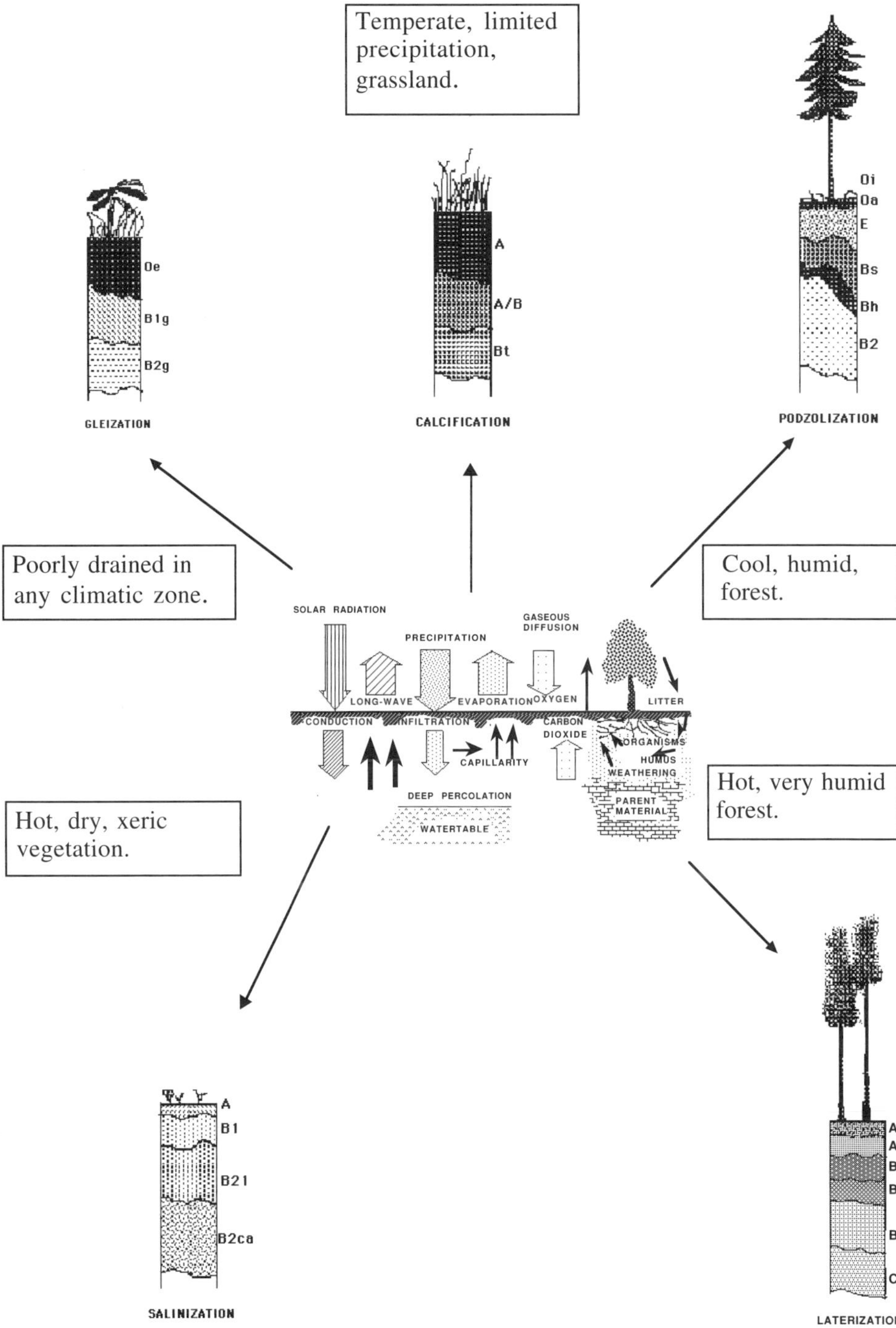

Fig. 4.2 The formation of different profiles under various climatic and vegetative regimes.

6. Moderate agricultural value requiring fertilizer and liming for intensive production; good forest soil for adapted species.
7. Designed mainly for an acidiphilous plant palette typical of eastern United States based on sandy texture.

Calcification (Mollisols and Alfisols)

1. Occurs under warm to hot, semiarid climate with grassland or savanna as covering vegetation. Leaching occurs only during precipitation events and soluble compounds are moved only to the subsoil with little removal from the entire profile. This horizon may be grayish in color. There is insufficient rainfall for complete removal as in podzolization or laterization. May occur in moist regions on highly calcareous parent material, but with the forest competing with the grassland. Fire is a definite factor. Where rainfall is adequate for a forest, the degree of leaching is greater and the depth to accumulated carbonates is greater than under grass.
2. Large amounts of organic matter recycled annually as grass dies back to root collar. Little accumulates on the soil surface. Organic horizons tend to be seasonal. Decomposition is rapid and complete with incorporation into surface mineral soil.
3. Very wide range of organisms present, especially bacteria. Earthworms, ants, and so on are abundant. Endomycorrhizas and nitrogen-fixing nodules are abundant. It's the best example of a fully diversified and active soil microcosm.
4. High level of energy absorbed in biomass with some accumulating in the incorporated humus of the topsoil, but most of it is dissipated on decomposition.
5. Soil is neutral to alkaline with high amounts of soluble nutrients present, especially calcium, magnesium, and (relatively) phosphorus. Free carbonates may accumulate in the subsoil; pH increases with depth in the soil.
6. Some of the most productive soils in the world. Major cause for concern is erosion from intensive agriculture.
7. Can support a wide range of plant palettes except eracaceous ones, but have the disadvantage of soil structure design problems and maintenance.

Laterization (Oxisols and some Ultisols)

1. Hot, wet climate with forest vegetation, including rainforest as typical. Occurs on a variety of parent materials.
2. Biomass is accumulated in the living portions of the ecosystem. Organic matter is decomposed extremely rapidly so that there is no accumulation

on the soil surface. Most roots exist at or near the soil surface to obtain nutrients before they are leached away by intense rainfall.

3. All components of the organism population are present. The invertebrates tend to be of giant size compared to temperate-region invertebrates.
4. The amount of energy cycled in the ecosystem is small since much is retained in the living biomass; however, the rate of cycling is very rapid.
5. Even though leaching and solution are intense, the soil pH remains relatively high because of the rapid release of bases in the organic matter cycle and those weathered from minerals of the parent material. Chemical processes are sufficiently intense to remove nearly all the soil silica. Bright red colors dominate due to the highly oxidizing soil environment.
6. Have been used in the past for *slash and burn* agriculture by natives with little detrimental effects. With increased intensity of farming, ecosystem nutrients are no longer preserved in forest cover, and they are quickly leached away.
7. Almost requires a permanent storied plant palette. Even ground cover should be protected by an overstory canopy. Obviously, the plant palette would be tropical.

Salinization (Aridisols)

1. Occurs mainly in arid regions with some to no rainfall. Soils are generally formed by physical processes with chemical processes occurring only during precipitation events. Soils tend to be sandy or skeletal (stony to rocky). Many soils exhibit a surface layer of gravel called *desert pavement.* The covering vegetation is sparse succulent or coniferous or woody species extremely efficient in the utilization of water. Some of the latter are green only during the spring of the year. Competition for water and nutrients is extremely intense.
2. Due to the sparse, scattered vegetation, little organic matter is cycled in this ecosystem. No organic matter accumulates on the surface and little is incorporated into the topsoil; thus, the latter are light colored (*ochric*). The remainder of the profile tends to be pale in color. Much of the litter may be blown away.
3. The organism population and activity are not abundant with the very little available energy source. Some of the organisms are especially adapted to these harsh conditions, forming a crust on the soil surface where they are able to take advantage of dew that condenses during the cool nights.
4. The cycling of energy is at a very low level and at a slow rate.

5. Soils tend to be very alkaline with accumulation of potassium, sodium, sulfur, calcium, and perhaps magnesium. Plants must be adapted to a high concentration of soluble salts.
6. Irrigation with freshwater is nearly always required if used for agriculture. The soils are productive, if irrigated, but many have been degraded by overintensive use.
7. The best plan here is to use native vegetation. Irrigation of landscape planting is being discouraged in many areas.

Gleization (Many Histosols and Aquic or Hydric (Wetland) Members of Other Orders)

1. Covering vegetation is composed of species adapted to poor aeration (flooding) of their roots or the roots occur exclusively on the surface. Poor drainage may be caused by climate (wet) and/or landscape position (floodplain, depressions, long footslopes, etc.). They can occur in most any climatic zone if a watertable exists near the surface.
2. Organic matter cycling is slow with most of the organic matter accumulated in the living biomass or mainly as *cumulose* materials in thick layers over mineral material that may be flooded.
3. Though anaerobic microorganisms dominate the gleyed soil, aerobic microorganisms exist in the upper portions where periodic aerobic conditions may occur. Generally, reducing conditions exist where the organisms gain their energy from inorganic reduction of oxidized compounds. These compounds include iron, manganese, copper, sulfur and nitrogen, and others. Yellow, gray, and bluish-green colors predominate in the gleyed horizons. The lowest portion may be nearly blue in some wetland situations.
4. The rate of energy cycling is very slow due mainly to the reduction conditions. Low-energy compounds are produced such as fatty acids, esters, and alcohols. Most of the energy is stored in the cumulose material.
5. The soil pH is usually strongly acid and nutrients are not present in abundance or in soluble form. Micronutrients may be present in toxic concentrations. Much of the nitrogen is volatilized (denitrification) by anaerobic bacteria.
6. Only specially adapted plants survive here. If drained and properly fertilized, it can become very productive muckland. However, subsidence and oxidation of the muck degrade the soil for agricultural use and are eventually abandoned, reverting to wetlands.
7. Design alternatives are limited to wetland use. Most are now regulated by federal and state laws unless they are constructed wetlands.

DESIGNING SOIL

The Topsoil Problem and the Need for Designed Soil

The former solution of covering the site material with topsoil and planting in that layer is no longer appropriate for two major reasons. The first is that true topsoil has become a scarce and expensive material with a high level of off-site environmental impact. Topsoil from other projects is no longer an acceptable practice. In most cases, the topsoil should remain for use on the original site rather than playing musical chairs with the topsoil among several different projects. The second and more important reason is spreading topsoil or any material as a layer over unknown subbase material is an unwise practice. The entire soil profile should be designed to accommodate the existing conditions and the planting plan and its palette. Opportunities for design alternatives may arise from information derived from the intensive sampling and analyses. The use of existing site materials as a component of a designed mix should be explored together with sources of recycled materials such as composted organic wastes and mineral materials. All of these actions would significantly reduce the environmental impact of restoration activities.

The Designed Soil Horizons in a Soil Profile

The horizons of a natural profile have the master designations of A, topsoil; B, subsoil; and C, parent material. Each may be subdivided into additional horizons. These represent either transitional conditions in the vertical gradient or the existence of some dominant pedogenic factor.

Topsoil. Designed soil requires a topsoil horizon. This horizon provides the most favorable growth conditions for roots. It should have the following characteristics:

(a) Low density and high porosity but not so light as to subside, or 1.2 to 1.4 Mg/m^3
(b) Moderate to high fertility
(c) High infiltration rate of at least 2 inches per hour
(d) Moderate to high organic matter content not exceeding 15% by weight
(e) Appropriate thickness of at least 4 to 6 inches.

Note that the organic matter content is not to exceed 15% by weight. The percentage is equivalent to a volume of approximately 33% or one-third on a proportional basis, depending on the particle density of the organic material. Many sources of organic matter have a particle density of about 0.22 Mg/m^3. In this case as in clay content, "too much of a good thing will kill you!"

Proportions of organic matter greater than one-third are very prone to settlement and eventual consolidation. The topsoil, even as a surface horizon, will not drain and becomes waterlogged, causing mortality of nearly all plants thereon. A flower pot or greenhouse mix is not appropriate for landscape applications simply because the physical conditions and requirements for drainage are entirely different in the two cases.

Particle Packing. To understand the critical volume of organic matter amendment to prevent settlement or compaction of a designed soil mix, the concept of particle packing must be examined.

A uniform sand (poorly graded) used as the basic matrix of a designed soil is said to have a single-grained structure. That is, the particles behave as individual entities; there is little aggregation without amendment and further structure development over time by soil organism activity, root activity, and the physical processes of wetting and drying, and freezing and thawing (Craul, 1992). It is primarily the initial physical condition of the designed soil that must concern soil designers. It is hoped the physical condition will improve over time by natural aggregation.

The first case of packing is that of minimal density and maximum porosity, which is *cubic* (open) packing. Each particle is in contact with six neighbor particles on the three orthogonal axes and has a *coordination number* of 6. The bulk density is $\pi\rho_s/6$ (where ρ_s is the density of the solid particles, usually taken as 2.65 Mg/m^3), and the porosity is 47.6% regardless of the diameter of the spheres. Bulk density is about 1.39 Mg/m^3. However, six-coordinate packing does not occur in nature and is unrealistic. The more realistic natural packing would probably be *pyramidal* or more likely *tetrahedral.* Both of the latter have the same density and porosity. By comparison, the tetrahedral packing is 12-coordinate, the bulk density is $\pi/3\sqrt{2}$ or 1.97 Mg/m^3, and the porosity is 25.95%. An intermediate packing of a cubic tetrahedral gives a porosity of 39.54%. This packing type is also likely unrealistic in nature.

If the sand is not uniform in size (diameter), having a range in particle diameters, the packing does not conform to the coordination described before. Depending on the array of particle diameters, coordination numbers of 30 or more and porosities less than 20% are possible. There is the tendency for self-compaction or compaction from external forces when the sand is well graded, especially when containing critical percentages of fine and very fine sand. Because the material is more dense, there is less pore space for occupation by organic matter amendment and greater settlement hazard results.

Sand particles are not perfect spheres in nature. They may range from rounded polygons to sharply angular. In the latter case, the orientation of the particles determines the packing, which is difficult to predict. Therefore, the discussion here sets out some model information that is very useful in the absence of other information, but is not entirely accurate for natural conditions.

The implication of organic matter amendment to a uniform sand, especially if most of its particle diameters are less than those of the sand matrix, is that the volume of the amendment should be no greater than the porosity of the packing type. Thus, organic matter amendment should not exceed 30% to 35% as maximum values. A more conservative value for the prevention of settlement or compaction would be between 20% to 25%.

It must be remembered that organic matter decomposes over time and may be reduced in the designed soil profile if not replenished by amendment incorporation or mulching. Addition of organic matter over time is better than an initial excessive amount. Greenhouse potting soil is an entirely different situation and the recommended proportions are not applicable to urban soil design.

Subsoil. This horizon provides bulk soil volume for additional waterholding capacity, nutrients, and strength for plant support. It should have the following characteristics:

(a) Moderate density of 1.4 to 1.6 Mg/m^3
(b) At least moderate fertility
(c) Moderate permeability of at least 1 in./hr
(d) A low organic matter content of about 2% to 3% by weight
(e) Appropriate thickness with a minimum of 12 in.

Parent Material. This horizon is included when it is desired to have additional rooting volume and soil strength. It is included when a specific layer is required for enhanced drainage of the profile, particularly in conjunction with underdrainage. It is also used as a filter to prevent clogging of any drainage structures. It should have the following minimum characteristics:

(a) High density but coupled with high porosity, usually a coarse to very coarse sand
(b) Low fertility
(c) Rapid permeability
(d) Organic matter content not required
(e) Minimum thickness determined by drainage requirements

How these characteristics are obtained is fully described in Chapter 5.

Generally, designed soils would have medium bulk density with a moderately high to high coefficient of permeability, k (Tables 2.5 and 2.6 in Chapter 2). Texture could range from silt loam, loam, sandy clay loam, sandy loam, and loamy sand to sand. To prevent excesses of compaction susceptibility and drainage, the silt loam and sand texture are promptly rejected. To be acceptable, the silt loam and loam would need to have a high percentage of water-stable aggregates (Craul, 1992) and the sand would have to be in the medium range with some organic matter content.

"Fitting" Horizons Together. As suggested earlier, omitting some transitional horizons in a designed soil may be desirable to reduce costs but creates abrupt changes in the properties between the horizons resulting in potential interfaces and the development of piping (Craul, 1992; Holtz and Kovacs, 1981). Some natural soil profiles have very abrupt changes and exhibit the problems associated with interfaces. A good example is the impeded drainage and root penetration that occurs immediately above a fragipan horizon (Bx).

The Interfaces. The presence of the transitional horizons and the subtle vertical gradient of texture, structure, and density within a natural horizon prevent the development of the interfaces. A natural soil profile may exhibit a gradual gradient from "top to bottom." Compensating factors also contribute to the gradual gradient. For example, a change to finer texture and lower organic matter content in the B horizon may be offset by an increase in the degree of favorable structure development. It is nearly impossible to create the subtle gradient within a horizon in a designed soil unless it is thick and with expenditure of a great deal of money and labor. A reasonable gradient in texture, pore-space distribution, and density is most desirable in a designed soil profile.

Abrupt changes in texture, structure or density, or all of them, from one horizon to another, can create an interface that may cause a sharp reduction in the flow of water. Perched water above the interface is a common result. The strongest example is the retention of water in the soil above a gravel layer even when not separated by filter fabric (Fig. 4.3a). The atmosphere exerts pressure upward against the water at the bottom of the soil in the gravel layer. The distinct change from small-sized pores of the soil to very large pores of the gravel "breaks" the water column. Soil water cannot enter the gravel layer until the soil becomes saturated to a hydraulic height above the interface exceeding atmospheric pressure. This height may amount to at least several inches. In some cases, the entire horizon must be saturated before the water flows into the gravel. Obviously, any roots growing in the saturated portion are subjected to anaerobic conditions.

Sharp contrast in bulk densities between horizons have the same effect as texture differences described before. Again, the water is impeded (perhaps root elongation as well) in the lower, denser horizon and the upper, less dense horizon becomes saturated. Not only do the plant roots suffer from the lack of oxygen, but on a sloping soil, the loading weight of the water can cause failure at the interface, leading to slumping or rotational failure.

Virtually the same situation occurs with a coarse-textured horizon over a fine-textured horizon (Fig. 4.3b). The permeability of the coarse horizon is greater than that for the fine horizon. Water flows rapidly through the coarse horizon but becomes somewhat impeded on reaching the fine horizon. The water then "backs up" in the coarse horizon and may become saturated if

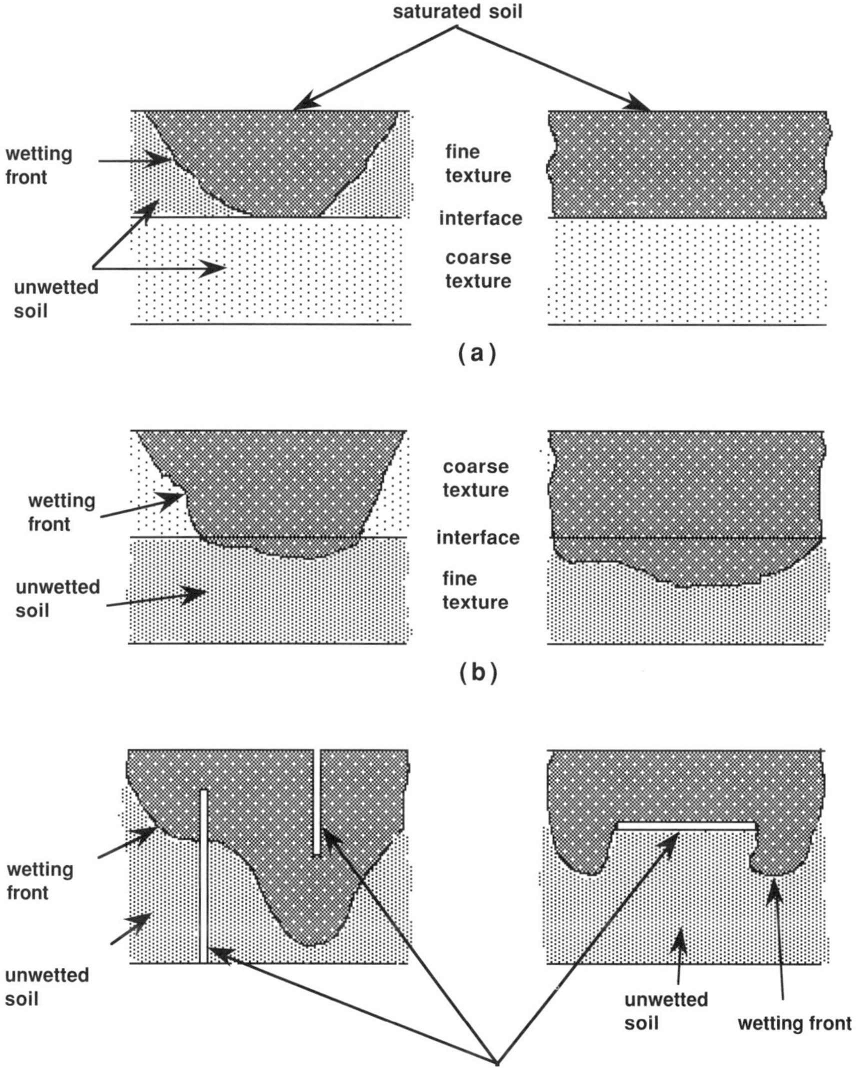

Fig. 4.3 Soil interfaces. (Adapted from W. H. Gardner, Washington State University.)

sufficient water is applied. The fine horizon becomes wetted slowly, and if insufficient water is applied at the surface, may not become wetted and remains relatively dry. This condition is illustrated in Fig. 4.4.

Figure 4.3c illustrates the effects of vertical interfaces connected and unconnected with the surface and a buried horizontal layer of coarse material on the movement of the wetting front.

An interface 24 to 36 inches (60 to 90 cm) or greater below the surface would cause much less poor growth conditions than one close to the surface such as 18 inches (45 cm). The abruptness of the interface, that is, the contrast

Fig. 4.4. Incomplete wetting of the soil profile in a raised planting of Bradford pear at the John F. Kennedy Center, Washington, DC.

between the physical properties of the adjacent horizons, also determines its effects: the greater the contrast, the greater the interface effect.

Interface Prevention. The creation of interfaces may be prevented in two ways. The first is to design adjacent horizons with only small differences in physical properties on either side of the interface. As a theoretical example only, a topsoil may be a medium to fine sandy loam and the next lower horizon a loam, but certainly no finer. The topsoil should have only a moderate amount of organic matter, probably less than 10% by weight and the lower one only about 4%. The next lower horizon would become coarser. The concept of minimum differences is applied to every interface except for the drainage layer, which may be at a sufficient depth so as to not to pose a problem to drainage and rooting with perched water above it.

The second method is accomplished during installation of the soil. Rototilling between horizons mixes the soil of two horizons so that a transitional layer is created rather than the interface formed when one soil material is laid on top of another without mixing. The specific procedure is described later in the section on "Construction and Installation."

There is little danger that significant interfaces will be formed if both of these procedures are followed. These procedures have been applied to both the Boston Commons Parade project. The specifications for these projects are provided in Chapter 12.

The "Piping" Factor. It is the migration of soil fines (potentially silts and clays and, sometimes, very fine sand) caused by water flow through the profile from one horizon to another or from the soil to a drain. Piping is somewhat similar to *stripping* that occurs with clay coats (cutans) in pedological processes. Either process may cause drastic changes in texture within the profile and may lead to clogging of fine pores in the lower horizons or filter fabric over drains. In dam and embankment engineering, piping may cause actual channels to form when the soil is saturated in earthen structures leading to eventual failure. The textures and the sequence of horizons must be carefully planned to prevent piping in a designed profile.

Soils are most susceptible to piping when the coefficient of permeability, k, is 10^{-4} cm/sec (Holtz and Kovacs, 1981). This value is also the approximate boundary between pervious and poorly drained soil under low hydraulic gradients. A k of 1.0 cm/sec is the approximate boundary between laminar and turbulent flow, and a $k < 10^{-7}$ cm/sec represents the permeability of impervious soils.

An empirical equation using D_{10} (the particle diameter in mm of the 10% weight passing the sieves from the particle size distribution curve of ASTM D-422-63), the "effective grain size," may be applied to *clean* sands (with less than 5% passing the No. 200 sieve), with D_{10} between 0.1 and 3.0 mm is

$$k = CD_{10}^2$$

where k is in cm/sec, D is in mm, and C varies from 0.4 to 1.2 with an average of 1. Holtz and Kovacs (1981) show k values for clean sands to be in the range of 1.0 to 10^{-3} cm/sec, and very fine sands and other common soil textures from 10^{-3} to 10^{-7} cm/sec. Critical conditions can occur when the hydraulic gradient approaches unity, leading to piping and erosion. Unfortunately, the gradient in many soils is near unity for most field conditions. The implication here is that the coefficient of permeability for each horizon in the profile should be greater than 10^{-4} cm/sec and should not greatly differ from one horizon to the next but should not be uniform (unity) either; thus, there should be differences in k from one horizon to the next (Fig. 4.5). The danger of piping occurs when a topsoil has a large organic matter content (probably near 50% or more) or other fines (very fine sand, silt and/or clay total of more than approximately 35%) coupled with a high permeability. Some protection against clogged drains and serious migration from a topsoil may exist with a subsoil having a lower coefficient of permeability and less fines. The intention is that the subsoil together with the parent material ho-

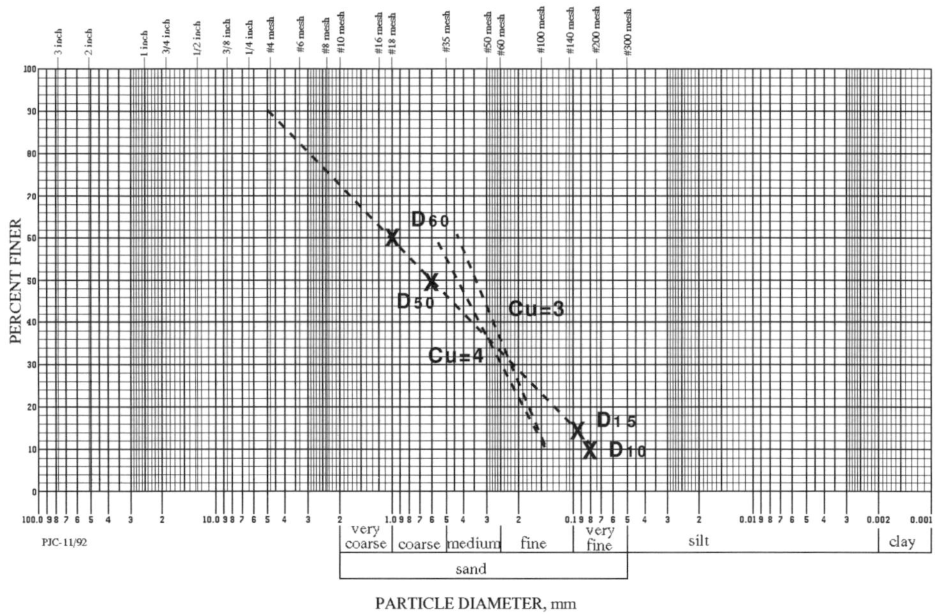

Fig. 4.5 The particle size distribution curve showing D_{10}, D_{15}, D_{50}, and D_{60} for a given soil material. Sieve sizes are listed across the top and diameters in mm are listed across the bottom. The y axis, or left vertical, is the percent of total weight passing the various sieves.

rizon or drainage layer act as a protective filter in addition to providing necessary additional rooting volume for water, nutrients, and mechanical stability.

Holtz and Kovacs (1981) provide four requirements for a protective filter:

1. The filter material should be more pervious than the material above it in the profile in order that no hydraulic pressure will build up to disrupt the filter and adjacent structures.
2. The voids of the inplace filter material must be small enough to prevent upper material particles from penetrating the filter and causing clogging and failure of the protective filter system.
3. The layer of the protective filter must be sufficiently thick to provide a good distribution of all particle sizes throughout the filter.
4. Filter material particles must be prevented from movement into the drainage pipes by providing sufficiently small slot openings or perforations, or additional coarser filter zones if necessary.

The last requirement could be fulfilled by some of the fabric filter materials that have been developed. The material is wrapped around the perforated PVC pipe or, as some engineers prefer, encase the pipe in medium to coarse gravel, which is further encased in the filter fabric.

Gradation requirements for protective filters are given in Table 4.3. The first ratio, R_{15}, ensures that the small particles of the material to be protected are prevented from passing through the pores of the filter; the second ratio, R_{50}, ensures that seepage forces within the filter are reasonably small. By the use of multilayers (horizons), the filtering effect of the soil is enhanced.

The X's in Fig. 4.5 designate the particle diameters for the soil finer (passing) at 10%, 15%, 50%, and 60%, which are 0.08, 0.095, 0.62, and 1.03 mm, respectively. To calculate the ratios defined in Table 4.3, the particle size distribution curve is plotted for each of the two materials (the filter and that which is to be protected), and the diameters are obtained manually and plugged into the equations. Iteration and reiteration of the particle size distribution curves may be necessary for "fitting" the two materials. For uniform filter material, the curve must begin and end as illustrated in Fig. 4.5 by the dotted line between the No. 4 and No. 200 mesh sieves, and the slope between the two points should be according to the $C_u = 3$ or $C_u = 4$ lines shown. A sigmoid curve within the rectangle delineated by the dotted diagonal is the general result. The various criteria for particle size distribution analysis are discussed in more detail in Chapter 5.

A brief example to the preceding discussion may be helpful toward clarification of the iteration process. Figure 4.6 shows the particle size distributions for a proposed planting soil and the coupled drainage-filter horizon. The solid curve in Fig. 4.6 is the planting soil and the dotted curve is the drainage-filter horizon. The relation between the two curves follows that of Table 4.3 for

TABLE 4.3 Gradation Requirements for Filter Materials

Filter Material Characteristics	R_{15}	R_{50}
Uniform grain size filters, $C_u = 3$ to 4	—	5 to 10
Graded filters, subrounded particles	12 to 40	12 to 58
Graded filters, angular particles	6 to 18	9 to 30

$$R_{15} = D_{15} \text{ of filter material}/D_{15} \text{ of material to be protected}$$

$$R_{50} = D_{50} \text{ of filter material}/D_{50} \text{ of material to be protected}$$

Note: Maximum size of the filter material should be less than 76 mm (3 in.). Use the minus No. 4 fraction of the base material for setting filter limits when the gravel content (plus No. 4) is more than 10%, and the fines (minus No. 200) are more than 10%. Filters must not have more than 5% minus No. 200 particles to prevent excessive movement of fines in the filter and into drainage pipes. The grain-size distribution curves of the filter and the base material should be approximately parallel in the range of finer sizes.

C_u is the coefficient of uniformity, which is a measure particle size gradation. The smaller the number, the more uniform the particle size or the smaller the range in particle size distribution.

Source: Holtz and Kovacs, 1981.

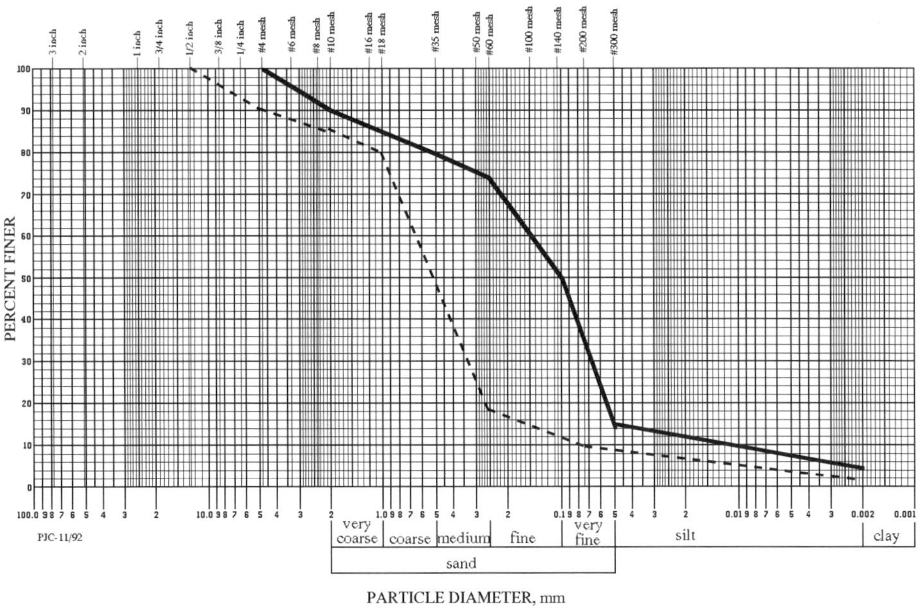

Fig. 4.6 Particle size distribution curves for a planting soil and its drainage-filter horizon.

the R_{50} ratio and the use of a uniform grain-size filter. The drainage area has an approximate $C_u = 3$.

Horizon Thicknesses. Horizons must have a minimum thickness to be effective in a designed soil. Topsoil thickness is usually most critical in a designed soil. A topsoil horizon ought to be at least 4 inches (10 cm) thick as an absolute minimum with 6 to 8 inches (15 to 20 cm) as being satisfactory and 12 to 18 inches (30 to 45 cm) as optimum. Some topsoil specifications may require a thickness of 24 inches (60 cm) (see Chapter 12).

The Topsoil. The topsoil thickness depends mostly on the plant palette composition and the use of the site segment, that is, bedding plants, ground cover, trees, turf area with heavy traffic, or a mixture of plants (Chapter 11). Generally, the topsoil should be thicker for bedding plants using annuals such as 12 to 16 inches (30 to 40 cm). Where perennials or woody shrubs are used, the topsoil may be thinner. Much the same recommendation holds for trees. Just because trees tend to have deeper roots does not mean that the topsoil must be as deep as the deepest roots. It has to be only about 6 inches (15 cm) thick to accommodate the needs of just the fine feeder roots as in the natural forest (Craul, 1992).

The Subsoil. Thickness is determined mainly by the amount of soil depth (volume) required to meet the waterholding capacity and anchoring requirements of the plant palette. The subsoil may be as thick as 24 to 36 inches (60 to 91 cm) in some instances where the need for irrigation is not desirable. On the other hand, it need not be thinner than 6 inches (15 cm); otherwise, the difference could be made up by topsoil with little, if any, additional cost since installation costs of an another horizon is omitted.

The Parent Material or Drainage Layer. The third soil horizon may also be used as "filler," but more often is designed as a drainage layer at the bottom of the profile. The thickness is normally at least 6 inches (15 cm) and may be 12 inches (30 cm) when filter design requires a thicker layer. Greater thickness is normally unnecessary and a thickness thinner than 6 inches (15 cm) reduces its drainage effectiveness.

Advantages and Disadvantages of the Designed Soil Profile. The designed soil profile is most advantageous to the deep-rooted plant species. The profile does the best job of meeting the growth requirements of these plants. It is an emulation of natural conditions. The topsoil provides an optimum medium for the fine roots and the subsoil provides additional amenities and the parent material or C horizon provides drainage and filter protection.

The disadvantages of the designed profile include the need for more intensive planning and testing of the materials and a more complex set of specifications; each horizon requires its own specification. The installation process, discussed later in this chapter, is more complex. The latter requires more day-by-day attention on-site by the landscape architect.

Obviously, the designed soil profile costs more. The costs include the complex design process, preparing elaborate specifications, the required varied source materials and their testing, more complex installation operations with the need for closer supervision and conformance. However, the payoff is the greater potential for a successful design with greater longevity.

The Homogeneous Soil

A homogeneous soil is one that is isotropic having no differences in three-dimensional space perhaps except for limited variation in depth. It is not differentiated into horizons. The texture, organic matter content, bulk density and permeability (drainage), and other characteristics are more or less uniform throughout. The drainage from the bottom of the profile is dependent on the material below it or the installed underdrainage.

Though the final design is simple, much care must be exercised in choosing the appropriate texture, organic matter content, and any structural amendments. It is in reality a compromise design (unless shallow—about 18 inches (45 cm) or less) because the upper portion must provide the attributes of a topsoil and the lower portion the attributes of a subsoil; this is difficult to do when the depth is greater than 18 to 20 inches (45 to 50 cm). A homogeneous soil 36 to 48 inches (91 to 122 cm) deep or deeper is totally inappropriate based on present-day soil technology.

Further, due to the inherent homogeneity, the soil is very dependent on the drainage characteristics of the material that lies below than with the designed soil profile, which has its own drainage layer. Of course, a drainage layer can be installed below a homogeneous soil if necessary, but then homogeneity is lost or at least reduced. Thus, underdrainage is very critical with a homogeneous soil design. This fact becomes more evident when the distribution of moisture in a homogeneous profile without drainage is examined. Harris (1983) provides an appropriate illustration (Fig. 4.7). It should be noted that the percentages at the bottom of each column indicates near saturation of the pore space. A drainage layer in contact below the columns would reduce the moisture percentage accordingly.

The criteria for the design of a homogeneous soil follow those for a topsoil with some minor modifications. The texture should be somewhat coarser and the organic matter content somewhat less than for the topsoil of a profile. Also, it may be more necessary to include a structural amendment to prevent settlement and compaction, particularly in the lower portion, as compared to the topsoil, which has less potential to do so.

Advantages and Disadvantages of the Homogeneous Profile. The design requires a single specification and is easy to install except for using lifts, which should be used in most installations. It is most appropriate for shallow installations and a simple plant palette.

It may not provide optimum soil conditions for all species in a diverse plant palette. A portion may do well and another portion may fail. Amendments usually must be added to the entire volume not just the surface horizon. The design, though simple, introduces some problems not inherent to a de-

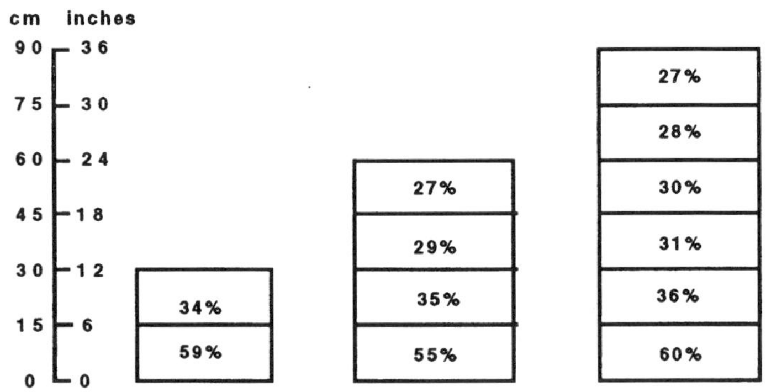

Fig. 4.7 Distribution of moisture in homogeneous soils of varying heights. (Redrawn from Harris, 1983.)

signed soil profile. The uniformity can cause potential piping problems because the hydraulic gradient will tend toward unity throughout the profile and fines can be transported directly into the drainage system. The physical uniformity creates greater potential for settlement or subsidence compared to the designed profile.

Profiles for Various Applications

It should be apparent from the foregoing discussion that the final designed profile becomes dependent on many factors. These factors include the *element situation* of the planting: open, curbside, covered plaza, or sidewalk; whether it is to receive traffic; the structure of the plant palette; on-grade or on-structure; irrigated or nonirrigated; and so on.

A wide array of profile applications are provided in Chapter 11. The chapter is a useful guide to landscape architects for the planning and designing of urban soil.

SUSTAINABLE SOIL

The definition of sustainable soil may be as follows:

> Sustainable soil is comprised entirely of recyclable products alone or in a mixture with derelict soil material, the latter of which is useless without supplementation. It contains little, if any, nonrenewable resources.

As yet, technology and economics have not yet permitted us to meet every criterion of sustainability. There is still opportunity to make use of derelict soil as part of the mix if it meets certain specifications; hence, the inclusion in the definition.

Some examples of sustainable components for use in a designed soil are (Craul, 1994b)

- Sand from river dredgings or sand pits or recycled ground glass
- Composted organic material derived from biosolids, selected municipal yard wastes, food-processing sludge, and so on
- Derelict soil material not otherwise useful such as selected mine tailings, basal glacial till, and so on
- Dehydrated washings from aggregate plants, certain smoke stack fly-ashes, and so on

Sustainable Soil Design

Various mixtures can be formulated that meet the biological requirements for plant growth, are resistant to compaction, and do not incur additional maintenance burden beyond that required by ordinary natural soil. The soil is constructed in layers, more properly horizons, to emulate the natural soil profile appropriate for the project objective.

When used as a mix in conjunction with mineral materials from other sources, a topsoil may be constructed. The mineral material must be specified to assure its usefulness and provide consistent data for prediction of the characteristics of the final mix. This portion of specification preparation is rather straightforward since this type of specification has been used by soil engineers for many years. However, the characteristics in the end product desired by the landscape architect and horticulturist are quite different from those desired by the engineer. Hence, the quantified engineering specifications differ from the quantified landscape specifications, even though the criteria list in the specifications are the same.

The formulation of a constructed topsoil must be accomplished with the consideration of many variables. These include the physical and chemical characteristics of the mix components and the matching of the mix to the landscape applications. The latter conditions are determined by topography, the soil physical and chemical characteristics and its drainage, the character of the bedrock or underlying surficial geologic material, and its depth.

The Mineral Component. The mineral component constitutes the primary fabric of the soil. It consists of the sand, silt, and clay particles. Organic matter is also a contributor to the soil fabric but is specified separately as an amendment. Here, the composition, size (diameter) distribution, pH, nutrient contents, cation exchange capacity, and source are included.

The mineral component may also include a subcomponent to enhance the structure of the mix. The subcomponent may consist of pumice, diatomaceous earth, sintered flyash, or expanded slate. The products increase the pore space of the mix. This in turn increases the waterholding capacity and drainage of the mix as well as its resistance to compaction (Craul, 1997).

The various properties specified for the mineral component include texture or particle size distribution, reaction (pH), organic matter content if a native soil is used in the mix, nutrient content, soluble salts, stone and debris content, and contaminant concentrations.

The Organic Component. In the past, shredded and screened peat moss has been a very satisfactory organic matter source. This material is now expensive and sources are declining because of the concern about the environmental impact of peat bog mining. Substitutes such as composted organic wastes, both agricultural and urban, have become available (Composting Council, 1996; Hornick et al., 1979). However, no specific standards of processing or manufacture have been promulgated, and a wide variety of materials have been marketed (Craul and Switzenbaum, 1996). Some materials have been given trade names with their composition as proprietary information. Some are documented by agricultural research station and university studies, but their composition remains proprietary. An organic component specification must be carefully and fully stated to ensure the acquisition of an appropriate organic material until standards are developed. The following elements are discussed with this necessity in mind.

The properties specified for the organic component include the carbon/nitrogen ratio, stability, maturity, odor, mineral and organic content, fineness, reaction (pH), cation exchange capacity, nutrient contents, heavy metal content, and potential pesticide residues (Craul and Switzenbaum, 1996).

The diversity of available organic materials creates a problem: Most producers have limited amounts in stock over a short time span. A project that requires large amounts in a short time, as during the installation stage of construction, may find inadequate amounts available from a single supplier and cannot easily go to another supplier to get similar material. Similar or nearly identical materials may be available from several suppliers, but the locations are dispersed and transportation costs become excessive. In some cases the budgeting for large transportation costs may be justified on the basis of obtaining sufficient quantities of the best material. Thus, several similar, not identical, materials from different suppliers may need to be specified to provide sufficient quantities to complete the project. Obviously, production of standard materials would eliminate this major problem in the future.

The Biological Component. Most of the time the biological component is totally ignored with the thinking that soil organisms can take care of themselves (Craul, 1997). Inoculation of relatively sterile material will occur over time. In many cases, this natural process does occur; in others, the failure or its prevention may contribute to the early demise of the plant palette by nutrient deficiencies or by direct pathologic effects. With the advent of sustainable soil, the soil organisms may be enhanced by the use of compost as the organic component. On the other hand, it is good professional practice to ascertain the biological status of the mix and to inoculate when necessary. Various organisms such as the mycorrhizae, beneficial bacteria (*Rhizobium* spp.), and so on, have been successfully inoculated into soil.

The translocation of a tree from the forest (or any other plant from its natural environment) by way of the nursery to the urban environment may leave out some important soil microorganisms. One of the most important to plant health is the mycorrhizae (Marx, 1997). Originally discovered only on tree roots on infertile soils, modern microscopic techniques have now discovered them on nearly all plants. The mycorrhizae are a group of fungi that live symbiotically on plant roots, supplying the host with enhanced amounts of nutrients, other plant growth compounds, and water because of their increased absorptive surface area and unique metabolic processes while deriving their carbohydrates from the host. Without mycorrhizae, a tree will use 100 times the photosynthate to produce the same size root system as it would with mycorrhizae present on the roots (Marx, 1997). Only about 15% of the total photosynthate produced by the tree is used to feed the mycorrhizae.

There are two types of mycorrhizae. The *ectomycorrhizae* (primarily *Pisolithus tinctorius*) were the first group identified since they occur on the outside of the roots as a continuous sheath and can be detected by the naked eye or by a pocket magnifier. They occur on over 2000 species and only on trees, including pine, oak, beech, birch, and spruce. The effects of the mycorrhizae are to increase drought tolerance, deter root diseases, and maintain root juvenility. The last is important because roots must continue to renew themselves to maintain the health of the tree. This group is reproduced sexually through aerial parts as puff balls and various mushrooms. The spores are disseminated by wind, birds, small animals, and so on. The second group is comprised by the *endomycorrhizae,* or more commonly, the VAM (Vesicular Arbuscular Mycorrhizae) group. This group occurs on agricultural, ornamental, and horticultural crops including vines, grasses, legumes, desert plants, berries, and tree species such as maple, apple, cherry, citrus, coffee, peach, pear, boxwood, dogwood, hawthorne, juniper, rose, and virburnum. This group reproduces by spores within the soil and is spread by soil fauna. VAM cannot be cultured in the laboratory as the *ectomycorrhizae;* they must be cultured on plant roots under sterile conditions then extracted for inoculation.

Inoculation of mycorrhizae may be done in the nursery on seedling stock, inoculation of the soil mix, or inoculation of the soil after planting. Combining the inoculate with fertilizer has been shown to be very beneficial to plant establishment and growth (Marx, 1997).

Because an urban soil or a designed sustainable soil may be relatively low in soil organism constituents and concentrations, it is advisable to consider the inoculation of the soil in most cases. Based on a review of the pertinent literature, it appears inoculation after planting is the most reliable procedure (Marx, 1991; Marx et al., 1997; Smiley et al., 1997). However, in the case of planting in paved areas, it appears that inoculation during soil mixing is the best procedure. Specifications and procedures are provided in Chapter 5 and the Appendix, respectively.

Testing and Analysis

Intensive testing and analysis and inspection of these materials are required because of the following:

- The chemical and physical composition of these materials are generally unknown and probably highly variable, and they are to be used in proximity to human activity and would exist on land potentially draining into water bodies.
- The material must be compatible to the objectives of the restoration as well as any plant palette involved in the design.

Therefore, detailed specifications must be provided and strict conformance enforced. Use of these materials cannot be haphazard. Standard tests are provided in the Appendix.

Costs of Sustainable Soil

Experience has indicated that we need to spend more money on soil. Some of this impact is caused by design considerations and some is caused by factors outside of the design and project realm and in the realm of economics. The ratio of plant palette costs to soil costs requires elucidation. In one notable planting failure, the ratio was about 7.5:1. The costs of specifying, obtaining, testing, and inspecting of soil should amount to a greater proportion of the project budget. The future ratio may approach 3:1, with the hope that the plant budget is not cut to reduce the ratio to that level.

Future of Sustainable Soils

Much data and information remain unknown about these materials when used as a sustainable soil. Empirical research and pilot studies should be undertaken to establish appropriate standards and guidelines and to develop cost-benefit data for wider, more comprehensive use by landscape architects, planners, public officials, and the producers of the materials (Craul, 1994a).

CONSTRUCTION AND INSTALLATION OF THE URBAN SOIL PROFILE

The importance of the construction and installation phase of soil design is greatly underestimated in many cases. The operation is more or less left to the whims of the landscape contractor, who normally will attempt to carry out the process by the most convenient and time-saving method. The choice does not always provide the proper results for a successful project. Specifications of soil installation must be carefully developed and supervised just as closely as is provided for the planting of stock; in fact, the two go "hand in hand." It has been the experience of the author that where a well-specified soil is carefully handled and installed, the degree of project success is greatly enhanced. Evidence of the enhanced success includes quick recovery from transplant shock with vigorously growing plants and virtually no mortality, appears almost immediately, and is prolonged over the time span of the project. The constant replacement of dead planting stock does not bode well for the planting and soil design. Obviously, something is wrong (*"there's a skunk somewhere in the woodpile!"*).

The details of the construction and installation phase depend very much on the design and type of soil that is to be installed.

Imported Topsoil

Though this process is not encouraged, there are circumstances when topsoil from an area undergoing development for which there is no topsoil use may be imported. Analyses, including texture, organic matter content, pH, and potential contaminants, are necessary before the use of the topsoil is contemplated. The analyses, especially texture, are needed to ascertain the compatibility of the topsoil to the on-site material that will become the base material. The textures should not be drastically different; that is, a sand should not be placed over a clay and vice versa.

The base or on-site material should also be analyzed for drainage, pH, texture, and contaminants. The common problem is the drainage of the on-site material, especially on restoration sites, which may be uniformly compacted on the one extreme or highly stratified with loose debris mixed with compacted lenses at the other extreme.

If it appears that the on-site material is acceptable as a base material, then the depth of imported topsoil applied is dependent on the amount available and the plant palette. As mentioned earlier, the topsoil should be at least 6 inches (15 cm) thick for most applications, but thicker for annual or perennial planting beds. If the on-site material is poorly drained, the drainage must be improved by creating water pathways through the compacted layers or by installing a drainage system. Augering holes through a compacted layer to a lower, better-drained portion of the profile and backfilling with gravel is a simple solution. Refer to Chapter 7 for more drainage details.

The effect of the topsoil is enhanced and a potential interface eliminated if the on-site base material is first loosened. Then a first 3-inch (8 cm) layer of topsoil is placed and tilled with the loosened on-site base material. The remainder of the topsoil is carefully placed.

Amendment of On-Site Material

In some cases, the on-site material may be made satisfactory for a planting medium with simple amendments. Medium to sandy texture, good drainage, adequate depth, absence of contaminants and debris, and pH extremes are all acceptable criteria. The usual deficiency is organic matter. Amendment and incorporation with composted organic material to form a topsoil meet this need.

The organic amendment process requires elucidation. As in the case of topsoil importation, the on-site material must first be loosened and then a 3-inch (8 cm) layer of organic material spread over the surface. This material is rototilled or disk harrowed until thoroughly mixed with the on-site material. A second 3-inch layer is spread over the surface and again thoroughly rototilled. Thorough mixing by rototilling is best accomplished by tilling in two directions at right angles to each other. This process has been successfully carried out by the National Park Service in the restoration of federal land used in construction siting for the Washington, DC, Metro System (J. C. Patterson, National Park Service, personal communication).

Fertilization with proper elements and amounts can overcome any nutrient deficiencies inherent in the on-site material. The soil mix is planted as required by conventional means.

Totally Designed Soil

A totally designed soil infers that the entire profile of several horizons or a homogeneous soil of several feet in thickness is designed and to be installed as such. The two situations permit alternative methods of construction and installation and are discussed separately. The designed soil is composed of various components and must be mixed.

Mixing. The mixing of components to obtain the desired soil composition may be accomplished by several methods, depending on the amount of soil required and the equipment available. The materials are best mixed when slightly moist without dust and certainly should not be mixed when wet. Continuous inspection of the product is required in all methods.

Concrete Mixer or Ball Mill. If small amounts of mixed soil are required, a standard concrete mixer or ball mill provides adequate mixing without a large amount of labor. Five rotations sometimes provide uniform mixing. The textures and moisture contents of the components determine the ease of mixing. Using a front-end loader has been found not to provide uniform mixing. Isolated cells of unmixed components are commonly found in the product and are cause for rejection.

Tub Mill or Mixer. The tub mill or mixer is a large machine with the tub being at least several feet in diameter. The larger mill is mounted on a trailer driven by a large diesel engine. The mill capacity may be several cubic yards. This equipment is used when large quantities of well-mixed materials are required (Couenberg, 1994). The uniformity is good.

Windrowing. For projects covering large areas or distances, such as roadsides, windrow mixing can be accomplished by a road grader or an angle-bulldozer. The separate soil components are placed side by side in windrows from the dump trucks. Windrowing with the grader of each component may be necessary, as it is placed to keep the windrows parallel to each other. The components are mixed by turning over the windrows one on to the other, first from one side then the other until a uniform mix is obtained. The skill of the operator determines the quality of the results. A large mixing area must be available, preferably on-site. This method has been employed by the author on one large project (107 acres or 43 ha) with satisfactory results (see Chapter 12).

The Soil Profile. The soil profile is best installed by horizon to make certain that the materials are in proper sequence and horizontally continuous. Wide variations in horizon thickness and presence of discontinuities are not acceptable for a designed soil.

Because of differential deposition of material and compaction in the spreading process, greater uniformity of *in-situ* material can be obtained by placing the soil in 4- to 6-inch (10 to 15 cm) *lifts* rather than dumping and spreading the entire amount of material at one time to obtain the correct horizon thickness. If horizon components need to be mixed on-site, spreading in layers then rototilling to mix the components one lift at a time facilitates uniform mixing and proper sequence. The process may be considered time-consuming, but the author successfully employed this technique on another project (see Chapter 12, the section on "Boston Commons Parade Garage"). It is especially important to mix the lifts between horizons, that is, the upper or last of the lower horizon and the first or lowest lift of the upper horizon. The operation eliminates the interface between the two horizons and forms a gradient between the two sets of characteristics. The profile is built up in vertical sequence.

The Homogeneous Soil. A homogeneous soil may be installed by at least two different methods. In one method, mixing of the components must be accomplished before installation.

Placement in Lifts. Obviously, a homogeneous soil will be thick, being at least 18 inches (45 cm) or more thick, and usually several feet. The material may be placed in 4- to 6-inch (10 to 15 cm) lifts as in the designed profile for the same reasons concerning the uniform distribution of materials and uniform compaction. If the material is mixed on-site, the components may be mixed as lifts, but the danger increases for nonuniform material within the profile more or less destroying the homogeneity of the profile. Attention to details and alert, continuous inspection are required.

Simultaneous Placement of Plants and Soil. In some situations, it may be advantageous to place the upper part of the homogeneous soil and the plants simultaneously. This process is recommended rather than installing the soil to final grade, then traveling over the final grade to excavate, and placing the planting stock. The major advantage to the process is that there is no traffic on the final grade material once it is in place, and it actually saves time. There are several conditions that must be met to be able to carry out the operation: (1) the soil must be filled to at least rough grade below the deepest root ball (usually the trees); (2) the soil at the rough grade must withstand both foot and vehicle traffic and not be compacted excessively; and (3) the remaining soil to be filled to final grade must be homogeneous. Deep tree root balls are placed directly on rough grade, whereas pedestals of compacted soil are built for placement of shrubs and small trees with smaller root balls (see Chapter 12, the section on "Battery Park City South Cove, New York"). The soil is filled to final grade around the root balls as they are placed. Mechanical tamping of the partially filled soil around the root balls to at least one-third to one-half the root ball height stabilizes the root balls. Ground cover, if part of the palette, is planted on the final grade as the areas are filled.

The process can be applied to a designed soil profile provided the preceding three conditions can be met.

Sequencing of the Installation

The installation of the soil, and the plants, too, may be accomplished in a way that protects both the soil and the plants. The concept is to start in the middle of the planting area and work outward so that the soil and planting stock are not disturbed or injured by subsequent operations. Running personnel and equipment back and forth over a site does no good and costs money and time. If it is a linear-shape area with limited access, start at the back end and work outward. If access is not a problem, begin in the middle and work toward both ends. The process may seem obvious and trivial, but it is amazing to see how many project operations do not take advantage of some common sense.

For the operation to be successfully completed, it is necessary to have all soil materials and planting stock ready to go.

Testing of Soil

Various tests should be conducted as the soil is placed and after completion of installation. They include mainly bulk density, permeability of the horizons, texture, organic matter content, and pH. More details are provided in Chapter 5 and the Appendix.

XERISCAPE DESIGN

Planting in xeric landscapes poses some challenges and risks. The first challenge is to minimize the need for irrigation. The choice of appropriate xeric plant species greatly reduces the irrigation requirement. A totally native xeric plant palette eliminates the need for irrigation (Fig. 4.8). On the other hand, there are irrigation designs that are more water-efficient than others and may be used where some irrigation water is required, even in humid temperate regions. It must be recognized that freshwater is becoming a highly competitive resource in many parts of the world and not necessarily confined to those semiarid and arid areas where water is naturally in short supply. Even humid areas experience water shortages due to intervals of drought or due to contamination of the water supply.

There are examples of deliberate xeriscapes created for special effects in humid temperate regions. Obviously, these require a different approach to planting and management than the natural xeriscapes and are not discussed here.

Fig. 4.8 A desert landscape for a residence, Scottsdale, AZ.

The Climatic Zone

Annual rainfall is generally less than 15 inches (380 mm) and may as little as 6 inches (152 mm). Summer air temperature exceeds 90°F (32°C) and relative humidity is low. Winter temperatures may be mild in southern portions and extremely cold in northern temperate areas. Soils may be frozen in winter in the northern areas. With many clear days, solar radiation is very high. Xeriscapes are usually quite windy. Hence, the high potential evapotranspiration levels.

The Plant Palette

To be sustainable, the plant palette must be one that contains moisture-efficient species and withstands intense radiation levels, wind, alkaline soil, and perhaps frequent frosts (Perry, 1992). The dominant species and cultivars include the various succulents (cacti, and so on), palms, the "hard" conifers (i.e., pinyon pine and juniper), evergreen and deciduous oaks, eucalyptus, and many others. Planting techniques for most of the species follow the conventional ones previously discussed. Palms are the exception and their planting is covered in Harris (1991).

The Arid Soil

The soils of the semiarid and desert areas exhibit *aridic* and *xeric* moisture regimes. The aridic moisture regime is defined as being dry more than half

Fig. 4.9 Desert pavement as a mulch, Scottsdale, AZ.

the time in most years and have insufficient moisture to grow crops without irrigation (Buol et al., 1989). The xeric soils are dry in summers and moist in winters. Stored soil moisture must be adequate to carry crops into the spring or early summer. Otherwise, irrigation is required.

The xeriscape soils are characterized by a thin A horizon with low organic matter content and covered by a desert pavement of loose gravel or stones. Wind erosion removes the soil fines. There are no surface organic horizons. The soil texture tends to be sandy. The horizon has a weak granular or crumb structure. The subsoil has prismatic or columnar structure that forms from deep cracking as the soil dries in summer. A layer of free carbonates occurs near the surface on upland sites. The soil reaction is usually alkaline to strongly alkaline (pH > 7 and approaching 8 or 9, and perhaps 10).

The use of desert gravel as a mulch rather than an organic one is recommended for xeric planting (Fig. 4.9). Not only does it provide the effects of mulch, but better fits the aesthetic quality of the natural xeric landscape. It is also erosion-resistant under the heavy thunderstorms typical of the region.

5

SOIL SPECIFICATIONS

The soil must be specified, the source investigated and the material tested and analyzed, handled and installed, and maintained just as carefully as the plant palette and its stock.

—P. J. Craul in a presentation of soil specifications for the J. Paul Getty Fine Arts Center, Los Angeles, June 22, 1994.

INTRODUCTION

There are two opposing considerations in the development of soil specifications for use on a planting site. The one consideration is to match the specified soil to the requirements of the elements within the plant palette. A simple plant palette normally indicates a simple soil specification. A diverse plant palette should dictate a diverse soil specification, though this latter procedure has seldom been used in the past as an attempt to keep it simple—the second consideration. A simplified specification facilitates conformance by the contractor as well as inspection and acceptance by the landscape architect. It also simplifies the soil mixing, if required, and increases the ease of the off- and on-site handling of the soil materials; hence, costs are greatly reduced. If a diverse plant palette is involved in the project design and a single, simple specification is employed, then there is a possibility that the specified soil is a weak compromise as a planting medium for some elements of the plant palette. These elements may not grow satisfactorily, or may fail; hence, this results in a variable success rate over the project.

Observation of the natural landscape shows variation in distribution, on a mesoscale of meters, of vegetation according to changes primarily in soil characteristics and associated topographic parameters. On a local or microscale, variation may be due to changes in competition between individual organisms for light, heat, water, and nutrients. Thus, species distribution within natural plant communities will change according to this variation in soil. Plant communities will themselves change when the soil changes are strongly significant to species growth over longer distances. The conclusion derived from this observation is that different plant species are adapted to different soil conditions, perhaps with overlap for some species. Conversely, many species do not occur on the same soil. Horticulturists and landscape architects can learn many lessons from these phenomena that verge into the discipline of basic plant ecology and eventually to landscape ecology.

In the past, there has been a prevalent attitude toward soil by many practitioners so that we can specify something to meet our needs; but keep it simple. Inputs with fertilization and irrigation will overcome any deficiencies in the soil characteristics. These maintenance operations require input of energy and materials that may not be available over the design life of the project. Manual labor is required in most cases, which is expensive. The design eventually deteriorates. The horticulturist or arborist is called to perform the task of determining the cause of death. By common opinion among practicing arborists, foresters, horticulturists, entomologists, pathologists, agronomists, and soil scientists (the last to be called!), about 80% of the urban vegetation deterioration and mortality may be ascribed primarily to problems in or with the soil! Further, there is greater emphasis nowadays on reducing the use of water for nonessential irrigation and decreasing chemical input to soils. These goals can be achieved by sound soil design gained through thorough and specific soil specifications accomplished early in the design phase of the project.

Standard Specifications

In some areas of the country, standard construction specifications are promulgated by various regulating agencies or professional groups. Soil specifications are usually included in these standards. Unfortunately, the soil specifications may be be stated from the engineering viewpoint and are inadequate for landscape architecture applications. Two examples from opposite ends of the United States are given here.

Standard Specifications for Public Works Construction, 1991 Edition, for Southern California. (Published by BNI Publications, Inc.)

Section 212-1, paragraphs 212-1.1.1 through 212.1.1.2 provide the following topsoil specification.

212-1.1 Topsoil

212-1.1.1 General. Topsoil shall be designated as Class A (imported), Class B (selected), or Class C (unclassified) as specified herein. The engineer shall determine the suitability of topsoil prior to use. Topsoil shall be transported from the source to its position unless stockpiling is specified.

212-1.1.2 Class A Topsoil. Class A topsoil shall be from a source outside the limits of the project selected by the contractor and in compliance with the requirements specified herein. The engineer may make such inspections and perform such tests as deemed necessary to determine that the material meets the requirements.

At least 15 days before scheduled use, the proposed source of topsoil must be submitted to the engineer for approval. The contractor shall submit a written request for approval which shall be accompanied by a written report of a testing agency registered by the State for agricultural soil evaluation which states that the proposed source complies with these specifications. Class A topsoil shall have the same relative composition and structure, a friable sandy loam character, and be free of roots, clods, and stones larger than 1 inch in greatest dimension, pockets of coarse sand, noxious weeds, sticks, brush, and other litter. It shall not be infested with nematodes or other undesirable insects and plant disease organisms.

Class A topsoil shall meet the following additional requirements:

(1) Gradation Limits. Sand, 50–80%, clay 20% maximum, and silt, 30% maximum. The sand, clay, and silt gradation limits shall be as defined in ASTM D 422.
(2) Permeability Rate. Not less than 0.5 inch per hour nor more than 2 inches per hour when tested in accordance with ASTM D 2434, Calif. Test 220, or other approved methods.
(3) Agricultural Suitability. The topsoil shall be suitable to sustain the growth of the plants specified.

Commonwealth of Massachusetts Department of Public Works, *Standard Specifications for Highways and Bridges,* 1988 Edition.

Section Ml—Soils and Borrow Materials, with Subsections Ml.05.0, M1.06.0, M1.06.l, and Ml.07.0 as follows.

M1.05.O Loam Borrow

Loam Borrow shall consist of a fertile, friable, natural topsoil typical of the locality, without admixture of subsoil, refuse or other foreign materials, and shall be obtained from a well-drained arable site. It shall be such a mixture of sand, silt, and clay particles as exhibit sandy and clayey properties in and about equal proportions. It shall be reasonably free of stumps, roots, heavy or stiff clay, stones larger than 1 inch in diameter, lumps, coarse sand, noxious weeds, sticks brush, or other litter.

Prior to stripping, the loam shall have demonstrated by the occurrence upon it of healthy crops, grass, or other vegetative growth that is reasonably well drained and that it does not contain toxic amounts of either acid or alkaline elements.

The loam shall contain not less than 4% not more than 20% organic matter as determined by the loss on ignition of oven dry samples. Test samples shall be oven-dried to a constant weight at a tempearture of 230° ± 9°F.

M1.06.0 Peat Borrow

Peat Borrow shall consist of specified material obtained from the locality of the project or commercially from other sources to be used as detailed or as directed by the engineer to fit adequately the intended purpose.

Peat Borrow shall conform to the following requirements:

1. Organic matter content	25% minimum
2. Acidity	pH 4.0–7.0
3. Water-absorbing capacity	250% minimum
4. Soluble salt index	100% maximum

M1.06.1 Processed Planting Material
Processed Planting Material shall consist of suitable organic soil containing a reasonable amount of fibrous material procured from within the project limits or elsewhere and mixed with a suitable type of subsoil, resulting in a homogeneous material free from hard lumps, brush, or litter satisfactory as a substitute for natural loam and capable of supporting plant growth.

This material shall conform to the following requirements:

Soluble salt index*	100% maximum
Acidity	pH 4.0–7.0
Organic content†	10–20%

	Percent by Weight Passing Through	
Sieve Designation	Minimum	Maximum
1/2	100	
No. 50§	45	74
No. 200	20	40

§This requirement shall be waived when the amount of fibrous material precludes an acceptable test procedure.

M1.07.O Topsoil and Plantable Soil Borrow
Topsoil and Plantable Soil Borrow shall consist of fertile, friable, natural topsoil, reasonably free of stumps, roots, stiff, clay, stones larger than 1-inch diameter, noxious weeds, sticks, brush, or other litter.

Prior to stripping the topsoil from the construction project, it shall have demonstrated by the occurrence upon it of healthy crops, grass, or other vegetative growth, that it is reasonably well drained and capable of supporting plant growth. Material classified as Topsoil can only be obtained within the project limits.

These specifications are not fully satisfactory for several reasons. The gradation limits are not sufficiently specific, topsoil may not be topsoil, "friable and well drained" is inadequately defined, and material source is not well specified. Further, no standards reviewed by the author address the design of a soil profile. The standards tend to assume the vertical homogeneity of the entire material. These points are now discussed in detail.

*As determined with a type RDl5 SoluBridge electrical conducfivity tester at a dilution ratio of I part oven-dried material to 5 parts of distilled water (by weight).

†Organic content will be determined by the loss on ignition of oven-dried samples. Test samples shall be oven-dried to a constant weight at a temperature of 230° ± 9°F.

There is a very wide range in the particle sizes of the five fractions of the USDA Classification (Table 5.3) within the sand separate. These classes, ranging from very coarse to very fine (2 to 0.05 mm) have significant effect on rooting and plant growth and on susceptibility to passive or active compaction on installation. The specification must be more specific in the particle size distribution of the soil, especially in the sand fractions; hence, the sieve sizes and percentage passing each sieve should be specified for every soil mix.

Soil structure is one of the physical soil properties that strongly influences its characteristics of porosity and degree of aeration, drainage, bulk density, and waterholding capacity. In-place, undisturbed soils will have this natural structure to a greater or lesser degree. Disturbed soils will almost always have greatly decreased degree of structure (Potter, et al., 1988). Sandy soils do not have a high degree of structure, but the large average pore size tends to compensate for the loss of the large pores (the macropores) that exist in a well-developed natural structure. Therefore, sandy textures are nearly always specified to compensate for the loss of soil structure in the excavation, loading, hauling, stockpiling, and spreading processes involved in all soil construction operations. The addition of organic matter and inorganic porous materials, such as agricultural pumice, also provide additional porosity and "lightens" the soil.

The standard specifications as written do not ensure uniformity. A "friable sandy loam" can embrace a wide range of soil properties and conditions that may not be satisfactory for use in a particular project. The term "friable" specifically refers to the physical condition of a soil at a given moisture content, usually at field capacity (see the Glossary). Unfortunately, friability varies widely among different soils, even within sandy loams. No one likes to discuss problems, but an example must be given. The 4.2-acre Enid A. Haupt Garden of the Smithsonian Institution, Washington, DC, was installed in 1987 with prime Victorian garden specimens obtained through a countrywide search. These specimens included 50-year-old boxwood and large magnolias and a pair of 47-foot ginko trees among other large specimens. The cost of the planting is about \$3 million. The estimated cost of the "natural friable soil," spread 2 to 6 feet deep, is about \$200,000. The specimens exhibited symptoms of severe stress the first growing season. Upon investigation by an expert team comprised of soil science and agronomic practitioners and researchers, the soil was found to be compacted, poorly drained, and anaerobic. The material is probably a mixture of dredged material and a boggy organic soil. Smithsonian horticulturists have been busy every year since to improve the soil by mechanical aeration, replacement of some of the bedding soil in the parterre, and other ameliorating operations. Some of the prime specimens have had to be replaced.

Agricultural suitability is a nebulous term unless the criteria are defined and evaluated for the proposed palette. It may only indicate the relative suitability of the growth of field crops. Generally, the criteria should include the soil reaction (pH), plant nutrient concentrations, organic matter content, soluble salt content, and concentrations of natural chemicals that preclude successful plant growth, perhaps with an evaluation of the suitable irrigation water to be applied, if required. Obviously, the commendable purpose of the criteria is to prevent absolute "junk" being used as soil on a project.

The soil solid phase or component, which includes both inorganic and organic materials, has profound influence on other properties and characteristics, so it more than forms the matrix of the soil. It determines inherent fertility and, partially, waterholding capacity, degree of drainage, and density. Texture is the soil science term that refers to the proportion of sand, silt, and clay particles (also called the particle size distribution) in the soil material (see the Glossary). The other soil property that influences these latter characteristics is structure, which is defined as the degree and pattern of aggregation of the solid particles (see the Glossary). Since structure is usually destroyed or greatly decreased in disturbed soil, some compensating factors must be applied. Compensation is gained by additions of organic matter or artificial amendments such as sintered fly ash, expanded clay, and pumice.

The soil design best begins with specifying the texture. Sometimes it is easiest to start with a single texture as a basic matrix and then modify it with amendments to fit the various soil requirements of the plantings in the construction of the planter soil profile. This may serve well for shallow to moderately deep and uniform soil profile depths over small areas. Rooftop applications may involve this plan. It would be unwise to specify a single texture for all plantings with great variation in soil depth and volume relations coupled together with a plant palette diversity over a large area or under widely different microclimates.

The intention is to emulate the "topsoil," "subsoil," and "parent material" of the natural soil profile with changes in the basic matrix together with different amendments when the profile is 18 inches (45 cm) or thicker. Textural changes are accomplished with additions of coarse sand or clay (or zeolite) materials, depending on the direction of the textural change. Waterholding capacity, bulk density, and drainage are modified accordingly. The use of sintered fly ash, pumice, and ceramic products such as Isolite or Turface are used to further the modification of the latter soil characteristics with or without texture change.

Soil amendment with organic matter of some type will also provide for the modifications described before in addition to contribution to the fertility status. However, the addition of an excessive amount of organic matter may be required to attain an effective modification. An excessive amount of organic matter (usually more than 30% on a volume basis) causes several problems. The soil is prone to settlement, causing waterholding capacity to increase to the point where the soil remains wet for prolonged periods after rainfall or irrigation. This condition decreases the movement of gases into and out of the soil, creating unfavorable aeration conditions. Decomposition by soil organisms may be actually inhibited. Many natural forest soils have only 4 to 5% by weight of organic matter content (about 10% on a volume basis) coupled with a well-developed structure. The addition of organic matter increases the fertility of the urban soil. Over time, the development of natural structure is enhanced that has been initially destroyed to varying degrees in these disturbed soils. To overcome the problem of excessive organic matter content, inorganic materials that lighten the soil are employed as amendments to the basic matrix.

Fertility and soil reaction (pH) can be adjusted during the mixing process to provide a broad range of nutrient availability. Local adjustment for individual plantings in a diverse palette can be made after installation and applied according to the needs of each plant palette either by the contractor or the client in routine maintenance. The latter needs to be specified in a postinstallation maintenance section.

WRITING SOIL SPECIFICATIONS

A great amount of time and thought is given to the plant palette on most projects undertaken by landscape architects and horticulturists. The plants are chosen on the basis of color, textural tone, size, time of blooming, and their availability and cost. They are specified by scientific name and cultivar, diameter or caliper, container size, and so on. The soil is usually simply specified as fertile, well-drained, friable natural topsoil, and the number of cubic yards needed with a cost estimate. Attempts are usually made to get the soil as cheaply as possible in order to allocate more money to the plant palette. Thus, to the soil scientist, the degree of specification and cost allocation applied to the two major components of a project are grossly unequal. And yet, most landscape architects will agree to the importance of a good soil as a rooting medium necessary for sustained success of the project The writing of soil specifications to meet the needs of the landscape architect is discussed in detail. Some of the terms are the same used by engineers and some are used specifically by soil scientists. An attempt is made to bridge the gap among these three groups of professionals so that they can communicate intelligently with each other. In this way, a greater appreciation may develop of the needs of each group for any project, and therefore facilitate the design and installation of a project successful for all three professional groups.

We can design and specifiy a soil that will meet the requirements of the plant palette and also overcome the problems presented by the urban soil conditions. The question arises as to what are the characteristics of an acceptable or satisfactory urban soil. The following elements are specified to meet the desired characteristics listed in Table 5.1. It should be understood that in many cases, the resulting characteristics from the specifications may be compromises of these criteria necessitated by the nature of their design and project site conditions. A compromise must not diverge greatly from the desired criteria; otherwise, unfavorable characteristics will result. Good scientific judgment must be employed. It is here that many persons create prob-

TABLE 5.1 Desired Soil Characteristics for a Nonlimiting Soil in Tree Planting

Soil Properties or Characteristics	Desired Criteria
Texture	Sandy loam to silty clay loam satisfactory, provided it has favorable structure; should not have extremes of sand or clay, which have limitations
Structure	Granular, crumb, or fine subangular blocky desirable; loose or single-grained and massive have limitations
Bulk density and porosity	Values between 1.1 and 1.4 Mg/m^3 desirable; should not exceed 1.60 Mg/m^3. Macro-porosity should be at least 15% by volume
Plant-available waterholding capacity	15 to 25% by volume
Infiltration	From 2 to 10 cm/ph
Drainage	Well drained to moderately well drained; no redoximorphic features (mottling) within 50 cm of the surface
Organic matter content	At least 1 to 5% by weight
Soil organisms	There should be ample evidence of soil organism activity such as earthworm casts, algal blooms, ant middens, etc., and physical disintegration of litter.
pH	From 5.0 to 7.5; extremes should be avoided.
Cation-exchange capacity	From 5 to 25 $cmol^+/kg$ (meq/100 g)
Nutrient content	Normal contents of nitrogen, phosphorus, and potassium for most tree species; no severely limited nutrients or imbalance. Applies to micronutrients as well
Soluble salts	Less than 200 ppm with caution at 600 ppm. Better measure of sodium effects under deicing salt impact is sodium absorption ratio (SAR) and exchangeable sodium percentage (ESP)
Contaminants	Should contain no anthropeic materials or be known to contain toxic substances

Sources: Craul, 1985, 1992.

lems for themselves and their client by not appreciating the limit to a compromise. If one criterion is compromised slightly, no harm may arise. If one criterion is greatly compromised, such as texture, or several criteria are moderately compromised, serious problems caused by interactions may result. It is evident that the criteria of Table 5.1 become the elements of the soil specification.

THE MINERAL COMPONENT

The mineral component constitutes the primary fabric of the soil. It consists of the sand, silt, and clay particles. Organic matter is also a contributor to the soil fabric but is specified separately as an amendment. Here, the composition, size (diameter) distribution, pH, nutrient contents, cation exchange capacity, and source are included. Organic matter content is specified if a natural soil is used. If a manufactured soil is designed, without any natural soil used as a component, the organic matter is treated as an amendment.

Texture

Texture is selected first because it has influence on many other physical soil properties such as bulk density, drainage and waterholding capacity, resistance to compaction, and inherent fertility (Table 5.2). It may be designated in several ways: USDA texture classes, Unified Soil Classification System (USCS) class, and the particle size distribution. There are advantages and disadvanatges to each method.

The USDA texture classes are important because most plant growth studies of interest to landscape architects, especially agricultural, silvicultural, and horticultural studies, have been based on these classes when texture was an experimental factor. The particle size classes applied to the texture classes are given in Table 5.3. The disadvantages are mainly semantic rather than technical. These arise because the engineering and architectural professions prefer

TABLE 5.2 Soil Characteristics Influenced by texture

	Texture Class			
Characteristic	Sand	Loam	Silt Loam	Clay
Identification in moist state	Loose	Cohesive	Shows fingerprint	Shiny streak
Permeability	Excessive	Good	Fair	Fair to poor
Available water	Low	Medium	High	Medium to high
Tillability	Easy	Easy	Moderate	Difficult
Runoff potential	Low	Low to medium	High	Medium to high
Water detachability	High	Medium	Medium	Low
Transportability	Low	Medium	High	High
Wind erodibility	High	Medium	Low	Low
Nutrient capacity	Low	Medium	Medium	High
Compactability	Low	Medium	Medium to high	High

Source: Craul, 1992.

TABLE 5.3 The USDA Particle Size Classes

Soil Separate	Diameter Range, mm	U.S. Standard Sieve Size (No.)
Very coarse sand	2.00 to 1.00	10
Coarse sand	1.00 to 0.50	18
Medium sand	0.50 to 0.25	35
Fine sand	0.25 to 0.10	140
Very fine sand	0.10 to 0.05	300
Silt	0.05 to 0.002	—
Clay	< 0.002	—

the USCS, which has been employed as a particle gradient classification for interpreting and predicting soil mechanical properties in the field and laboratory.

The texture classes are made up of various proportions (percentages) of the separates of sands, silt, and clay. There are twelve classes and are illustrated in the familiar texture triangle of Fig. 5.1. Ranges in the separate percentages exist. This condition does not permit the precise definition of a particle size distribution, but acts as a reliable starting point in the choice of texture for the specification. When a texture class is chosen, information on waterholding capacity and drainage, and so on, can be derived from Table 5.4 or other sources.

Because of much documented engineering data and research results from standardized tests, the USCS is very reliable for characterizing soil mechan-

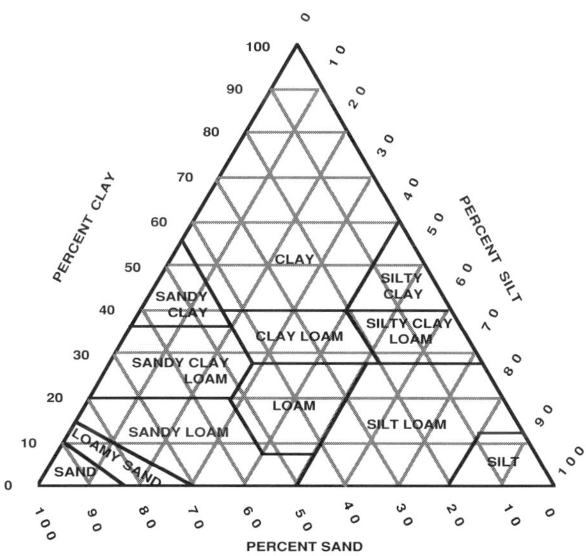

Fig. 5.1 The USDA texture triangle. A newer version is a right triangle.

TABLE 5.4 Waterholding Capacities of Several Soils and Soil Mixes

	Total Storage Capacity		Macropore Space	Plant-Available Capacity	
Texture Class	(%)	(in./ft)	(%)	(%)	(in./ft)
Coarse sand	24.4	2.93	17.7	6.7	0.80
Coarse sandy loam	24.5	2.94	15.8	8.7	1.04
Sand	32.3	3.87	29.0	13.3	1.59
Loamy sand	37.0	4.44	26.9	10.1	1.21
Loamy fine sand	32.6	3.91	27.2	5.4	0.65
Sandy loam	30.9	3.71	18.6	12.3	1.48
Fine sandy loam	36.6	4.39	23.5	13.1	1.57
Very fine sandy loam	32.7	3.92	21.0	11.7	1.40
Loam	30.0	3.60	14.4	15.6	1.87
Silt loam	31.3	3.75	11.4	19.9	2.39
Sandy clay loam	25.3	3.04	13.4	11.9	1.43
Clay loam	25.7	3.08	13.0	12.7	1.52
Silty clay loam	23.3	2.80	8.4	14.9	1.79
Sandy clay	19.4	2.33	11.6	7.8	0.94
Silty clay	21.4	2.57	9.1	12.3	1.48
Clay	18.8	2.25	7.3	11.5	1.38
25% peat to 75% sand	51.0	6.12	6.0	—	—
50% peat to 50% sand	60.0	7.20	10.0	—	—
75% peat to 25% sand	71.0	8.52	11.0	—	—
100% peat	80.0	9.6	15.0	—	—

Source: Kays and Patterson, 1982; sand-peat mixes from Juncker and Madison, 1967.

ical properties (Fig. 5.2 and Table 5.5). It differentiates soil into four classes: (1) coarse-grained, (2) fine-grained, (3) organic soils, and (4) peat. However, its objective and purpose are very different from those of the USDA classification. The former, based on texture and plasticity of soil, is to clearly predict the mechanical properties of soil for use as structural material, whereas the latter is based on texture of mineral soils only and indirectly on chemical activity. It is used to predict its plant growth response as a rooting medium. Thus, the test and prediction parameters in each case are generally diametrically opposed with little apparent compatibility. There are several compatible parameters that may be used by landscape architects in writing soil specifications. One of these is the particle size distribution. In this case, it doesn't make any difference what sieve sizes for sand fractionation are used in a stack, as long as the represented range in each class including those for silt and clay, as well as the total range (from at least No. 4 mesh to 0.002 mm particle diameter), are about equal for the two classifications. When the particle size distribution is plotted on a five-cycle semi-log graph, as in Fig. 5.3, interpretation may be derived for either classification system. The results for the graph are obtained through the particle size distribution analysis,

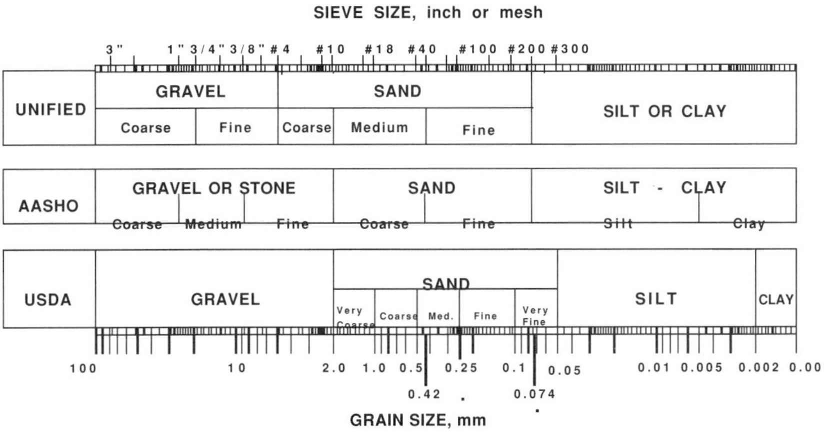

Fig. 5.2 Particle size class comparisons with sieve sizes for three classifications. (Adapted from Buol et al., 1989.)

ASTM D-422-63, in which sand fractions are determined by dry sieving, and silt and clay determined by hydrometer readings in a soil-water suspension.

Another widely used particle size classification system is that of the American Association of State Highway Transportation Officials (AASHO). It has some similarities to the Unified Soil Classification System, but also some significant differences in the size ranges. A good comparision is given in Das (1979). The size class limits are compared in Fig. 5.2. Here, we will use only USDA and USCS systems.

Particle Size Distribution

Rather than simply placing percentage data by sieve size into a table, a better understanding is gained by the plotting of a distribution curve. Several examples are illustrated in Fig. 5.3. Further, various analyses can be applied to

TABLE 5.5 The Particle Size Classes for Three Different Classifications

Grain Size	USDA	USCS	AASHTO
Gravel	3 in. to 2 mm	3 in. to 5 mm	3 in. to 2 mm
Sand	2.00 to 0.05 mm	5 to 0.074 mm	2 to 0.074 mm
Silt	0.05 to 0.002 mm	< 0.074 mm	0.074 to 0.005 mm
Clay	< 0.002 mm	< 0.074 mm	< 0.005 mm
Organic[1]	Not classified	Not classified	Not classified
Peat[1]	Not classified	Not classified	Not classified

[1]Peat and muck particle sizes are not classified, but organic soils based on content are included in the classification schemes for each system.

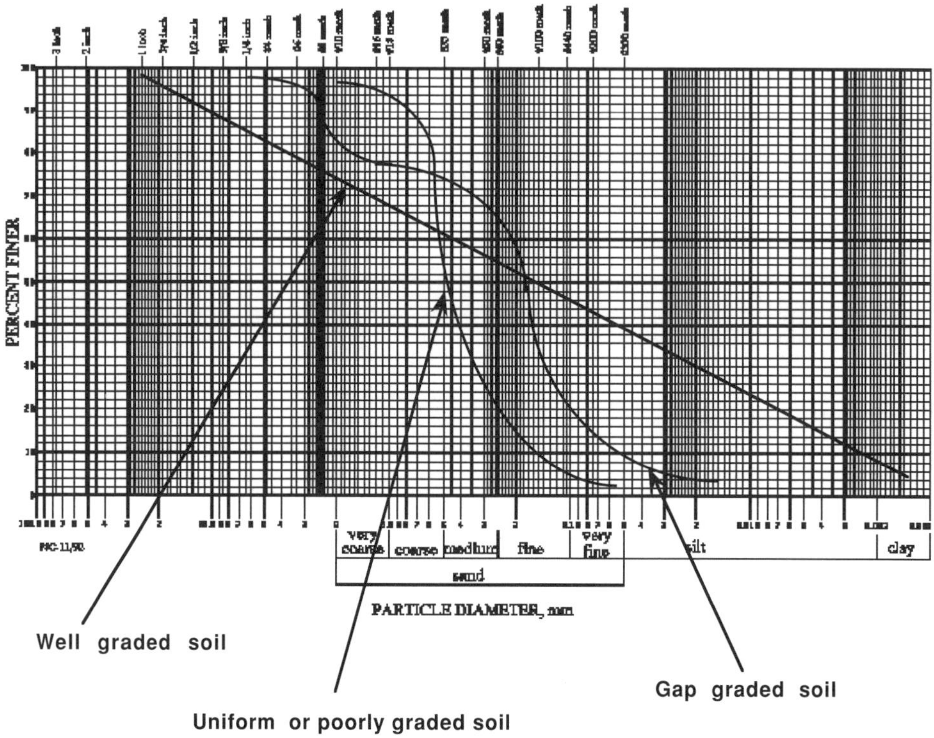

Fig. 5.3 The particle size distribution graph with plotted soil grades. Various sieve sizes are given above the graph.

the curves that better define them for the purpose of more precise description and specification. Conformance to specifications is also enhanced.

Three different graded soils are illustrated in Fig. 5.3. A well-graded soil is one with a wide range of particle sizes represented in the distribution. This soil has the desirable engineering characteristic of the relative fit of smaller particles in the spaces between the larger particles. The soil can be easily compacted to maximum density for engineering subgrades and bases. Unfortunately, a natural soil with such a distribution may be self-compacting if vibrational or settlement forces are applied to it. The uniform or poorly graded soil exhibits a narrow range of particle sizes so that only several classes are represented in the distribution. This soil tends to have uniform pore spaces and are not generally filled with smaller particles. The soil is usually loose and tends to be resistant to compaction, especially if the soil contains little fine and very fine sand and silt. The gap-graded soil may exhibit characteristics of either the well-graded or uniform graded soil depending on the average particle diameter in the distribution.

In writing soil specifications for landscape design, a self-compacting, that is, well-graded, soil should be avoided where settlement or vibrational forces are present. If these forces are absent, a well-graded soil with organic matter amendment is appropriate especially in planting beds for biennials and annuals. This soil would not be suitable for permanent plantings with surface traffic.

Engineers apply several values derived from formulas to numerically describe the particle size distribution curves and are useful in soil specifications for landscape architects (Holtz and Kovacs, 1981). These are the coefficient of uniformity, C_u, and the coefficient of curvature, C_c. The coefficient of uniformity is given by the formula

$$C_u = D_{60}/D_{10}$$

where D_{60} represents the point on the curve with 60% of the material is finer (passing) than this diameter. In like manner, D_{10} represents the point on the curve with 10% of the material finer (passing) than this diameter. For the well-graded, poorly graded, and gap-graded soils in Fig. 5.2 the C_u values are 64.2, 3.53, and 3.33, respectively. This value represents only two points on the curve. The coefficient of curvature represents three points on the curve and gives additional weight to D_{30} by squaring it. The formula is

$$C_c = (D_{30})^2/(D_{60} \times D_{10})$$

The coefficients of curvature for the three curves in Fig. 5.2 are 0.127, 1.36, and 1.38, respectively. Both parameters have nearly the same value for both the poorly graded and the gap-graded soils even though the two curves are different. When specifying sand as a mix component, C_u should be less than 3 (Couenberg, 1994).

Mean Particle Diameter

Mean particle diameter is another parameter used to partially define a mineral material. It tends to place laterally the curve on the graph, indicating its relative coarseness or fineness. It is widely used in hydraulic engineering and fluid mechanics. It is defined as

$$d = \left[\left(\sum_{i=2}^{n} \frac{m_i}{d_i} \right) + \frac{3}{2} \times \frac{m_1}{d_1} \right]^{-1}$$

where d = mean diameter of soil particles
m_i = ratio of the weight of a size-class fraction to the weight of the whole soil sample
d_i = mean diameter of particle size in a fraction class
m_1 = ratio of weight of smallest size class to weight of whole sample

d_1 = the upper limit of the smallest size class (0.002 mm in USDA system)

The d_i value is calculated from

$$d_i = \left(\frac{1}{d_i'} + \frac{2}{d_i' + d_i''} + \frac{1}{d_i''}\right)^{-1}$$

The value must be used with other parameters to fully describe the curve; one point does not describe a curve.

Another value used in the field of petrology (an area of geology and mineralogy) is the *phi* value of Krumbein and Pettijohn (1938).

$$d = 2^{-\phi}$$

where d is the particle diameter in mm, and $\phi = (\ln d)/(\ln 2)$. Some ϕ values for the USDA class limits are given in Table 5.6.

McCammon (1962) uses the ϕ value in his eight-point equation to describe mean particle diameter. The equation is

$$\sigma_\phi = \frac{1}{9.1}(\phi_{70} + \phi_{80} + \phi_{90} + \phi_{97} - \phi_3 - \phi_{10} - \phi_{20} - \phi_{30})$$

It is more complex and requires greater effort for solution than the C_u and C_c values. A programmable hand calculator or desktop computer can be used effectively. The process moves closer to the objective of describing a curve with one value, but the visual advantage of a plotted curve is clear.

An Example

It was decided that the planting soil for a project should have the particle size distribution envelope given in Table 5.7 and illustrated in Fig. 5.4. It provides

TABLE 5.6 Phi (ϕ) Values for USDA System Class Limits

Class Limit, mm	Phi Value
2	−1
1	0
0.5	+1
0.25	+2
0.10	+3.3
0.05	+4.3
0.002	+9.0

TABLE 5.7 The Particle Size Distribution of the Planting Soil

U.S. Sieve Size, No.	Percent Finer, by Weight
4	100
10	81 to 93
18	66 to 87
35	52 to 82
60	35 to 66
140	12 to 34
300	5 to 18
Silt	4 to 13
Clay	1 to 5

a high infiltration rate but low waterholding capacity. However, the deep profile, 4 feet, compensates for the moisture deficiency.

The material available on-site is a silty glacial till residuum from the subsoil of a New England soil with a well-graded particle size distribution, as illustrated in Fig. 5.5. The characteristics of a silty, well-graded soil indicate that the soil is self-compacting and must be amended to correct the problem. A coarse sand is found that is mixed with the silty till to partially overcome

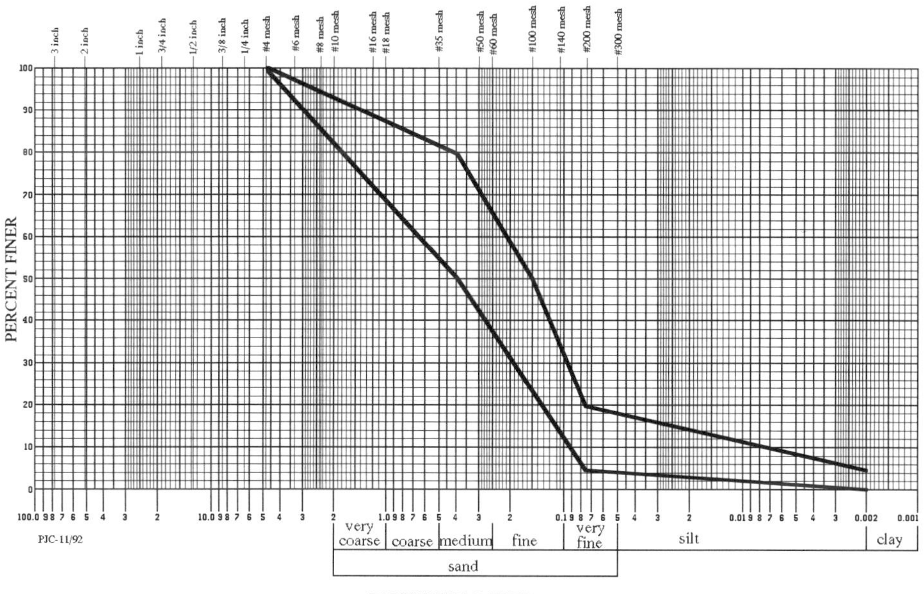

Fig. 5.4 Planting soil particle size distribution.

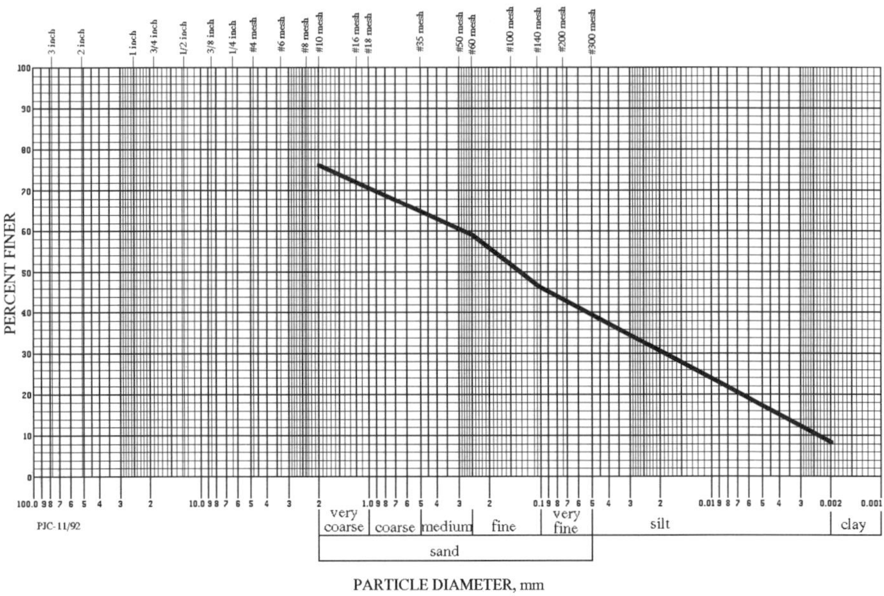

Fig. 5.5 The glacial till particle size distribution.

the self-compaction. The particle size distribution of the coarse sand is shown in Fig. 5.6.

A trial prediction of a 2:1 sand:till mix is constructed. The procedure is outlined in Table 5.8 and the resulting curve is shown in Fig. 5.7. Tests on samples of the actual mix indicate that the predicted curve is correct. However, the client rejected the 2:1 mix as too expensive and used a 1:1 mixture. The material is presently self-compacting, as predicted. The planting soil is further amended by the addition of composted sewage sludge. This addition will further contribute to the reduction of the self-compacting character of the soil.

In some geographic areas, sand is derived from the weathering of limestone or calcareous shales, as contrasted to sandstone, granite, and so on, parent material. The fine and very fine sand fractions in the limestone-derived soil may be comprised of up to 65% calcium carbonate. The presence of this material acts as a cementing agent and can cause self-cementation leading to severe compaction. To eliminate the problem, the sand must be specified as "double-washed" to remove the calcium carbonate. "Mortar sand" is another term applied to washed sand.

THE STRUCTURE COMPONENT

This component is included when the designed soil requires additional pore space to increase resistance to compaction or reduce the soil weight to meet

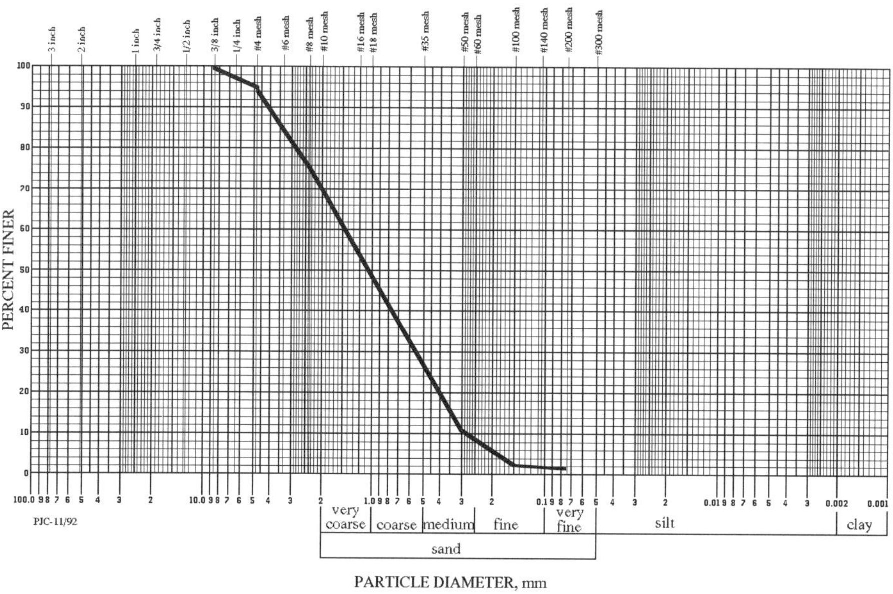

Fig. 5.6 Particle size distribution of the coarse sand.

rooftop loading constraints. The waterholding capacity may be increased. Rooting depth and root volumes are significantly increased. The component is not required in a soil that has a high degree of water-stable aggregates as a residual effect of a well-structured natural soil as part of the mix or a soil that will be used as bedding for annuals or perennials and frequently cultivated. Organic matter is the best amendment in the latter case.

The purpose of the structure amendment is to provide additional pore space to the designed soil that is lost during the handling and mixing of the soil components and make it less susceptible to compaction. To be effective, the material should be nontoxic or noncorrosive with some level of cation-exchange capacity, very porous with internal pore space containing plant-available water, structurally stable over time, resistant to compression (rigid), and does not self-compact in bulk. Usually, one treatment persists for a long time. Many products, both old and new, meet these requirements to varying degrees. Each one must be evaluated on the basis of all of the stated criteria.

Type

There are various types of structure-enhancing materials other than organic matter itself. There are essentially five categories: naturally occurring materials (mined), processed natural materials, by-products, and two of synthetic materials. The materials are listed in Table 5.9. The listing is not inclusive; only the more commonly available materials are given.

TABLE 5.8 Calculation of a Predicted Mixture Particle Size Distribution Curve

Component and Proportion Multiplier	Soil Separates, % by Weight: Coarse Fragments, #10	Very Coarse Sand, #18	Coarse Sand, #35	Medium Sand, #60	Fine Sand, #140	Very Fine, #300	Silt	Clay, 0.002 mm
1 Part till (× 1)	24	5	5	7	12	6	32	9
2 Parts sand (× 2)	0	33	38	17	9	0	3	0
Sum	24	71	81	41	30	6	38	9
Mixture (total divided by 3)	8.0	23.7	27.0	13.7	10.0	2.0	12.6	3.0
Percent passing 100	92	68.3	41.3	27.6	17.6	15.6	5[1]	2[1]

[1]Determined by hydrometer.

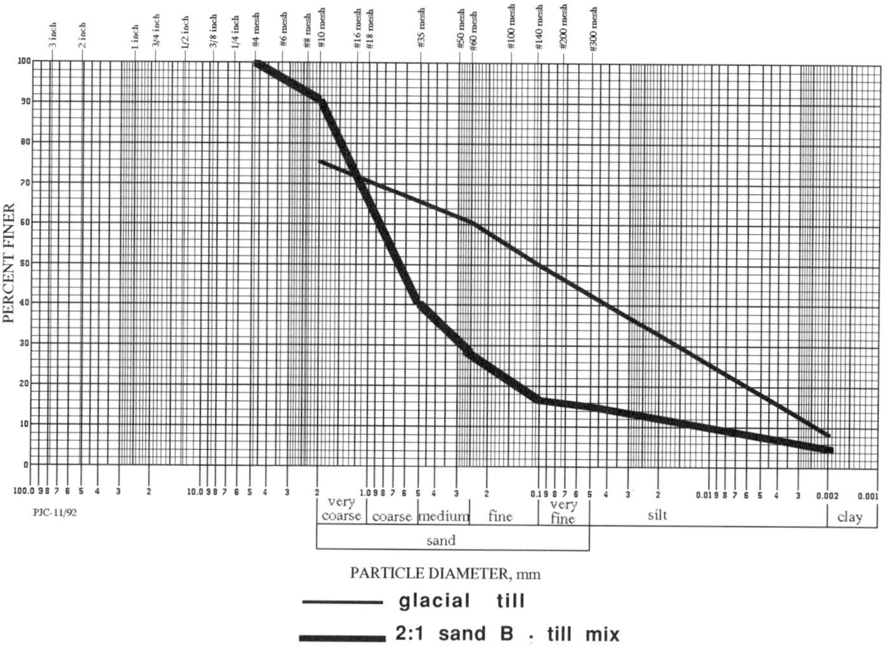

Fig. 5.7 The 2:1 sand:till mix for the planting soil.

TABLE 5.9 Structure-Enhancing Materials: A Partial Listing

Type	Material Examples
Natural-occurring material	Vermiculite
	Pumice
Natural-processed material	Expanded vermiculite
	Expanded clay
	Expanded shale
	Diatomaceous earth
Processed by-products	Sintered flyash
	Steel mill slag
Synthetic product	Perlite[1]
Synthetic agent	VAPAM[1]

[1]A trade name. The trade names of the various products are used here for clarity and discussion purposes only and does not constitute endorsement or outright rejection.

Description

Many of the materials are marketed under various trade names and will be used here for clarity. The reader's attention is called to the footnote of Table 5.9.

Vermiculite. A natural material derived from geologic deposits. Its light weight and platy structure lighten the soil and contribute to increased rooting depth, better drainage, and improved tilth. The expanded form, heated to very high temperatures, is the most commonly marketed product The particles do break down over a long time period and may orient themselves horizontally under compactive forces and potentially contribute to soil compaction. The material is used best in greenhouse and nursery mixes rather than in field applications.

Pumice. Pumice is a volcanic rock originally containing large amounts of gas. As the rock was thrown from the volcano, the molten magma cooled around the bubbles to form a very porous, lightweight rock. The gas eventually escaped as the rock cooled and fractured. The material may occur in a range of sizes from huge rocks to fine gravel and smaller. Soil formed in this material is very porous and a poor conductor of heat; thus, it has a very high surface temperature, which is lethal to many native plants. The material is usually marketed as pumice. The smaller sizes do not exhibit uniformity of porosity and have little waterholding capacity. The major benefit is the lightening of the soil and partial prevention of compaction. The demand for the larger rocks as landscape elements has increased its cost. It is not commonly specified as a soil amendment at present.

Expanded or Calcined Clay. Natural clay is heated and sintered (or calcined) at high temperatures. The material is then crushed and screened to desired size. Internal pore space and size are variable. Examples are TURFACE® and SOLITE®.

Expanded Shale. Natural shale is heated and sintered at high temperatures much the same as expanded clay. Certain shales are amenable to expansion and lightening by this process. The treated material is crushed and screened to desired size. An example is RIDGELITE®, which is offered in a variety of particle size ranges.

Diatomaceous Earth. Diatomaceous earth is derived from natural terrestial deposits of ancient marine diatoms. The diatoms were one-celled animals that inhabited shallow seas in large numbers in early geologic times. Their bodies, composed of hardened calcium phosphates with a very high percentage of silica, persist in large deposits today. The material is extremely porous, much like coral. Some manufacturers claim that the internal pore space contains plant-available water. The diatom earth is fired at high temperatures and later crushed and screened to size. Uniform sizes or mixtures of sizes (particle size ranges) are offered. Porous ceramics may be manufacturered from a mixture of inorganic minerals and extruded. There is a wide variation in the properties of the products. Each one needs to be evaluated for the desired use. Diatom earth products are ISOLITE® and AXIS®.

Sintered Flyash. The material is derived as a by-product of smoke stack scrubbing. The particulates of smokestack air are washed out with water and certain wetting agents and then dried. It is then fired at high temperatures to sinter it. As with the other materials, it is crushed and screened. The material generally has a high pH.

Steel Mill Slag. A by-product from the manufacture of steel from iron ore, it is primarily the residue of impurities lapped from the surface of the molten iron. It contains high amounts of silica, calcium carbonate, and other substances. The pH is usually around neutral (pH = 7). It is porous, much like pumice with pores formed from gas bubbles. It is crushed and screened for use. The material tends to vary according to the smelting process employed or the type of steel produced. With the demise of the steel industry in the United States, it is not as common as formerly. It is very little used in landscape applications.

Perlite. Perlite is a synthetic material with a plastic base. It is extremely light and enhances the waterholding capacity and drainage of the mix. It is not completely rigid and can be easily crushed. It is very dusty and blows very easily in the field. It is best used as a component of greenhouse and horticultural potting soil in which it is generally mixed with sand, loam soil, and, sometimes, vermiculite.

VAPAM. This material is somewhat different from the others discussed here. It is a chemical agent (organic compound) that is applied to the soil to enhance the cementing of primary particles into compound aggregates. The result is an increase in the natural aggregation of the soil, its primary use and benefit. Reduction in bulk density and some increase in compaction resistance has been exhibited in various studies and practical applications. It has been widely used in golf course applications. It takes some time for the material to be effective as it must react with the soil.

Proportion in the Mix

The proportion of the mix may be as high as 50% (volume basis) if the loading constraints require it. However, experience (J. C. Patterson, personal communication, unpublished data, 1997) shows that a 60% mixture caused very dry conditions, with reduced rooting depth and volume. A maximum in the mixture of 30 to 40% seems to be satisfactory for most growing conditions. Where traffic or loading is not a problem, the mixture may be 10 to 15% or perhaps not used at all. Manufacturers provide their own recommended mixtures for various applications.

Particle Size

Most processors supply the materials in various uniform particle size classes or provide a particle size range. Generally, the particle size and/or its range is dictated by the soil horizon in which it is to be mixed, the planned use of the soil (traffic or no traffic, etc.?), and the plant palette. Manufacturers provide particle size recommendations for various applications.

If the soil is to be used for bedding without traffic, the particle size may be uniform and small, 2 to 5 mm. With traffic, as for a turf area, the proportion should increase as well as the size. The maximum size may increase to 15 mm (1.5 cm) with a particle size range down to 5 mm. In the subsoil horizon, it is advantageous to have a uniform size of about 20 mm (2.0 cm) or perhaps larger, with the proportion sufficient (probably between 40 and 50%) for near contact between the particles if under traffic; otherwise, the proportion should be between 30 to 40%. The drainage layer may contain structure materials to lighten the overall weight Here, the particles may be large, 20 to 30 mm (2 to 3 cm), and make up to 50% by volume if necessary. In some special situations, the proportion may be greater than 50% if there is sufficient satisfactory rooting depth (at least 20 to 24 inches or 50 to 60 cm) above the drainage layer.

Evaluation

Each user of structure amendments must evaluate each product to find the best one for the specific application. For example, pH, density, cation-exchange capacity, porosity and pore size, particle size (texture), chemical composition, stability (rigidity), and electrical conductivity all should be examined. For example, a rooftop application would require one with a very low density, whereas a turf area with traffic would require one with strong, persistent rigidity. Where fertility is a concern together with compaction resistance or lightness, then a product with high cation exchange capacity and rigidity should be selected. More details on proportions in mixes and particle sizes are given in Chapter 12.

THE ORGANIC COMPONENT

Organic matter is a significant major component of the soil. It provides the source of energy for soil organisms, especially the bacteria, algae, and fungi. It supplies most of the natural soil nitrogen, half of the phosphorus, a major portion of the sulfur, and many other organic compounds essential to healthy plant growth. It contributes significantly to the cation-exchange capacity of the soil, which acts as a reservoir for plant-available nutrient storage. It also enhances the development of soil structure so necessary for water movement and storage, gaseous diffusion, and root penetration and extension. Therefore, most designed soils require an organic matter component. Part of the content may be supplied in natural soil as part of the mix. However, it is likely that most soil material available for future projects will have low organic matter contents or none at all; supplementaton is required.

In the past, shredded and screened peat moss has been a very satisfactory organic matter source. This material is now expensive and sources are declining because of the concern about the environmental impact of peat bog mining. Substitutes in the form of composted organic wastes, both agricultural and urban, have become available. However, no specific standards of processing or manufacture have been promulgated, and a wide variety of materials have been marketed. Some materials have been given trade names with their composition as proprietory information, and claims are given of near miraculous results when applied to growing plants. Little independent documentation exists for some of the claims. Some almost verge on the "snake oil" claims of old. Other claims are valid, being documented by agricultural research station and university studies. A soil specification must be carefully and fully stated to ensure the acquisition of an appropriate organic component until standards are developed. The following elements are discussed with this necessity in mind.

Type

There is a wide variety of organic materials available or specifically produced for use as soil amendments. These may be placed into several general classes (Table 5.10).

When used as a mix in conjunction with mineral materials (even a soil) from other sources, a topsoil may be constructed. This procedure is appropriately applied to landscape projects when topsoil is a desired design feature but sufficient amounts are not available or its use may have severe environmental impact, thus nonconforming to a sustainability requirement. Further, the mineral material must be specified to ensure its usefulness and provide consistent data for prediction of the characteristics of the final mix. This portion of specification preparation is rather straightforward since this type of specification has been used by soil engineers for many years. However, the charactersitics in the end product desired by the landscape architect and horticulturist are quite different from those desired by the engineer. Hence, the quantified engineering specifications are quite different from the landscape specifications, even though the criteria in the specifications are the same.

The formulation of a constructed topsoil must be accomplished with the consideration of many variables. These include the physical and chemical

TABLE 5.10 Classes of Organic Materials

Plant Residues
woodchips—hardwood, softwood, lumber, shavings
bark
sawdust
shredded leaves
municipal mix
Agricultural Manures
horse
cow/steer
sheep
hog
poultry
Sewage Sludge
pellets
dried cakes
granulated
Composted Sewage Sludge
Organic Wastes from Processing
brewer's waste
food-processing residues (sludges)
wood pulp

characteristics of the mix components and the matching of the mix to the landscape applications. The latter conditions are determined by topography, the soil physical and chemical characteristics and its drainage, character of the bedrock or underlying surficial geologic material and its depth.

There is a wide variety of organic materials available or specifically produced for use as soil amendments. In most instances, the material has been composted before put in use, but not always. In fact, the composting process has such beneficial effect on most organic materials that it is almost a universal step in their preparation for recycled use. Therefore, in the following discussion, it is assumed that the composted material is being considered (de Bartoldi, 1987). Some remarks will be made concerning the uncomposted or raw materials when an especially detrimental condition exists.

The plant residues cover a wide range of source materials or feedstocks. Chipped wood and bark, shredded leaves, and sawdust probably are the most frequently composted materials. When in uncontrolled or unsorted mix, the material is called a "municipal mix." If properly composted and screened, the material can be a very beneficial material. However, many products are very coarse (unscreened) for a mix component. In addition, plastic fragments, usually the black plastic bags used for disposal, and other debris make them undesirable as well. Debagging at the collection point would do much to make the material more acceptable. Wood chips, bark, and shavings used alone are best for mulching materials rather than as a mix component. Sawdust is very raw material and has limited application when used alone. Each of these products have an extreme carbon:nitrogen ratio and do not decompose very rapidly—an advantage for mulch, but a disadvantage for use in a mix without thorough composting. Peat moss has been frequently used in the past as a mix to produce a designed topsoil.

Agricultural manures usually require composting for use in urban areas to reduce the carbon:nitrogen ratio, odor, and pathogens. Farmers may use it raw, but with less restrictions. They are a very beneficial supplement to the plant residues. If composted together and finely screened, the product is very desirable as a mix component or used alone. This product is some of the potting soils sold in garden stores. The preferred manure is horse and cow manure, sheep and hog manure less so. The difference is the bedding material that is usually contained in the cow and horse manure that the sheep and hog manure usually lacks. The bedding material is straw, hay, sawdust, and sometimes shredded newspaper. With the new soy-base printing inks, newspaper is safe. Poultry manure is considered "hot" and is best mixed with other manures for composting. Other manures are minor in extent but locally may be important, such as from zoos and special animal farms.

If we are dealing exclusively with composted biosolids, then the primary specification is that it must meet EPA Part 503 Grade A "exceptional quality" requirements. All of the following specifications are framed in that concept. The public and our clients must be assured of being risk-free and protected.

The following individual specifications together give a scientifically sound basis for a material that is safe to use, beneficial to the plants to be grown on it, and subject to strict conformance. Experience by the author indicates that the set of specifications is feasible and attainable.

Carbon:Nitrogen Ratio

The carbon:nitrogen ratio of plant tissue and farm manure ranges from 20:1 to 30:1 and increases to 100:1 in stems of cereal crops and reaches 400:1 in sawdust. Thus, woody material has a wide or large C:N ratio. The C:N ratio of soil organisms is quite narrow being about 4:1 to 9:1. Mixtures of organic materials are expected to have a wide range of C/N ratios. In all cases of compost processing, the C:N ratio narrows and the stability of the material increases (Hirai et al., 1983).

The importance of the carbon:nitrogen ratio to the ease and rate of organic residue decomposition is well known to soil biologists and soil scientists. Generally, the narrower or lower the ratio, the greater the ease of decomposition. Those organic materials with a narrow C:N ratio will decompose faster since there is more nitrogen available to the decomposing microorganisms. When the compost is applied to the soil, the C:N ratio should be less than 33:1 (Brady, 1990). Below this threshold, a surplus of nitrogen is available to plant roots as the material is decomposed further. Above it, there is a nitrogen deficiency for the associated vegetation and fertilization is required. For most applications of compost, it is best for the C:N ratio to be below 33: 1. Some farmers prefer to apply manures with a higher C:N ratio because the material has a higher total nitrogen content. However, they are not as concerned with the environmental factors of odor and exposure as would be the case in urban areas or along highways.

The C:N ratio is sometimes used as a measure of compost stability. A better indicator of stability is necessary, as suggested by Zucconi and colleagues (1981) because the carbon:nitrogen ratio is influenced by many other factors not related to the curing process. The reporting of the C:N ratio of the initial material and that of the processed material is a good measure of the efficiency of the composting process. It also provides a means to detect if nitrogen fertilizer was added to the compost to improve the final C:N ratio.

Stability

Stability refers to the degree or stage of compost decomposition (Composting Council, 1996). As discussed under the C:N ratio, the organic matter becomes more stable as decomposition progresses. Carbon dioxide, heat, and water vapor are produced and nitrogen and oxygen consumed with a successively more stable substrate as residue. A "cured" compost is one that has reached a stage in which little further decomposition occurs. This product becomes the most favorable material for use as a soil organic amendment. If the compost is unstable or too raw and not handled properly, plant growth inhibition may occur. The various degrees of compost stability as used by various authors may be described as "raw" for fresh feedstock; "very unstable" for compost under initial, active decomposition, "unstable" for advancing stages of decomposition; "stable" for compost in the later or "curing" stage of decomposition; and "very stable" for compost that has reached the final stage of decomposition with little further evolution of carbon dioxide, heat, and water. The most suitable compost for use in planting media is stable or very stable.

One indicator of stability is by the measurement of heat evolution from a moist sample, called the self-heating test (Zucconi et al., 1981). A second indicator is the measurement of carbon dioxide evolution from a sample as produced by the microorganisms on oxygen uptake. In some instances, both indicators may be used for specifications.

Maturity

This criterion is a measure of the acceptability of the compost for a planting medium (Composting Council, 1996). The indicator is a growth-screening procedure that includes seed germination tests coupled with subsequent plant growth and survival (Riffaldi et al., 1986; Saviozzi et al., 1989). The tests may include pot tests and root elongation measurements. They are used to detect the presence of volatile fatty acids, alcohols, soluble salts, heavy metals, or ammonia. These substances may cause delayed seed germination or death, and seedling damage or death. Because volatile fatty acids and alcohols are produced under anaerobic conditions, stored compost in anaerobic conditions subsequently may not be mature. Stored compost should be kept in small piles and protected from the weather. Harada and colleagues et al. (1980) suggest using the carbon:nitrogen ratio as a measure of maturity, but Zucconi and colleagues (1981) question this method.

Odor

Odor is a descriptive rather than a quantitative factor, but is valuable to the evaluation of compost stability. Generally, stable compost should produce no unpleasant odor. It may have a slight to moderate musty odor caused by its moist condition. The musty odor should not be overwhelming. This indicates a compost that is too moist and may be in the initial stages of unfavorable anaerobic decomposition.

Compost that produces an odor of rotten eggs (hydrogen sulfide), the bitterness of ammonia, or other strong odor is not sufficiently cured and has a low degree of stability and is not mature. It certainly is not "stable" or "very stable," as the evolution of these gases is produced by the microorganisms continuing the decomposition process.

Mineral/Organic Content and Fineness (Particle Size)

Compost includes some mineral material besides organic matter. The amount present and its particle size distribution are an important part of the specifications for determining usefulness. The minimum amount of organic matter on a dry weight basis must also be included. Much of the weight of compost is absorbed water. Many composting processes involve the use of wood chips or other woody products as absorbants. Generally, these woody materials are screened out subsequent to composting. Some remain in the compost product. The amount and fineness of the coarse residue must be part of the specifications.

1. *Organic Content.* The organic content may vary from 30 to 70% with the moisture content varying from 30 to 60% although some standards allow as low as 10%. The desirable percentages would be greater than 50% organic matter and a moisture content of 40 to 50%. The bulk density should range from 800 to 1000 pounds per cubic yard. Particle size distribution of the organic material is necessary only for potting soil.

2. *Mineral Content.* The mineral content may range from 0 to 1.5% and reported as ash residue on a dry weight basis. Ash contents greater than this range may indicate the presence of other mineral impurities.
3. *Coarseness.* The size and amount of coarse material contained in the compost are related to the screen size used by the processor. Some compost is specified to pass a 1-inch (25 mm) screen and others pass 1/2-inch (13 mm) and smaller screens. The coarseness permitted depends on the use of the compost. Topdressing of turf should be very fine and specified to pass a 1/4-inch (6 mm) screen. For general application and as a mixture component, the 1/2-inch size limit is appropriate. This limit may vary according to the drainage of the site, with well-drained sites permiting finer compost than sites with restricted drainage. The 1-inch limit is too coarse for grass planting. Coarser limits are appropriate for erosion control measures.

Reaction (pH)

As the organic material is decomposed, the pH undergoes change together with the C:N ratio. High C:N ratio materials are usually acid having a pH less than 7.0, usually in the range of 4.5 to 6.0. As the C:N ratio narrows during the composting, the pH tends to rise. However, it seldom exceeds 8.0 on completion and may be in the range of 6.0 to 7.0. A final pH of less than 6.0 is not acceptable for the stabilized compost. Sludge slaked with lime would have a higher pH value, in the range of 8 to 11 or even 12.

Cation-Exchange Capacity

The cation-exchange capacity (CEC) is a measure of the nutrient-holding (adsorption) power of the material and is related to potential fertility. It is also related to the state of decomposition, tending to increase as the material is progressively decomposed. However, it is not a measure of maturity or stability. Again, too many other factors influence CEC, but the data provide a basis for comparative purposes and to estimate the potential CEC of a compost-amended mixture. It does not need to be a criterion for every compost specification.

Nutrient Contents

The nutrient contents of compost are low compared to commercial fertilizer materials (Table 5.11). It is best not to consider compost a primary fertilizer, but it does contribute to the nutrient content of a mixture or as an amendment. It provides a more valuable contribution to the soil physical properties. Compost nutrient content varies according to the feedstock and process used. Ranges in nutrient content are appropriate for specifications.

TABLE 5.11 Plant Nutrients in Composted Organic Materials, Dry Weight Basis

Macronutrients	Minimum Admissible Levels, %
Nitrogen	0.6
Phosphorus (P_2O_5)	0.5
Potassium (K_2O)	0.3
Calcium (CaO)	2.0
($CaCO_3$)	3.0
Magnesium (MgO)	0.3

Source: Zucconi and de Bertoldi, 1989.

Heavy Metals

Limits to heavy-metal contents in compost are governed by federal and state regulations. Some state standards are more stringent than the federal standards. The federal limits are provided in Table 5.12.

Standard Compost Specification Example

The following are proposed specifications for compost quality for its procurement and use in amending soils on transportation projects. They are based on a review of the literature, the response of a survey conducted among compost vendors in Massachusetts, and the personal experience of the author.

TABLE 5.12 Concentration Limits for Heavy Metals in Grade A (Exceptional Quality Sewage Sludge) Compost

Element	Concentration Limit, mg/kg dry wt. basis
Arsenic	41
Cadmium	39
Chromium	1200
Copper	1500
Lead	300
Mercury	17
Molybdenum	18
Nickel	1420
Selenium	36
Zinc	2800

Source: EPA Part 503, Table 3.

General

The compost shall be a stable, humuslike material produced from the aerobic decomposition of organic residues. The residues may include biosolids as well as yard wastes, agricultural wastes, and source-separated municipal solid waste.

Color

The compost shall be of a dark brown to black color and be capable of supporting plant growth with appropriate management practices in conjunction with addition of fertilizers and other amendments as applicable.

Moisture

The compost moisture content shall be such that no visible free water or dust is produced when handling it.

Carbon:Nitrogen Ratio

The ratio shall be in the range of 10:1 to 25:1. The reported C:N ratios before and after processing are to be preferred together with total carbon.

Degree of Stability

Stability must be assessed by either a CO_2 evolution test or a reheating test. Protocols for each are specified by CONEG (Coalition of Northeast Governors, 1996). For the CO_2 test, the compost respiration shall be no more that 5 mg CO_2-C/g BVS day. For reheating using the Dewar self-heating test, the maximum heat rise should be no greater than 20°C above room temperature (20 to 25°C).

Odor

The composted material should not have an unpleasant odor.

Pathogens/Metals/Vector Attraction Reduction

As per 40 CFR Part 503 rule (EPA CFR, Part 503 Regulations, Table 3, page 9392, Vol. 58, No. 32, Friday, Feb. 19, 1993).

Organic Content

The compost should contain at least 40% organic matter (dry weight).

Particle Size

One hundred percent of the material should pass a 1/2-inch (13 mm or smaller) screen. Debris such as metal, glass, plastic, wood (other than residual chips), asphalt, and masonry shall not be visible and should not exceed 1% dry weight.

pH

The pH shall between 5.5 to 8.0.

Salinity

Soluble salts should not exceed 4.0 mmhos/cm (dS/m) or 2560 ppm salt.

Nutrient Content

The mineral content of the compost must be given including: nitrogen, phosphorus, potassium, calcium, magnesium, sodium, and micronutrients including iron, copper, boron, manganese, and molybdenum.

Availability and Costs

The diversity of available organic materials creates a problem: Most producers have limited amounts in stock over a short time span. A project that requires large amounts in a short time, as during the installation stage of construction, may find inadequate amounts available from a single supplier and cannot easily go to another supplier to get similar material. Similar or nearly identical materials may be available from several suppliers but the locations are dispersed and transportation costs become excessive. In some cases, the budgeting for large transportation costs may be justified on the basis of obtaining sufficient quantities of the best material. Thus, several similar, not identical, materials from different suppliers may need to be specified to provide sufficient quantities to complete the project. Obviously, production of materials that meet a universally accepted standard would eliminate this major problem in the future.

Conformance

Conformance to standards must be enforced to ensure that the desired characteristics and quality of compost are obtained, otherwise project success could be jeopardized. The process involves submission of processor's specifications based on independent analyses, inspection at the source, with sampling and analysis on delivery to the site. Visual and olfactory inspection may satisfy transient inspection reinforced with subsequent laboratory analyses. The two characteristics most likely to be troublesome in consistence and conformance from batch to batch or load to load are particle size (woodchip content) and stability. It is extremely important for consistency in these two characteristics for most landscape applications.

Amount of Organic Matter in the Mix

The percentage of organic matter in the final mix should be in the range of 1 to about 8% by weight, with the most generally recommended amount as 5% by weight, which is about 10% by volume. At least 3% organic matter by weight is required to attain a bulk density of less than 1.45 Mg/m^3 (McCoy, 1992), and 3.5% is required for available water content of more than 0.2 m^3/m^3. For golf course applications, which have high pedestrian traffic, McCoy (1992) does not recommend organic matter content greater than 3.5% by weight (about 7% by volume). There have been soil failures (severe settlement with subsequent compaction and waterlogging) observed by the author when the organic matter component has been as great as 50% on a volume basis. The maximum proportion specified should never exceed 30% by volume for landscape projects and then used only for the topsoil for bedding plants, with 20% a less risky maximum. The organic matter content for the subsoil horizons is greatly reduced or the amendment may even be omitted.

THE BIOLOGICAL COMPONENT

The importance of soil organisms to the well-being of landscape plants is becoming better recognized. Most perople are familiar with the role of organisms in the decay or composting of organic residues as well as the importance of earthworms to soil tilling, but few carry the concept to living plants. Importance of the organisms to crops, especially the nitrogen-fixing bacteria (*Rhizobium* spp.), have been long recognized by agronomists and farmers (Brady, 1990), and the mycorrhizae (primarily *Pisolithus tinctorius,* or *Pt*) for tree crops by foresters and nurserymen (Pritchett and Fisher, 1987; Wilde, 1958). Marx (1991) has carried out a long series of mycorrhizal studies, culminating in material suitable for injection into mycorrhiza-poor soils with reasonable success (Marx et al., 1997).

The conceptual development of the sustainable soil (Chapter 4) requires attention to the microorganism status of the designed soil. Thus, a biological component specification is required. When, where, and how to use biological inoculation are covered in Chapter 4.

The questions to be answered for inoculate specification are as follows:

1. What type of organisms (mycorrhizae) are present on the existing site plants and how well are they developed?
2. If the site is a new planting, what is the conposition of the plant palette?

As discussed in Chapter 4, there are two different types of mycorrhizae, each appropiate for a separate group of plants. If the plant palette consists mostly of oaks and pines, and so on, the *Pt* ectomycorrhiza is appropriate. On the other hand, if the plant palette for a element of the project design is a mixed one, then a combination ecto-endo innoculum may be applied. If the palette is bedding plants or other horticultural material, the VAM inoculum is appropriate. One manufacturer has a separate inoculum for palms and associated plants.

ELEMENTS APPLIED TO ALL COMPONENTS OF THE SPECIFICATIONS

The following elements are each stated under each soil component in the specification for clarity. For redundancy reasons here, they are given separately from the components. Appropriate references to specific components are made where necessary. Every writer of soil specifications should make certain that all elements are covered for each component.

Reaction (pH)

The reaction (pH) of the soil material or mix component should approximately match that of the specified final pH for the project. In humid temperate regions, the pH should range from 6.0 to 7.5. Most plants can tolerate pH in this range with some exceptions. Otherwise, addtional expense and effort is required for pH adjustment. Further, an initial extreme low or high pH value in humid temperate regions may indicate possible nutrient deficiencies or chemical toxicities. In arid or semiarid regions, pH values will be alkaline. In these areas, the plant palette should be adjusted accordingly rather than the soil.

Adjustments. If the soil pH is not in the desired range, it usually can be adjusted by the application of various amendments. These amendments should be applied to the soil either during the mixing of the soil or broadcast on the surface and tilled into the soil. If vegetation exists on the soil, broadcasting is still an effective operation. Mulch should be removed to increase the effectiveness of the broadcast material.

Raising pH. Ground limestone is the most effective material for raising soil pH. It is slow acting and does not burn the vegetation even as a surface dressing if not applied in exorbitant amounts. Split applications in subsequent years are recommended when a large adjustment is required. The effect of one application usually lasts about 5 years depending on soil properties and the rainfall of the area. Other materials used as liming agents are burnt lime, CaO, and hydrated lime, $Ca(OH)_2$. Both of these materials are more effective than ground limestone on a per unit weight basis, but they have the disadvantages of greater expense and greater potential for burning. Most state agricultural experiment stations have developed tables for use in each state. It

TABLE 5.13 Lime Requirement for Soils with Given Initial pH, to Raise to pH 7.0

	pH					
Soil Type	4.5	5.0	5.5	6.0	6.5	7.0
	←——lb/1000 ft^2——→					
Sands	78	64	37	18	8	—
Sandy loams	165	138	78	41	18	—
Silt loams	266	207	122	69	32	—
Clay loams	321	271	161	88	39	—

Source: Generalized data from Michigan, Missouri and New York.

is best to use the applicable table rather than a generalized one. A generalized table for ground limestone is provided here for illustrative purposes only.

An Example. A silt loam soil has an initial pH of 5.5. The specified pH level is 7.0. By using Table 5.13, the amount to be applied is 122 pounds per 1000 square feet. If the ground limestone is added during the mixing process, then it should be mixed on a volume basis for only the upper 3 inches of soil.

Lowering pH. Lowering pH may be more common in semiarid and arid regions than in humid regions. The practice is increasing for urban areas where the pH is anthropogeically raised above native soils. Addition of elemental sulfur is the most effective method (Craul, 1992).

THE CONCERN FOR SUSTAINABILITY IN SOIL SPECIFICATIONS

The term *sustainability* may be defined two ways in the context of soil specifications. The first definition is applied to the reduction of on-site and off-site (mainly the latter) environmental impact by the proposed project. The second definition is applied to the ease and cost of maintaining the project upon completion, especially in minimizing the need for resource inputs of water, fertilizers, high-energy-consuming materials, electricity, and so on. Landscape architects should be deeply concerned with both definitions.

This concern can be manifested in the use of recyclable materials as substitutes to natural soil and other natural materials as amendments, such as peat moss. This topic has been discussed more thoroughly in Chapter 4 on sustainable soil design.

6

ROOFTOP APPLICATIONS

I'll huff and puff, and blow your house down!

INTRODUCTION

Increased sensitivity to and interest in the greening of rooftops require that some attention be given to the formulation of rooftop soils and the associated planting techniques (Fig. 6.1).

Several design criteria are present in rooftop plantings that do not apply to other urban planting situations. These criteria are architectural loading constraints, windfirmness of shrubs and trees, and special installation and sequencing requirements. An irrigation system is usually required due to limited soil volume. The design requires special care because of the limited soil volume and drainage in most rooftop applications.

LOADING AND SPACE CONSTRAINTS

In most cases, the building architects will provide for the additional weight of the rooftop garden, but the allowance is generally less than what weight would be required to provide for adequate soil depth and volume and moisture content under the exposure conditions. A rooftop garden frequently has greater exposure to sunlight and wind but perhaps lower temperatures. Reflected heat from the building facade may also become a problem. These environmental conditions normally require large amounts of rooting volume to provide for the high evapotranspirational stress. The limited space of the rooftop also causes the plants to be enclosed in small spaces or containers. Thus, the flexibility present on ground-level plantings does not exist on the rooftop. A further constraint is caused by the rooftop itself. The rooftop must

Fig. 6.1 Rooftop garden with trees and shrubs around a swimming pool over a parking garage in Needham, Massachusetts.

be waterproof, but at the same time carry off rainwater and irrigation water from the garden. The soil profile does not have the advantage of a connection to the ground watertable through the subsoil, parent material, and bedrock. The rooftop tends to perch the water in the soil profile, giving rise to drainage conditions similarly found in soils with fragipans. The rooftop garden soil must be designed to be well- or self-draining so that water may reach the drainage outlets provided in the structure. The rooftop planting behaves very much like a container planting if the planting stock is not already placed in containers. As a result, the soil water retention in the containerlike situation must be assessed with respect to its contribution to the loading of the roof. An illustration of the distribution of water within a containerized soil profile is shown in Fig. 6.2.

By using the examples shown in Fig. 6.2 and the weight of water at 62.4 lb/ft^3 (1000 Kg/m^3), the total water weight per square foot of roof surface for each example is 28.6, 45.5, and 66.1 lb/ft^2 (458, 729, and 1059 Kg/m^3), respectively. The soil weight contribution is dependent on the bulk density of the individual layers, if different. If homogeous soil is assumed with a bulk density of 1.45 Mg/m^3, the total surface loading without vegetation is 119.1, 180.96, and 271.4 lb/ft^2 (1908, 2899, and 4348 Kg/m^3), respectively. The loading limit is frequently specified by the building engineer as an average weight per cubic foot. The average weight for each example then becomes 119.1, 90.5, and 90.5 lb/ft^3 (1908, 1450, and 1450 Kg/m^3), respectively. Thus, the shallow soil has a higher weight per cubic foot simply due to its

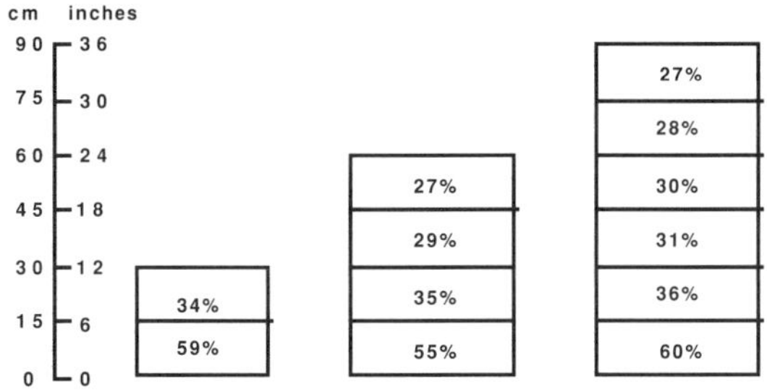

Fig. 6.2 Distribution of water in containers. Percentage of water on a volume basis. (Adapted from Harris, 1983.)

high moisture content. The additional soil volume with lower moisture contents of the upper layers reduces the average cubic foot weight in the other examples. The average value does not generally remain constant with additional soil depth, but will increase as soil depth increases.

It should be kept in mind that the soil weight of ball and burlap or container trees and shrubs is very heavy to that specified for general rooftop application. Table 6.1 provides root ball size and weight data for some common root ball

TABLE 6.1 Size, Depth, and Weight of Root Balls for Trees

Tree Diameter		Ball Diameter		Ball Depth		Approximate Weight of Ball and Tree			
in.	cm	in.	cm	in.	cm	lb	kg	U.S. Tons	Metric Tons
—	—	10	25	8	20	34	15	—	—
—	—	12	30	9	23	55	25	—	—
—	—	14	35	11	28	91	42	—	—
—	—	16	40	12	30	124	56	—	—
—	—	18	45	14	35	193	88	—	—
—	—	20	50	15	38	254	115	—	—
—	—	24	60	16	41	392	178	—	—
—	—	28	70	19	48	624	283	—	—
—	—	32	80	20	50	867	394	—	—
—	—	36	90	22	55	1216	552	—	—
—	—	42	105	25	63	1877	853	—	—
5	13	48	120	29	74	—	—	1.5	1.3
6	15	60	150	32	81	—	—	2.4	2.2
7	18	72	180	34	86	—	—	3.7	3.4
8	20	84	210	36	96	—	—	5.4	4.9
10	25	96	240	38	96	—	—	7.4	6.8
11	28	108	280	40	100	—	—	9.9	9.0
12	31	120	300	40	100	—	—	12.2	11.1

Sources: Himelick, 1981, and Watson and Himelick, 1997.

sizes. As a result of the additional loading, it is necessary to carefully consider the location and arrangement of larger plants and trees across the rooftop. These topics are discussed later in the chapter.

Weight reduction can be achieved by planting bare root stock on the roof or in lightweight containers (molded plastic or fiberglass) with lightening amendments to the soil. The lightened soil may significantly reduce the rooftop loading. However, the procedure has the disadvantage that the plants are not as windfirm as those planted in heavier, unamended soil. Staking or some other type of protection is required. In all cases, the root system should be well established before exposed to the potential winds of a rooftop application.

LIGHTENING AMENDMENTS

The designed soil must reflect the preceding constraints by incorporating special components or features. The soil character becomes more of a potting mix application rather than a so-called street or field soil application. However, there is a limit to lightening of the soil. There must be sufficient soil weight to hold the root plate in place as a resistance to windthrow, especially for the larger shrubs and trees. Toppling of these plants is common on rooftops when the force of winds are not considered in the planning stages of rooftop design.

Estimating Bulk Density from Texture and Amendment Content

In order to ensure compliance with the roof loading constraints it is necessary to obtain an estimate of the net dry bulk density of the designed soil. Rawls (1983) provides one method, which is

$$\rho = \frac{100}{\left(\dfrac{X}{\rho_o}\right) + \left(\dfrac{100 - X}{\rho_m}\right)}$$

where ρ is the soil bulk density (Mg/m^3), X is the percent by weight of organic matter, ρ_o is the average bulk density of the organic matter (0.224 Mg/m^3), and ρ_m is the bulk density of the mineral matter (Mg/m^3). Other components can be added to the equation provided the bulk densities and percentages by weight are known. For example,

$$\rho = \frac{100}{\left(\dfrac{X}{\rho_o}\right) + \left(\dfrac{100 - X}{\rho_m}\right) + \left(\dfrac{100 - (X + Y)}{\rho_a}\right)}$$

where ρ_a is the bulk density of the amendment, and Y is the percent by weight of the amendment. In the case of rooftop soil applications, the amendment is normally a lightening agent. However, the author has found that the equations significantly underestimate the bulk density of a mixture material.

Another possible method of calculation to obtain net dry bulk density by the author is that of direct component proportion. This method has been used to obtain approximate bulk density of mixes on several projects, though it, too, tends to underestimate bulk density for some mixtures. For example, a mixture is to be 75% sand (ρ = 1.55 Mg/m^3), 15% lightening agent (ρ = 0.60 Mg/m^3), and 10% organic matter (ρ = 0.22 Mg/m^3). Then:

Sand:	0.75 × 1.55 Mg/m^3 =	1.162
Lightener:	0.15 × 0.60 Mg/m^3 =	0.090
Organic matter:	0.10 × 0.22 Mg/m^3 =	0.022
Weighted bulk density:		1.274 Mg/m^3

Once the mixture has been chosen, actual samples should be mixed and packed to the appropriate Proctor value and the dry bulk density determined. Modification may be required to adjust the bulk density to the desired value. Iterative testing should be done for every mixture employed in the soil design.

Several soil engineering and soil physics texts examined by the author do not provide any methods for estimating bulk density other than for application to assumed homogeneous material and regular packing of uniform spheres. No methods are given for the determination of bulk density of mixtures of particles. The only reference that provides guidance to the calculation of random packing of nonuniform particles is by Wise (1952). The author is now pursuing application of the methodology to soil mixtures.

Lightening Materials

Some of these materials were described as amendments for the designed soil structure component in Chapter 5. The descriptions are repeated here for those materials that are used primarily as lightening agents. The physical characteristics of each are described in more detail to provide a basis for their evaluation in use as lightening components to the designed rooftop soil. It goes without saying that their amendment to the soil also contributes to the structure component as well as lessening the soil weight itself.

Peat Moss. A very light material that is soft and easily compressed. In shredded form, it must be mixed as a very large proportion in order to have a significant lightening effect. The overabundance of the peat moss then leads to settlement and compaction, especially in large containers or extensive rooftop planters, creating potential waterlogging, which is difficult to repair. The author no longer includes peat moss in any soil amendment specifications even though it may be favored by many landscape architects and organic gardeners. The bulk density of shredded peat moss is about 0.27 Mg/m^3 (Table 6.2).

Composts. Compost derived from organic materials will have a lightening effect when mixed with mineral soil material and the resulting distribution of the mix. However, the lightening effect will vary depending on the particle size distribution of the compost. Some composts are very coarse, containing a large proportion of wood chips and/or other coarse organic fragments. The large fragments dispersed throughout the mix will tend to lighten the soil more than a screened compost consisting primarily of fine particles less then 1/4 inch (6 mm) or so. A large proportion of wood chips or coarse fragments is not desirable in most designed soil mixes. Additional amendment of another material is required to achieve the load limit.

Diatomaceous Earth. This material is derived from hydrated silica mineral formed from the remains of diatom sea organisms. The material is calcined at high temperatures and then crushed and screened to various sizes and mixtures of sizes. Some manufacturers extrude the material to form tubular granules, increasing the lightening and porosity effects. Commercial examples of this material are AXIS® and ISOLITE® (Table 6.2).

Sintered Flyash. The slurry from limekiln or coal-fired generator plant smokestack scrubbers is dried and then fired at high temperature to form sintered flyash. It is lightweight and relatively hard. However, it usually has an alkaline pH that limits its horticultural use. It is seldom specified in landscape architecture projects.

TABLE 6.2 Properties of Various Lightening Materials Used in Soil Mixes[1]

Property	Axis®	Compost[2]	Isolite®	Peat moss	Profile®	Ridgelite®	Turface®
CEC, cmols/kg	27.0	—	1.2 to 1.9	—	33.6	—	28.8
Bulk wt., lb/cy	972	850	850	456	938	1161	900
Bulk density, Mg/m^3	0.60	0.22	0.50	0.27	0.60	0.63 to 0.69	0.54
Particle density, Mg/m^3	2.20	—	2.27	1.71	2.50	1.16 to 1.28	2.55

[1]Commerical names are used as examples and only for comparison purposes.
[2]Composted biosolids as processed by Allgro Corp., Merrimack, NH.
Sources: Data from various manufacturers information sheets.

Expanded Shale. Shale is calcined to form crystallized expanded stable particles. The coarse aggregates are screened to various sizes. The material has been used extensively in golf green mixtures. An example is RIDGELITE® (Table 6.2).

Pumice. This material is of volcanic origin, having been molten material filled with large bubbles of gas. The gasses escaped on cooling leaving behind large circular pores. The material is extremely light and can be crushed into finer particles. The material increases the overall porosity of the soil but the water retention of pumice itself is not high when crushed because of the shape and size of the pores.

Fired or Calcined Clays. Montmorillonite clay is fired at high temperatures (>1300°F). It is a hard porous material. It is then crushed and sieved to various sizes (textures) for different applications. An example is TURFACE® (Table 6.2).

Styrofoam. This material may be produced in several forms as sheets, cast blocks, or granules (pellets). The usual form utilized in landscape projects is the sheet or block form. The reason for this preference is that the cast styrofoam is more stable and resistant to compression from the force of soil weight applied to it from above. The granules or pellets are not favored as a mix component because they are subject to compression under soil loading, reducing their beneficial effect of soil lightening. Rigid sheets of extruded styrofoam were employed to lighten the loading and create a better distribution of weight over the roof area in the Boston Commons garage (Fig. 6.3). Styrofoam was also used in the Battery Park City South Cove design to lighten the loading over the steel overhang of the promenade. Cast styrofoam blocks were used in the J. Paul Getty Fine Arts Center plantings that were on-structure. Severe weight constraints existed. Placement of the cast styrofoam blocks had to be "double-staggered" in order to reduce loading over the span and yet not interfere with the required soil rooting depth for medium to large trees (Fig. 6.4.)

Load Arrangement

The use of styrofoam and its arrangement on the rooftop suggests that the proper arrangement of the load on the rooftop may overcome some of the weight limitations imposed by the building architecture. The most obvious solution is the placement of the heavy components of the plant design over the vertical columns or roof supports (Fig. 6.4), particularly where a thicker soil profile is required, or containerized trees must be placed. In the latter case, the weights of the tree and root ball (Table 6.1) and the container must be considered.

Fig. 6.3 Styrofoam sheets placed over the rubber membrane on the Boston Commons Garage.

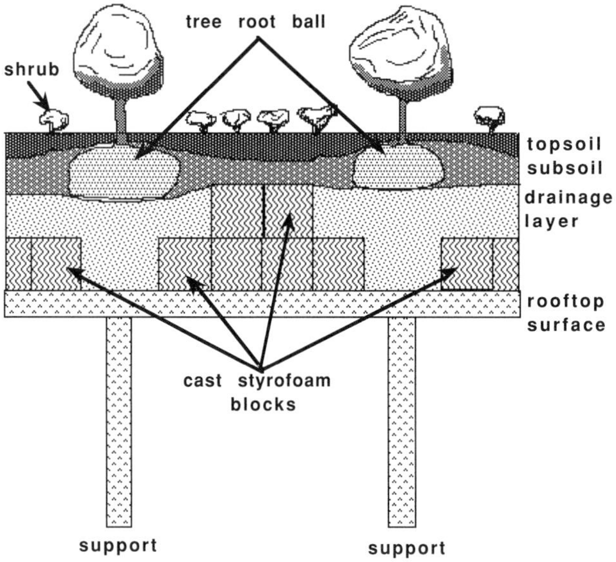

Fig. 6.4 The cross-section of a planter showing cast styrofoam block arrangement, J. Paul Getty Fine Arts Center, Los Angeles.

In cases when it is desired to have a heavy load placed on a span rather than over a support column, a cantilever base can be used to redistribute the load to the support columns (Fig. 6.5). However, it must be remembered that the cantilever base must be strong enough to support its load and it becomes a component of the total load. It is possible that the total load may exceed that for the support column.

TREE ARCHITECTURE

The architecture of a tree is the shape and form of its trunk, how the branches are attached and arranged around the trunk, and the general outline of the crown, which includes the fine branches and foliage. The several forms are illustrated in Fig. 6.6. The forms are part of the genetic design and also of chance. It is a challenge to distinguish the genetically determined form and structure from environmental damage and phenotypic plasticity (Tomlinson, 1982). The structure of a mature to aged tree is a visible historic record of an organism, the integration of a lifetime of continued growth, organogenesis, organ death, and active structural responses to a changing environment. Obviously, the form and structure of a young tree recently transplanted have a crown history reflective of its treatment in the nursery and its response to transplanting, but defined by the genetic form.

The various crown shapes present different resistance to winds. Open, coarsely branched crowns exhibit less wind resistance than fine, closely branched compact crowns. High clustered branching produces greater bending force on the stem than a uniformily distributed branching along the stem. Whether the tree was subjected to winds in the nursery also determines its resistance to wind damage after transplanting. The oscillating movement of

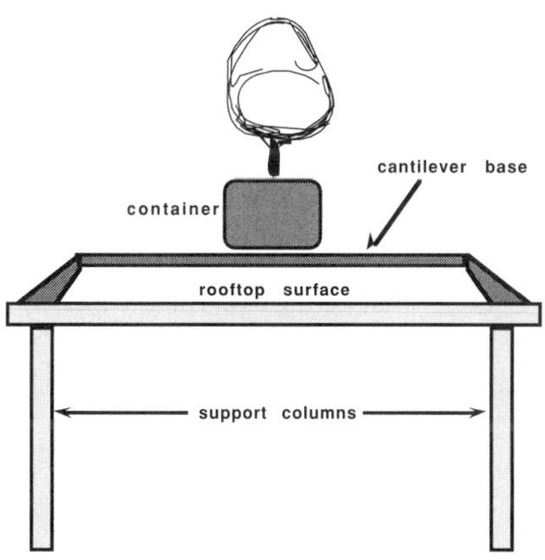

Fig. 6.5 A cantilever base to redistribute a containerized tree load to support columns.

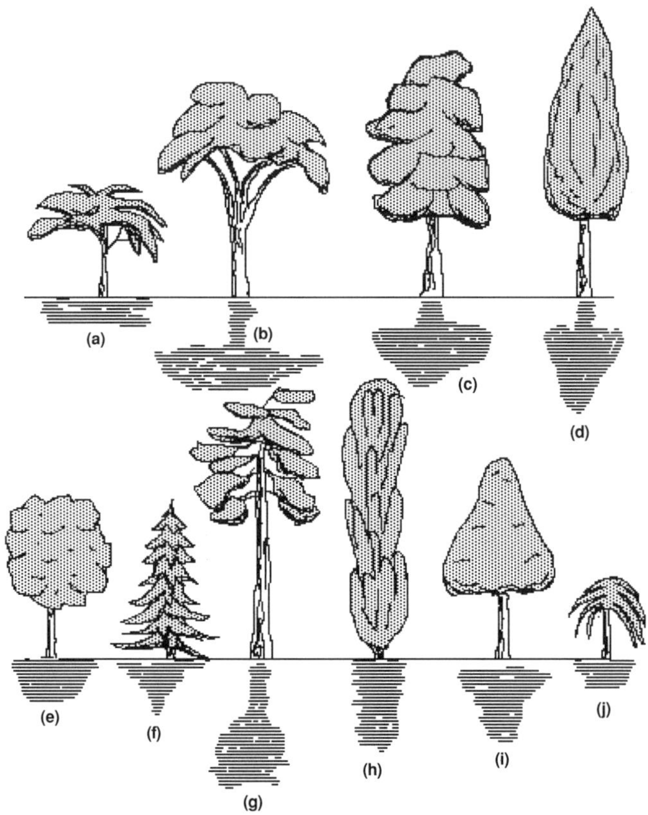

Fig. 6.6 The forms of tree architecture. (Craul, 1992.)

the stem reinforces the development of a strong stem base during juvenile growth (Harris, 1991). If the tree is not subjected to wind or is firmly confined by strong staking, a weak stem base develops. Harris cites examples of trees falling over when the staking is removed. Some flexibility in the stem is desirable, but the mechanical system of the tree must be able to support itself.

THE EFFECTS OF WIND ON TREES

It should be apparent from previous discussion that wind has effects on trees. Wind passing through the crown will tend to lower foliage temperature, but at the same time transports moisture from the boundary layer over the leaf surfaces. The transpiration rate may be increased until the stomates close. Calm wind allows the boundary layer to thicken, reducing the transpiration rate, but increases the crown temperature if in full sunlight. If the temperature is raised to a critical level, the leaves may wilt or become scorched. On the other hand, transpiration may increase by 20 to 40% in moderate winds (Chaney, 1997). Further, the CO_2 level in the crown may decrease under calm air as compared to the concentration outside the crown. Reduction in photosynthesis results. A wind of about 12 mph (27 m/sec) is required to maintain the crown CO_2 level at atmospheric concentration. Species intolerant to wind will close their stomates such as rhododendrons (*Rhododendron sp.*) so that moisture loss is less than that in calm air. Species tolerant of wind (*Larix sp.* and *Alnus sp.*) will maintain open stomates and continue photosynthesis and transpiration. Closure of stomates may completely inhibit photosynthesis without eliminating water loss because the leaf cuticle of many trees is more permeable to water than to CO_2. Species intolerant of wind include most of the Japanese maples, paw paw, camellias, flowering and Chinese dogwoods, most evergreen hollies, magnolias, sourwood, sequoia and redwood, yews, arborvitae, hemlocks, most viburnums (Flint, 1997).

The shaping of tree crowns by wind on mountains and coast lines is very obvious to most observers. The windward branches are bent and stunted by mechanical effects of wind and by the dessication of buds and shoots. Those branches on the leeward side are elongated by the wind "training" them and the normal growth of the buds permits branch elongation. An asymmetrical crown results with a sometimes pleasing appearance. The desire for symmetrical traditional shade trees for formal rooftop applications seems to be the usual case, but the author has seen some excellent alpine trees used on rooftops with great aesthetic appeal.

The type of leaves determines the influence of wind on the foliage itself in the form of drag (Vogel, 1996). Broad leaves, such as maples, can be easily folded, creased, or torn by winds, especially when they are immature. Many leaves will simply curl and form a cone on the wind. This reaction minimizes the drag of the wind and reduces the turning force on the bole (Vogel, 1996). Mature leaves appear to be more resistant to shredding, but all leaves will shred above a critical wind velocity. There are even differences within the maples. Norway maple (*Acer platanoides,* L.) has very tough leaves, whereas red maple (*Acer rubrum,* L.) leaves will curl and tear in moderate wind. Even the pinnate leaves will curl up along the rachis to form a hollow cylinder to reduce drag. The strength of the petiole varies among species as well. Long petioles and protruding lobes at the point of connection between the leaf and petiole strengthen the arrangement. Short petioles have grooves or ridges or a flattened cross-section that allows them to twist in the wind. Aspens and cottonwoods have flexible petioles and flex in the wind giving the pleasant flickering appearance of the foliage in a light breeze. Thus, flexibility within the structure of the tree makes it more resistant to the destructive forces of wind.

The stem or bole of the tree also exhibits the effects of wind by its growth response. Trees exposed to wind movement will have shorter trunks with thicker bases and greater taper than those of trees grown within a copse or forest stand. The cross-section of the trunk will be elliptical in shape with the longer axis parallel to the wind direction. This growth response is an attempt by the tree to strengthen the trunk against the force of the prevailing wind. But as Vogel (1996) points out, most trunks are nearly circular in cross-section to provide greater resistance to twisting loads. As mentioned earlier, the turning force on the trunk is the cause of a trunk snapping or breaking. The taller the tree, the more susceptible the tree is to breakage because the turning force moment is higher in the bole.

Windfirmness Considerations

The mechanical effects of winds, especially those of gale force, on trees is well known to the general public. The injury may involve blowdown, shear-off of the trunk, limb breakage, stripping of foliage, or simple bending over of the trunk. Blowdown is nearly always fatal (there are examples where the tree was uprighted, dislodged roots and soil replaced, and the tree survived) because the tree suffers excessive damage. Trunk shear is always fatal. Damage by limb breakage may be repaired if not severe, and stripping of foliage is usually not fatal to most healthy deciduous trees. Stripping of needles from conifers, though not common, is nearly always fatal.

Though this discussion may seem out of place here, windfirmness is a primary concern for rooftop applications even though it is fully realized that it is applicable to any tree planting. Foresters and arborists have studied the windfirmness of trees for years, especially in plantations and old growth forests (Coutts, 1983).

The windfirmness and resistance to breakage of a tree is dependent on the architecture and aerodynamic parameters of its crown and foliage, strength of its wood, its root system pattern, and the physical aspects of the soil in which it grows, as discussed earlier. To fully appreciate the concept of windfirmness, it is necessary to discuss in some detail the components that contribute to windfirmness.

Based on observations made after a recent hurricane in Charlotte and Raleigh, North Carolina, winds of 50 to 55 mph (80 to 88 km/h) seem to be the threshold for windthrow of ground-surface trees. The apparent cause was that the soil itself was unable to hold the tree down primarily due to saturated soil (more than two-thirds of the trees observed). At this point, the soil exhibits little friction among its particles and aggregates and between these and the tree roots. Roots can then slip from the soil, and if anchoring sinker roots are not present, the tree falls. It was found (Smiley, 1997) that 33% of the trees had reduced root systems caused by fill being placed over the root systems during adjacent construction. Root rot was a factor in 8% of the observations. Some trees subsequently diagnosed as susceptible to windthrow were actually protected by adjacent upwind trees; isolated trees are high risk for windthrow if the root systems have been reduced or impaired for some reason.

The Root System

The support and anchorage functions of root systems receive little attention from plant physiologists. The lack of information on the subject creates a tree planting design situation with insufficient information.

The reaction of trees to wind stresses varies with age and maturity. Juvenile trees will tend to bend at the stem base and are toppled without breakage or uprooting. If unattended, the tree will continue to grow in the prostrate form and become permanently bowed. A more mature tree will have a rigid stem. When wind stress is applied, the tree becomes uprooted or windthrown. In some instances of turbulent wind, the stem may be broken or sheared off at some distance above the stem base. In other cases, the stem may be delaminated. The toppling and windthrow may be the result of previous unstable conditions in the soil.

Wide variation in stability is exhibited between different sites (soils) and species. The ideal root system for any condition of anchorage is not known because root morphology tends to be dominated by site conditions, so simple direct observation by excavation is not helpful. Vogel (1996) suggests four distinct schemes of root anchorage. Combinations are possible and an individual tree may use several different schemes as it progresses through its life cycle. The four schemes are compressive buttressing, tensile buttressing, taprooting, and diagonal guying.

Compressive Buttressing. The most common scheme in many angiosperms and gymnosperms of temperate regions is the development by the tree of a stiff or rigid wide base so that the pivot point or axis of turning is well aside the center of the trunk and root mass (Fig. 6.7). Compression loading occurs

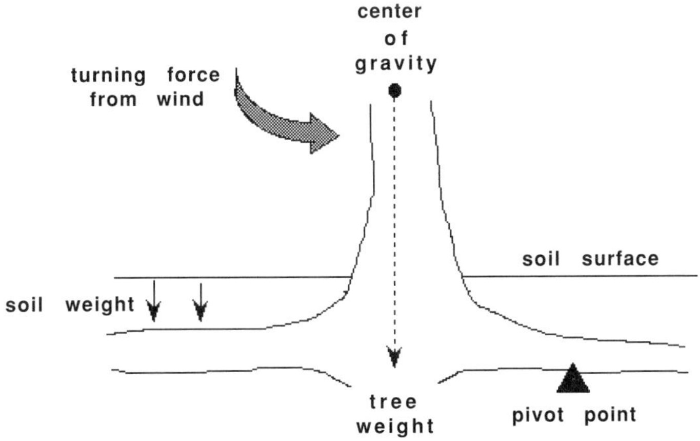

Fig. 6.7 Factors in compressive buttressing of roots. (After Vogel, 1996.)

on the leeward side with the pivot point displaced laterally depending on the root and soil strength. Uplift force against the weight of soil occurs on the windward side. Thus, to topple or throw the tree, work must be done against the soil loading and the soil and root tensile strength on the windward side together with the compression bending of the roots at the pivot point on the leeward side. The stronger the roots and soil and the deeper the roots, the less the danger of windthrow. The effectiveness of the arrangement is increased if the trunk is stiff, minimizing downward drift of the trunk center of gravity in the wind.

Tensile Buttressing. This type of buttressing occurs mainly in tropical rain forest trees, but is included here to provide a full range of anchorage schemes (Fig. 6.8). The buttresses common to tropical trees are very strong in reaction to tensile forces but very weak by compressive forces. Their narrow width does not provide for compressive strength. Since the buttresses completely surround the base of the trunk, tensile support is provided in every direction and compression forces are reduced accordingly. Stabilization of a very stiff trunk is less important here. Vogel (1996) suggests that trees with tensile buttressing should not be planted in isolation.

Taprooting. The scheme depends on the compressive resistance of the soil. If a taprooted trunk is moved in one direction, the taproot moves in the opposite direction since the pivot point is at the center of the taproot and lateral root mass. The lateral roots must be lifted on the windward side and compressed on the leeward side (Fig. 6.9). Additional vertical sinker roots, common on longleaf pine (*Pinus palustris,* Mill.), will strengthen the system. Some authorities list the oaks as taprooted species. However, many lose the taproot characteristic as the laterals develop with advancing age. The system becomes more like the lateral, widespread system of many other angiosperms.

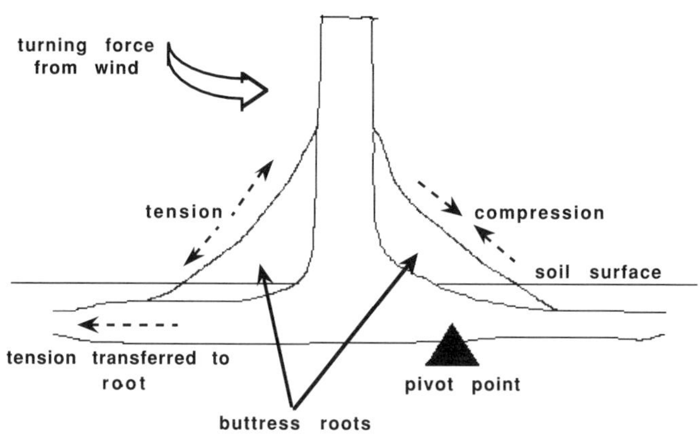

Fig. 6.8 Tensile buttressing of tree roots. (After Vogel, 1996.)

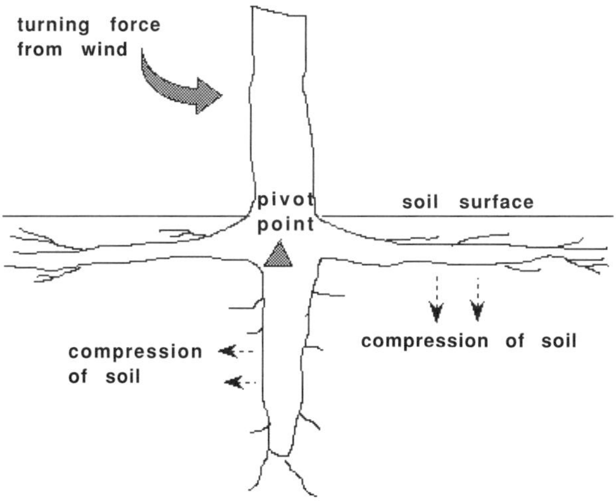

Fig. 6.9 The taprooting scheme of tree anchorage. (After Vogel, 1996.)

Diagonal Guying. This scheme occurs in bamboo and corn (Fig. 6.10). It certainly provides some interesting, though not widely applicable, details. The scheme is really a variation of tensile buttressing, with guying structures connected from the trunk to the lateral roots, which are tensile strengthened by a great mass of tangled roots. Guy structures also extend from the root mass downward to the toproot so that the guying is symmetrical around the stalk. Wall thickening of the stalk at the base and downward through the taproot provides the necessary stiffness.

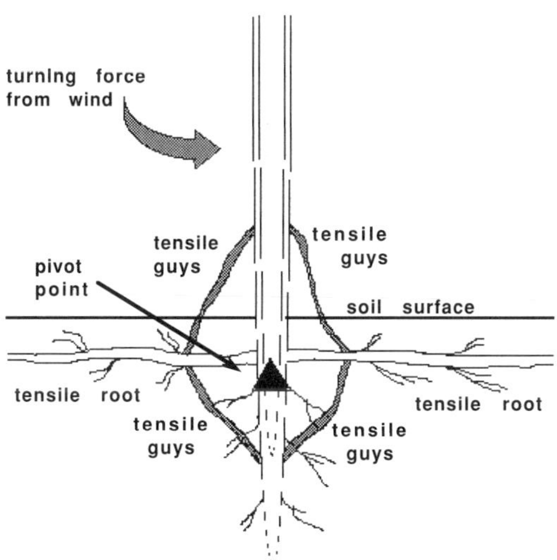

Fig. 6.10 The guying scheme of tree anchorage. (After Vogel, 1996.)

The Root Plate. From above there is differential reaction between the leeward and windward side of the tree root system. On the leeward side, the roots are subjected to bending and compressive forces against the bearing soil below. On the windward side, the roots are subjected to tensile and shearing forces as the root system tends to be uplifted against the overburden pressure of the covering soil. Main laterals lying near the surface stiffen the soil more efficiently on the windward rather than leeward side. Deeper horizontal roots located on the leeward side will have a reinforcing, stiffening effect on that side when they become under tension in the lower convex part of the bending root-soil plate (Fig. 6.11). Additional resistance to displacement arises from the mass of the root-soil plate and from soil tensile strength around its edge and beneath it.

The root may be considered as a circular beam. However, Nicoll and Ray (1996) have found that the cross-section of the main windward roots of Sitka spruce (*Picea sitchensis,* [Bong.] Carr.) differ from that of the leeward roots. The leeward roots have an upright T-beam cross-section, wherease the windward has an I-beam cross-section. These differences suggest the response of the root system to the differential stresses on either side of the tree. The differential thickening starts within the first 3 years of radial root growth. Wilson (1975) reports that white pine roots show the thickening on the top, and as the distal end is approached, the thickening shifts to the bottom side of the root. Circular beam stiffness is related to the fourth power of its diameter. Therefore, if a root forks into two equal-sized roots with the same total cross-sectional area of the initial root, the total stiffness is reduced by half (Coutts, 1983). Branching in the leeward roots reduces stiffness, so that the fulcrum of root-soil plate uplift occurs closer to the stem and stability is reduced. This condition suggests that a root system of a few thick lateral roots at the stem base (the framework roots of Lyford, 1980, in Craul, 1992) will be stiffer and more stable than one with many thin ones. Wide divergence between laterals also reduces the stiffness of the root system. The reduction

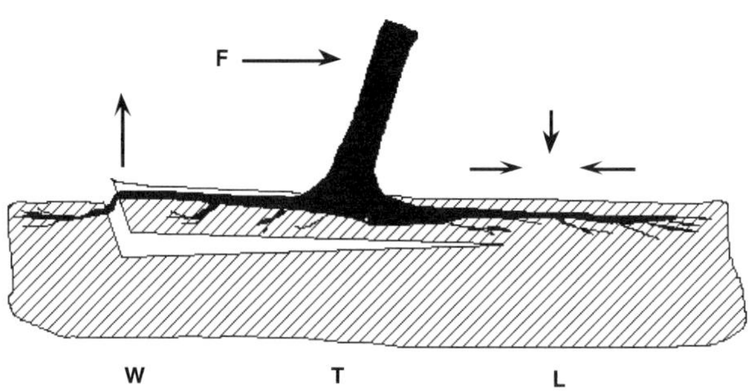

Fig. 6.11 The root-soil plate for a soil under wind stress from left to right. W is the location of the upward force and L the hinge of the root-soil plate with compression of the roots and the soil. T is the center of the tree stem and F is the force of wind. The roots to the left of W are under tension and tear as the plate is lifted. (After Coutts, 1983.)

in stiffness is significant when the angle between laterals exceeds 60 degrees. The windward side of the root system is under tension, and subdivision does not affect tensile strength unless the loading is applied to a single root rather than both or all.

There is danger that early and severe root pruning in the nursery will have detrimental effect on the development of thick laterals close to the root collar and the wide divergent (greater than 60° between forked laterals) laterals. Unfortunately, the goal in many nursery operations is to encourage the development of fine fibrous root systems to aid the survival and quick establishment of the transplanted stock. Whether these roots will eventually develop into thick laterals with little subdivision close to the stem base is open to question. The question requires elucidation by some very specific root research.

Some observed root strength values (Coutts, 1983) show that root failure occurs at 11 to 20% strain under loads of 15,000 to 63,000 kPa. Bark contributes little to strength and strength varies greatly among species. Soil tensile strength is 3 to 5 orders of magnitude less than the tensile strength of roots. The strain at failure is only about 1.8%. Variation occurs with soil moisture and organic matter contents, with both factors reducing the strength. Obviously, a saturated organic soil would have the lowest strength and hence the greatest windthrow hazard. Root slippage in saturated fine textured (clay) soils is the cause of failure for that group of soils. The soil properties are more important than root morphology in this case. However, as the soil dries from saturation, root morphology increases in importance and soil properties decrease. Where a large amount of much-branched root material exists, the root strength will exceed the tensile strength of the soil.

The root-soil plate is generally elliptical in shape, with the longer axis perpendicular to the wind direction. In cross-section, the plate tapers toward the edges. The depth and thickness of the root-soil plate is dependent on both root morphology and soil properties. The root biomass is not necessarily uniformily distributed throughout the root-soil plate. If the root biomass is clustered asymmetrically in the root-soil plate, trees may be easily thrown if the stress loading is applied in a direction other than the center of the mass. The various mechanical adaptations of the root system increase rigidity and move the hinge point away from the tree, thereby lengthening the lever arm and the tree's resistance to overturning.

Spatial Arrangement of the Trees

It may be a simple application of common sense, but proper arrangement of trees will greatly aid in the reduction of problems associated with windthrow hazard. Even the consideration of physical windbreaks on rooftops is appropriate and should not be overlooked in the planning stages of a rooftop design.

Large shrubs and trees should not be isolated on a rooftop. Either they should be placed in protected locations or be arranged in a copse or cluster so that common protection results. A cluster of three or four individuals makes a pleasing arrangement. Plants of low form such as the heath and alpine plants are more satisfactory for isolated locations. Where space is not at a premium, it may be beneficial to develop a living windbreak. Plants resistant to the effects of wind are placed in rows on the outside with the less tolerant species placed on the inside or leeward side. The first row may be short plants and each successive row would be of taller, less tolerant species. If single rows are used, it is best to align the row parallel to the prevailing wind rather than perpendicular to it. This method allows the first individual to protect the remainder of the row. The arrangement must also keep in mind the weight loading of the rooftop.

Implications to Rooftop Plantings

Because of the greater exposure to winds, the rooftop garden must provide more windfirmness for plants, especially small trees, than many on-ground plantings except for those planted in a wind-tunnel-effect situation. Greater soil depth and larger root balls compensate for the greater windthrow hazard. However, this requirement creates a paradox on the rooftop, where soil depth and volume are constrained by loading limitations. Since space is limited, there is danger that insufficient rooting area in soil is available so that the shrub or tree may develop the appropriate root anchoring scheme outlined earlier. It would appear that large flat planters or planting zones are preferable on rooftops. The purpose is to allow the pivot point of the root system to be an adequate distance away from the stem for most trees (the compressive buttressing). Otherwise, trees with tensile buttressing or taproots should be planted in the closed areas.

One solution to the problem is to plant only small trees with thin crowns or place them in relatively sheltered spots on the windward side or confine them to the leeward side of the garden, if this flexibility exists. Another solution is to place the trees in their deeper soil over the building support columns rather than over support spans. The span locations can be planted to low shrubs, herbs, and grasses. But, it is absolutely essential to be informed about the loading limits over the entire rooftop.

Most trees, regardless of size, are usually placed in planters or containers for rooftop applications. It is hoped that the containers have adequate volume to sustain root growth as the trees mature. Initially, the root system is mainly within the root ball and the stability of the tree is determined by the relative stability of the root ball itself within the container. If the backfill soil of the container does not stabilize the root ball, then the tree and its root ball is subject to windthrow. The tree and root ball can simply be turned or tilted in the container. The backfill soil must be removed, the tree and root ball straightened, and the backfill replaced. As the roots extend themselves into the backfill soil and the tree adds to its vegetative weight, the tree becomes stabilized by the entire soil volume and possibly the weight of the container. Above a threshold strong-wind velocity, the entire system of tree, soil, and container can be tilted or tipped over if the tree does not first break up, stripped of foliage and branches, and so on, in the wind gusts (Chaney, 1997). The event is possible since container weight and volume are limited in rooftop applications.

The shape of the container influences the windfirmness of the containerized plant as a unit. As suggested earlier, a broad flat container provides more lateral rooting space for the plant, enhancing its windfirmness. The container shape also provides great stability to the unit, as it is difficult for it to be overturned in high winds. Straight-sided cylindrical containers are more stable than inverted cone-shaped containers (the common flowerpot shape). Square or rectangular shapes are more stable than cylindrical ones of equal basal area. Some authorities suggest that perhaps an unstable container that tips over in high winds may better protect the containerized plant as compared to it being held at the vertical position. If this alternative is chosen, provision must be made in the design for the potenital tip-over so that the occurrence does not damage adjacent plantings or the structure.

ROOFTOP IRRIGATION

Rooftop gardens usually receive a high degree of postinstallation maintenance attention. The value of the planting requires necessary greater care in the function and operation of the irrigation system than in other planting situations. In many cases, the plants, especially if containerized, are watered by hand from a water can or hose.

The low-volume drip irrigation system is generally recommended for rooftop application where the area or number of plants is extensive. There are several reasons for the recommendation. Because of the increased potential evapotranspiration common on rooftops, the drip irrigation scheme would conserve more water than other methods. The required lightweight soils dry faster and require more frequent watering. The lightweight soils also are subjected to windy conditions and may blow easily when dry; thus, they must be kept moist as a preventive measure.

THE DESIGN AND SPECIFICATION OF ROOFTOP SOIL

The Soil Profile

Because of space and weight constraints, most rooftop soil profiles will be homogeneous. That is, they usually will not have designed horizons such as deeper, on-grade soils would have. There are exceptions, of course.

A contradiction arises in rooftop soil design. It is desirable to have the soil as light as possible to meet the roof-load limitation. And yet it cannot be too light as there may be insufficient weight to hold the plant root plate in place. Further, the addition of excess amounts of organic matter to lighten the soil may cause waterlogging and settlement as mentioned earlier. The structural lighteners meet this requirement. But, even this material should not comprise more than about 35% on a volume basis. Volumes between 40 to 50% or more of structural lighteners tends to make the soil mixture excessively drained or too dry, depending on the particle size distribution of the lightener. A compromise design that does not exceed the roof load limit and yet provides sufficient wieght for windfirmness is the primary objective. See Chapter 11 for more details on rooftop soil design.

Drainage. Either a well-drained soil must be designed for a rooftop or there must be sufficient depth to permit the inclusion of a deliberate drainage layer. Applying Davis and colleague's (1964) advice for a homogeneous soil without a drainage layer in a container, a homogeneous profile is appropriate for shallow rooftop soils, but the profile must be well drained to excessively drained to prevent the accumulation or perching of water, as suggested in the shallow soil examples of Fig. 6.1.

Project Weight. In summary, the design must address the combined weights of the moist soil, the plants and their root balls, and, if appropriate, the weight of any containers used and any other structures. If large trees are to be grown, and all other design requirements are met for them, their mature weights should be included in the total weight calculation. Tree biomass data are available in various forestry and orchard publications. Unfortunately, the data in the former publications are mainly for forest-grown trees rather than as individual open-grown trees. Orchard data are usually for trees of more appropriate size for rooftops, but not for traditional-size large trees.

Installation and Sequencing of Construction

The placement of the planting stock and soil requires more care on rooftops than elsewhere. On-site space is usually limited, access to the site may be only by crane or elevator, and the roof-loading limits preclude the use of mechanical equipment. Hence, much manual work is involved. Mixing of soil components is done off-site. The planting stock and soil are moved into place as needed. In most cases, the simultaneous placement of the planting stock and soil is the most efficient. Installation proceeds from the center outwards, or, depending on the location of the roof access, from the outer edge inwards toward the access door. If access is by crane or temporary construction elevator, the usual procedure is to begin at inside areas and work outward toward the access point. Worker safety is a major concern.

7

DRAINAGE TECHNIQUES

INTRODUCTION

Drainage deals with the movement of water over its surface and within it. Surface runoff is the usual term applied to the surface component of drainage, whereas permeability and hydraulic conductivity refer to the movement of water within the soil. Some authors use the term percolation to evaluate the rate of water movement in the soil. For specific definitions, see the Glossary.

The growth of many plants is adversely affected by the saturation of a significant portion of the rooting zone or ponded water on the soil surface (Soil Conservation Service, 1973). Poorly drained soils affect plant growth because of the following:

1. Wet soils have lower daytime temperatures than well-drained soils and warm up more slowly in the spring; hence, a shorter growing season.
2. Saturated soil contains little or no oxygen, inhibiting soil aerobic organism activity.
3. Some plant diseases and root parasites are favored.
4. Root penetration is limited by a high watertable.
5. Soil structure may not develop or may deteriorate.
6. There is concentration of salts and alkali in the root zone or at the surface if they are present in the soil and the water.
7. Wet areas within the project may prevent uniform growth of the plant palette or interfere with maintenance.

Surface drainage problems are generally encountered where the land is level or nearly so. Ponded water occurs in depressions or ridges and may retard natural runoff on sloping land. Low-capacity outlets may also slow the rate of runoff, causing backup and flooding behind the outlet. In some urban situations, there may be no obvious, accessible outlet. Physical obstructions such as curbing, walls, foundations, piping, and other infrastructure may also retard or inhibit surface runoff.

Subsurface drainage problems are related to the permeability of the soil or to the presence of physical obstructions below the surface. Soil profile drainage is determined by the *least permeable* horizon or layer in the soil (Craul, 1992). If a soil has a compacted, massive subsoil horizon (usually designated Bx), water may be perched above that layer at periods during the year, leading to saturated conditions in the upper profile. It may be totally disconnected from the watertable. A rooftop or other impermeable infrastructure may have the same effect as does a high watertable. High watertables are common on floodplains, low topographic depressions, and flat coastal terraces. Lateral subsurface flow of water from a higher location may also cause poor soil drainage in a lower topographic location. Upwelling of the water to the surface creates a seep.

The soil exhibits retentive capacity for water. The capacity is related to the texture, structure, and density of the soil (Craul, 1992). A portion of the water will flow in response to the force of gravity, thus draining the pore space that does not exert sufficient force to retain the water. The drained pore space is called the air-filled pore space. As the soil dries beyond this moisture content, due to water absorption by plant roots, evaporation from the soil surface, or other capillary forces, the proportion of air-filled pore space increases. It is the water contained in the macropores, the gravitational water, that is drained from the soil by a designed drainage system.

Drainage relates to the aeration of the soil as much as it does to water movement through the soil. Not only must water be able to infiltrate into the soil, be stored there for absorption by plant roots when required, and for excess amounts to drain away to the watertable below, but oxygen must be continually supplied to the roots and soil organisms through air-filled pore space (Craul, 1992). There are times and conditions when the pore space is completely filled with water, or saturation, and when the soil may be nearly air dry. There are periods during the year and instances in time when the amount of water applied to the soil exceeds its capacity to absorb and hold it (waterholding capacity), resulting in surface runoff. More specifically, the rate of application, or the intensity of the rainfall, exceeds the infiltration rate of the soil, causing surface runoff. The phenomenon is discussed in Chapter 10, "Project Erosion and Sedimentation Control." The soil must supply both oxygen and water to the plant roots; the soil should neither be saturated nor bone dry. In this chapter, techniques and designs to accomplish these objectives are discussed and evaluated (Fig. 7.1).

Even though drainage was discussed as part of the soil design process in Chapter 4, there are some facets of drainage that must be examined in more detail. Surface drainage may be addressed by soil surface shaping or topographic modication. Drainage of the soil profile is a more complex matter.

Fig. 7.1 Tree pit auger holes to a better-drained lower layer. The holes were backfilled with gravel, Lexington Avenue, New York City.

An internal artificial drainage system is normally required. The design and placement are critical to the success of the system.

An additional concern is the improvement of drainage and aeration for vegetation already established on a site. Some commercial products can aid the design and operation.

PRINCIPLES OF DRAINAGE

To better understand the design of drainage systems, it is important to be familar with the physical principles that govern water drainge of the soil. They will be examined on the basis of surface and subsurface drainage.

The drainage system is comprised of three functional parts (Fig. 7.2):

1. *The Collection System.* Small channels, rills, ditches, swales, buried pipe, and so on, through which flow occurs.
2. *The Disposal System.* The large channels or pipes that carry the water to the outlet. Sometimes the collection system may be connected directly to the outlet.
3. *The Outlet.* The end point of the immediate drainage system. It may be a road culvert, a roof drain, or an existing stream.

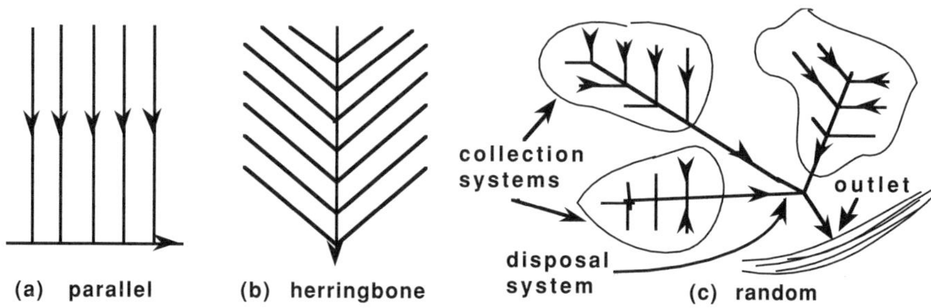

Fig. 7.2 Three types of drainage designs with the components.

Surface Drainage

Surface drainage involves the potential energy of water that exists when there is a difference in elevation. The elevation difference creates a hydraulic gradient that permits the water to move in response to gravity, or "water runs downhill." The water moves from a position of high potential to low potential. The potential energy is dissipated as kinetic energy when the water flows. It is a free-water-surface slope and may be temporarily confined in piping along its pathway.

The water-surface profile is the starting point in the design of the disposal system channels. The elevation of the outlet establishes the lower hydraulic control point. Control points along the gradient are critical low areas and the restrictions in the pathway such as culverts, bridges, and weirs. Within a project area, these features may be miniatures of those that occur on the natural landscape, but may restrict the flow sufficiently to have significant effect on the drainage of the project area. Serious impact on the growth and survival of the plant palette is possible if adequate surface drainage is not provided where it is necessary.

Where flow volume is large, Bernoulli's theorem is used to compute the hydraulic gradeline for steady-flow conditions. Losses of head due to friction are computed by an open-channel formula such as Manning's. Any hydrology textbook will explain these applications in detail.

Subsurface Drainage

Soil water occurs in the pore space of the soil. As such, its movement and retention is strongly dependent on the pore space geometry and volume. Pore space geometry includes the distribution of pore sizes (diameters), the shapes of the pores, the arrangement of pores, and their connectiveness with each other.

Pore space geometry is determined by the texture, type of structure, and density. Organic matter influences the structure and density. Coarse sandy soils have a preponderance of large pores with very few if any small pores, reflective of the large diameters of the particles. As the soil texture becomes finer, shifts occur in the pore sizes with increases in the proportion of smaller pores. The degree of shift depends on the type and degree of development of structure.

Granular structure will tend to have mostly large pores (macropores) with fewer medium (mesopores) and small pores (micropores). The macropore shapes will usually have concave surfaces, reflecting the convex surfaces of the granular peds. The points of contact between the peds will have irregular annular rings like coarse sand. The mesopores are the channels between ped (intrapedal) surfaces and may have regular (parallel) walls or be very irrregular. The micropores are usually the very small pores within the peds themselves and may be continuous through the ped or be close-ended. Subangular blocky to blocky structure will have less macropore space but more mesopore and micropore space. Platy structure has mostly mesopore and micropore space.

Dense soils by definition have less pore space than "light" soils. Very loose soils have a large volume of macropore space. A compacted clayey soil has a small amount of pore space mostly contained in micropores. Medium-textured soils with granular structure or well-developed subangular structure have a favorable distribution between the three pore size classes. Organic matter will tend to enhance the development of structure, especially the creation of macropore space, and increase the total pore space.

The Three Conditions of Water in the Soil. The definition of the three pore size classes discussed before depends on the reaction of the contained water to various forces acting on it. Saturated flow occurs mainly in the macropore space. Flow here is in response to the force of gravity and is referred to as *gravitational* water. This water contributes to the watertable below or may flow laterally as subsurface flow. Little water is retained in the macropore space unless the water is impeded in its downward movement and becomes perched. The soil then may remain saturated until downward or lateral movement occurs.

Water retained against the force of gravity within the mesopores and micropores is called the *capillary* water. This water moves in response to capillary or matric forces that exist between the soil particle surfaces and the water molecules, as well as the gradients created within the water "column" by water absorption of plant roots and evaporation from the soil surface.

The third class of water is in the form of water films that exist over the soil particle surfaces. It is the remaining water when the soil has reached wilting point. This water is not available to the plants and is called *hygroscopic* water.

The Watertable and the Capillary Fringe. In soils where a watertable exists, the water molecules at the surface are drawn upward in response to the capillary force of the soil (Fig. 7.3). The height of the capillary fringe is depen-

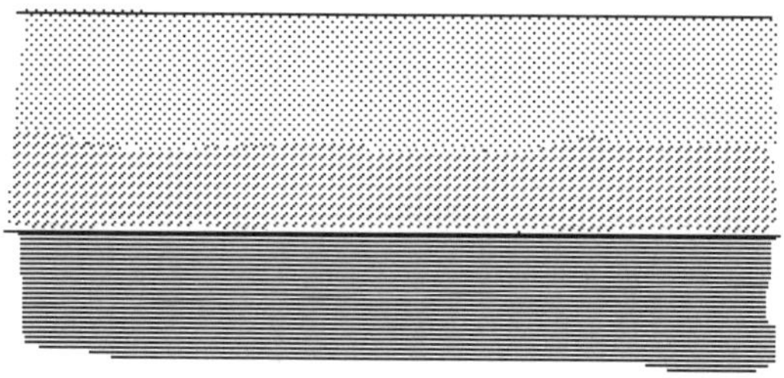

Fig. 7.3 The capillary fringe above a watertable.

dent on soil texture. The capillary fringe will be higher in clayey soils than in sandy soils (Holtz and Kovacs, 1981). In coarse sand, the height may be 1.5 to 6 inches (0.4 to 15 cm); fine sand, 15.6 to 138 inches (0.40 to 3.5 m); silt, 98 to 138 inches, or nearly a 12-foot maximum (2.5 to 3.5 m), and clay, equal to or greater than 394 inches, or 33 feet (> 10 m). The fringe is abrupt and obvious in sands and very diffuse and obscure in silts and clays.

The implication to drainage is that silty to clayey soils may require subsurface drainage even though the watertable may be below 6 to 8 feet (1.8 to 2.4 m). Sands, on the other hand, would not require subsurface drainage with a watertable at 10 feet (3 m) unless it is a dense fine sand.

Flow in the Saturated Zone. Soil drainage mainly deals with gravitational water, suggesting that soils to be drained are saturated in some portion that interferes with plant growth or certain land uses. The objective in drainage design is to remove the gravitational water. Saturated flow involves the phenomena of hydraulic head and hydraulic gradient.

The Hydraulic Head and Gradient. Saturated flow, whether in open channel or in confined space (the soil), occurs as a difference in total energy. The total energy is the sum of kinetic, pressure, and gravity components. Since saturated flow in the soil is low-velocity, the kinetic enegry is neglible, leaving the pressure and gravity (elevation) components. These components are illustrated in Fig. 7.4.

Hydraulic head (potential energy) is simply the difference in elevation of two points above a datum, usually the watertable depth. The hydraulic gradient is the difference in hydraulic head over a stated distance measured along the path of flow. Flow occurs in response to elevational or positional difference (running downhill), and from high pressure to low pressure. Drainage design takes advantage of the two phenomena. Hence, drain pipes must be placed below the upper boundary of the saturated zone in order for it to drain. The drain outlet must be open to the atmosphere so that the water may flow from high pressure (the saturated zone) to low pressure (the open pipe).

Flow Paths or Streamlines. Within the draining volume, there will be lines of "equipotential" head (Fig. 7.4). The flow of water will occur normal (per-

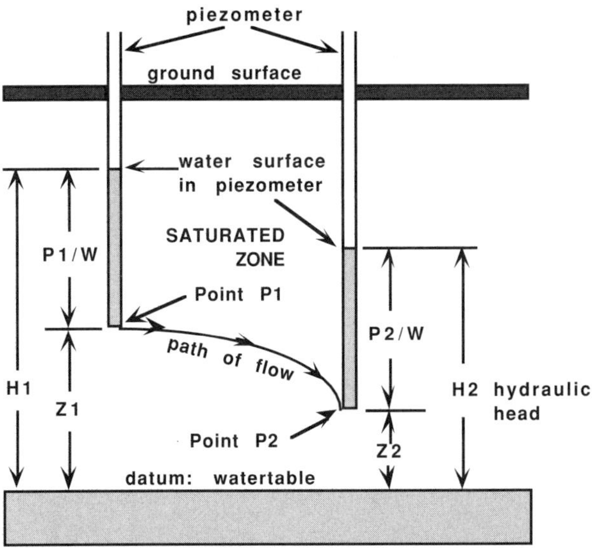

Fig. 7.4 The hydraulic head (H1 and H2) and hydraulic gradient (P1 and P2) showing the path of flow.

pendicular) to the lines of equipotential, provided that the soil is isotropic—having equal hydraulic conductivity in all directions. If the soil is stratified or anisotropic, the flow lines will be some vector between the normal of the equipotential line and the orientation of the stratification. The streamlines are predictable based on the knowledge of the equipotential lines and the hydraulic conductivities within the soil.

The Flow Nets. The vertical cross-section of flow paths to a drain within a saturated zone will exhibit a flow net (Fig. 7.5). The flow is the result of Darcy's law of saturated flow (Hillel, 1980). Topography, location, and quantity of the water source, stratigraphy, and drain locations set the boundary conditions for the flow net.

Figure 7.5 illustrates several conditions. The left flow net is for an isotropic soil and the right one for an anisotropic soil. The latter has a horizontal permeability that is 16 times its vertical permeability.

Knowledge of the flow net provides information for the design of the drainage system such as the number and placement of drain pipes to achieve a given drainage result.

Permeability and Hydraulic Conductivity. Some confusion may exist between the two terms. Permeability refers specifically to the capacity of the soil to transmit fluids (in this case, water) and is dependent only on the physical properties of the soil itself. It is also used as a descriptive term for the relative ease of a soil to transmit water. Hydraulic conductivity is based on the per-

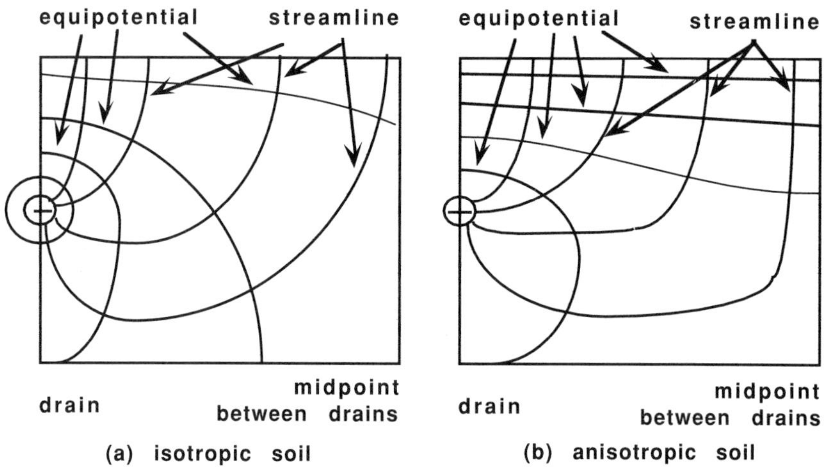

Fig. 7.5 The flow nets for (a) an isotropic soil and (b) an anisotropic soil.

meability of the soil and the physical properties of the transmitted water. The properties include viscosity, density, electrolyte concentration, and so on. Research has found that these properties may significantly influence hydraulic conductivity under certain conditions (Hillel, 1980).

ELEMENTS OF DRAINAGE DESIGN

The process of drainage design consists of several elements. These include the soil requirements for the plant palette, site investigations to determine existing drainage conditions, development of the design criteria for the project, and the plans and specifications for the drainage system.

Plant Palette Requirements

Various types and species of plants have different drainage (aeration) requirements. Some prefer well-drained soils, whereas others may require moist soils with a high watertable. Some plants may tolerate regular flooding of their root systems (many bottomland hardwood trees and shrubs), whereas others cannot survive even several hours of flooding (the Japanese cherries, *P. serrulata,* around the Tidal Basin in Washington, DC). The plant drainage requirement is based on (a) the maximum duration and frequency of surface ponding, if it occurs, (b) the maximum height of the watertable or the occurrence of a perched watertable during the dormant season, and (c) the minimum rate and depth at which the watertable must be lowered.

If drainage is complex, it sometimes is wise to accept drainage variability as an opportunity to have a more diverse plant palette compared to a uniformily drained site. It should also be remembered that the effects of a high watertable are carried well above the watertable depth due to the capillary fringe and fluctuations in watertable height during the different seasons of the year. The moisture content within the capillary fringe is well above field capacity. Thus, in some areas, hydrophilic species are favored and mesophilic species in other areas. The concept of diversity here is very similar to that for irrigation diversity of Chapter 8.

Site Drainage Conditions

The character of investigation for drainage needs and design depends on the project location. A project located on a natural landscape will entail a different investigative approach than one in a central city restoration area or a totally designed environment such as a rooftop. It is unlikely that a proposed project site presents uniform drainage conditions especially over large areas of natural landscape or in urban restoration areas.

Topography, especially on natural landscapes, is the first element to be investigated. It reveals surface slope, hydrography, and potential drainage outlets. It may also indicate the bedrock situation from the present landform. If the natural drainage is favorable, the drainage design may be very simple and inexpensive. If the natural drainage is unfavorable, the system may be more complex and costly.

For natural landscapes, an examination of the soil map will reveal drainage classes, depths to watertable and bedrock, and permeability of the soils. Sometimes it is necessary to carry out a special intensive soil survey (Order 1) to provide greater soil detail. Mapping at a scale of at least 1:12,000 or even 1:4000 is appropriate. Geologic maps and hydrographic information is very useful for the determination of impermeable rock formations or pervious unconsolidated deposits, watertable depths, presence of aquifers, and so on. Drilling logs are a source of subsurface materials, their sequence, and thickness. A vegetative survey will also provide information on soil drainage on relatively undisturbed sites. If on agricultural land, the farm plan on file in the local agricultural office (USDA) will provide information on existing drainage systems. Finally, on-site permeability and percolation tests may be required especially where drainage may be a serious problem. Installation of peziometers to determine seasonal watertable depth and flow provides even greater information for drainage design.

For restoration sites, existing data and information may be limited. The topography may not provide valid information if the constructional history of the site is complex. The soil material may not be natural and reflect the impact of urban development, being highly variable and complex across the site. Detailed site assessment is required as suggested in Chapter 2.

Rooftop drainage design requires an examination of the architectural drainage. Input by the landscape architect to the roof drainage design should be early. Many times the drainage provided by the architect and building engineer is not favorably arranged for soil drainage even though the engineer may have properly designed the drainage capacity. As the result of misplacement of rooftop drainage, or failure to provide a sloping surface, the engineer has greatly compromised the design alternatives available to the landscape architect. Valuable soil volume is then consumed in order to provide proper soil drainage. A rooftop application should always be considered as one with serious drainage problems to make certain that proper drainage receives adequate attention. The details for a rooftop drainage design will vary from project to project, but the principles of drainage hold everywhere.

Design Criteria

The design criteria are usually set by the landscape architect early in the planning process by the choice of plant palette and its layout. The two drainage design criteria are the rate of water removal from the area and the depth to which the watertable is to be lowered.

Generally, the water removal rate must include the precipitation amount since there is to be no additional water stored on the site. Precipitation data from the climatological records for the area will provide the necessary information. If irrigation is included in the design, the frequency and amount of irrigation must be determined. These data are then coupled or added to the drainage rate required to lower the watertable if the action is required.

Lowering of the watertable is dependent on the number of drainage pipes (actually, the interval between them) and the depth at which they are placed. It must be kept in mind that lowering the watertable to a desired depth is inexact. Many factors determine the efficiency of buried pipe drainage, and predictability is crude at best. However, the available empirical data remain useful for successful design in most cases.

Plans and Specifications

As with any other design projects, plans are made for layout, required equipment and supplies, construction sequence, costs, and inspection procedures. Then the specifications are written on the basis of the plans. For surface drainage, the specifications are more in the form of rough and final grade plans (see the following section), whereas the subsurface drainage involves specification of type and quantity of equipment and supplies, sequence of the work, the site trenching operation, installation of the system, backfilling, and finishing.

SURFACE DRAINAGE TECHNIQUES

Properly planned surface drainage systems eliminate ponding, prevent prolonged saturation, and accelerate flow to an outlet without siltation or erosion of soil. The system must collect the water from the source and carry it to the outlet without interfering with the basic landscape design. Sometimes, the landscape design can be modified to accommodate the surface drainage system, perhaps adding some aesthetic quality.

Swales and Open Ditches

Swales have a simple, shallow cross-section where the width is much greater than the depth. A common swale is only several inches (6 cm) to 1 foot (30 cm) deep and 2 to 3 feet (60 to 90 cm) wide (Fig. 7.6a). A swale is employed to intercept water from a source and direct it, such as from an upslope area around a planting, rather than to carry water directly to an outlet. In some cases, a swale may merge with a ditch at the point the capacity of the system must be significantly increased.

An open ditch has a deeper cross-section to carry greater flow of water (Fig. 7.6b). The cross-section is usually trapezoidal to take advantage of reduced friction and increased capacity as the water level rises in the ditch. The sloping sides reduce erosion potential and permit planting of vegetation, usually turf.

A permanent trench filled with rip-rap or large stones may be used as an interceptor where the surface flow is large. It is similar to those used in erosion and sediment control (Chapter 10). The trench then carries the water directly to the disposal system (Fig. 7.7). The trench may even serve as temporary storage to reduce peak flow during a storm.

Surface Drainage System Design

The arrangement of a system of swales and ditches provides a solution to complex surface drainage problems. Buried drain pipes may serve as a disposal system. The arrangement may be very flexible as long as no component carries more water than its design allows. For example, an open ditch should not flow into a swale.

The surface drainage may be laid out in parallel form and perpendicular to the contours on a very gently sloping site. The subdivision of the site into crowned segments with the collection in the direction of flow removes the water quickly from the site (Fig. 7.8). For steep slopes, the swales and ditches

Fig. 7.6 The cross-sections of (a) a swale and (b) an open ditch.

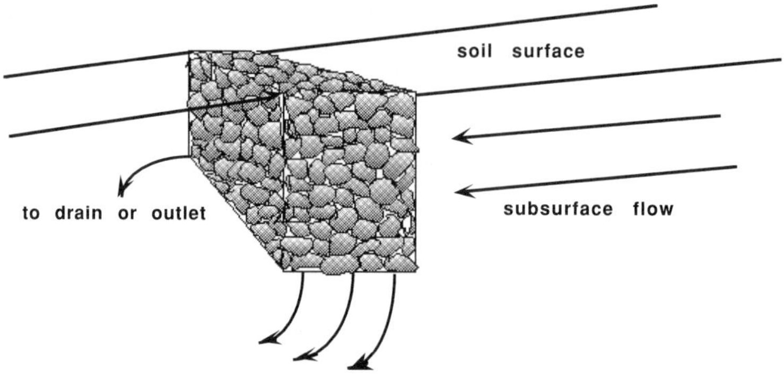

Fig. 7.7 An open trench filled with large stone as an interceptor.

need to be parallel with the contours (cross slope) or in diversion arrangement. The water is carried to points along the contour and into open ditches or into buried drain pipes.

A complex drainage system is comprised of swales, open trenches, and buried pipe drains as the disposal system to carry water to the outlet. Most surface drainage systems do not need to include all the components. A com-

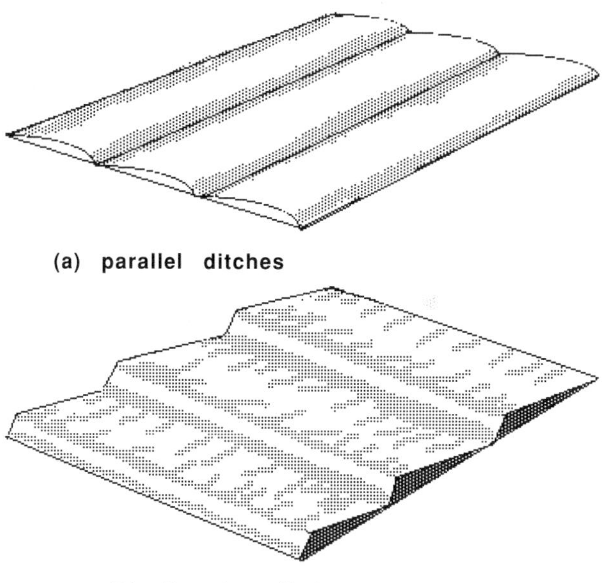

Fig. 7.8 The parallel arrangement of swales or ditches in two situations, (a) gentle slope with ditches parallel with the flow of water, and (b) steeper slope with ditches perpendicular to the flow of water.

bination of swales may be adequate in simple cases, especially if the soil is well drained and the watertable is deep.

Landforming

This process is the mechanical changing of the topography of the land surface. It includes such practices as land smoothing, land grading, and land leveling. The major requirement for its successful practice is uniform, very deep soil over broad areas. The technique is intensively used in the agricultural valleys of the western United States.

Land smoothing eliminates minor concave surfaces and enhances surface drainage over large areas. The resulting grade may be less than 1%. Land grading involves cutting, filling, and smoothing to a planned surface grade. Deliberate crowns may be created to aid surface runoff and open ditches may be constructed to collect the water. The arrangement is similar to those shown in Fig. 7.8 but on "larger" scale, covering tens of acres (tens of ha). The technique is best applied to very level, smooth areas where the surface drainage must be enhanced. Land bedding is a modification of land grading where the distance between the "rows" is short and the crown height is greater. Land leveling is a precise process where a very uniform grade and a very smooth surface are created to aid typical open flow irrigation of agricultural fields. In landscape design, it is primarily used for small areas within a project such as a turf area used for receptions, parties, and so on.

Landforming is also used in the shaping of land for urban development in many major areas of California with sometimes disastrous results (see your national TV news for the effects of El Niño!).

SUBSURFACE DRAINAGE TECHNIQUES

The primary purpose of subsurface drainage is to remove excess water from the soil profile and to lower the watertable. Rooting zone conditions are improved. The drained soil becomes capable of supporting a land use that previously was not possible. With the extensive regulations concerning wetlands that now exist throughout the United States, there is much less opportunity for the unregulated drainage of wetlands. However, there are situations where wet spots occur within a project area that are not subject to regulation but still need to be drained to enhance the success of the project. Further, drainage conditions external to a project may require the installation of subsurface drainage to protect the project from subsurface flow. Techniques given here will provide solutions to most of these potential problems.

Planning for subsurface drainage requires some degree of observation and measurement. These efforts include open-observation well hydrographs where water heights are measured and plotted over time to reveal watertable fluctuations. Profile flow patterns by plotting results of peziometer measurements over time to indicate flow patterns is another source of information. Contour maps of the watertable are obtained by well observations located on a grid pattern over the area. These can then be superimposed over topographic maps to give the relationship between surface configuration and the groundwater configuration.

Open Ditches

The use of open trenches is the simplest and cheapest technique of subsurface drainage (Fig. 7.9). They also serve as interceptors for surface drainage. However, there are disadvantages that must be recognized:

1. Because the ditch is open, it becomes breeding area for mosquitoes, muskrats, and snakes.
2. They consume a large amount of surface area when spaced for effective drainage.
3. They require continued attention to maintain flow and the stability of the sides.
4. They are obstructions to access over the area; bridges or culverts are required.
5. They must be deep enough to effectively lower the watertable; hence, they can be a hazard.

Open ditches are used most extensively for the drainage of agricultural and forest lands, especially in muck soils. Because of space limitations, they are seldom used in landscape architecture projects except perhaps on large development projects where space is not a limitation.

Buried Drains

The older buried drain consisted of tiles placed in trenches and backfilled. The joints, deliberately spaced apart to permit entry of water, were covered

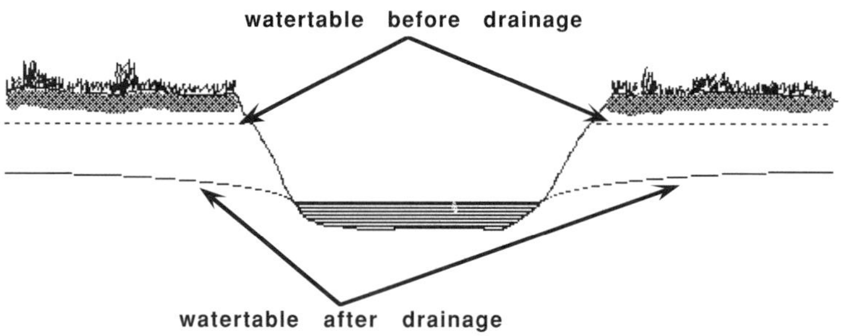

Fig. 7.9 The cross-section of an open ditch with its drainage flow net.

with tar paper to prevent entry of soil during backfilling and subsequent soil settlement. This was succeeded by "Orangeburg" pipe (bituminized fiber pipe), a rigid pipe with holes drilled in a single line along the length of a section. The holes were placed on the bottom during installation in the trench. Collars joined the pipe sections. The newest style is flexible plastic pipe with narrow slots in ridges around the circumference of the pipe. For most applications, the pipe may be placed automatically in the trench as a component of the trenching machine. Otherwise, the pipe is manually laid after the trench is dug by backhoe. Buried drain pipe is the most popular technique for subsurface drainage.

Design of Buried Drains. Several steps are involved in the process of buried drain design. The steps include determination of the following:

Drainage Coefficient. The rate of water removal to achieve the desired protection of the plant palette from excess subsurface (or surface water, too) is called the drainage coefficient. It is usually expressed in depth of water (in. or cm) to be removed in a specific time period of 24 hours in humid areas. The coefficient is strongly dependent on the soil permeability and the amount and pattern of precipitation. It may range from ⅜ to 1 ½ inches (0.95 to 3.8 cm) in 24 hours for areas that have a separate surface drainage system (Soil Conservation Service, 1973).

On arid area projects that are irrigated, the additional applied water must be taken into account in the subsurface drainage system design. A formula provides a means for obtaining the drainage coefficient for the situation. The formula, mainly for agricultural lands, contains parameters that are not applicable to landscape architecture projects. However, the drainage coefficient is approximately equal to ⅒ to ½ the amount of water applied (Luthin, 1957).

The simplest method for determining the drainage coefficient for an urban soil application is to consider the excess precipitation (the dormant season precipitation as well as amounts added by irrigation) together with the storage capacity of the soil profile, subtract the potential evapotranspiration amount, and calculate the approximate excess as the drainage coefficient.

Design Capacity. To carry off the excess water as determined by the drainage coefficient, the capacity of the drainage system must be calculated. This involves the determination of the depth and spacing of the buried drain system.

For parallel relief drains across the area the formula becomes (English units used as in original)

$$Q_r = \frac{qS(L + S/2)}{43{,}200}$$

where Q_r = relief drain discharge, cubic feet per second (cfs)
q = drainage coefficient, inches per hour
S = drain spacing, feet
L = drain length, feet

As an example, a drainage system has parallel drains 100 feet long and spaced at 5 feet. The drainage coefficient is 0.5 in./hr, then

$$Q_r = \frac{(0.5)(5)(100 + 5/2)}{43{,}200} = 0.0059 \text{ cfs}$$

The Q_r is a very small value, indicating the drainage system may be over-designed. The main drain design capacity would be (8)(0.0059) = 0.0472 cfs. The drain size (diameter) may be determined by (Luthin, 1957)

$$d = 0.892q^{0.375}A^{0.375}s^{-0.1875}$$

where d = inside diameter of pipe, in.
q = drainage coefficient, in./hr
A = drainage area, ac
s = hydraulic gradient, ft/ft

The minimum hydraulic gradient for 4-inch tile is 0.15% decreasing to 0.05% for 12-inch tile. A velocity of 1.5 ft/s is usually designated as the basis for minimum slope. A recommended minimum grade is 0.2% (Luthin, 1957), where some practitioners recommend no less than 1%. The next larger commerical size is selected if the calculation results in a fraction.

In humid areas, the depth and spacing of drains have been largely been determined by experience over many years from many field applications. Formulas have been derived from this experience (Luthin, 1957, Soil Conservation Service, 1973). Optimum drain depth for buried drain laterals is influenced by soil permeability, drain spacing, optimum or desired depth of the watertable, depth to impervious strata, outlet depth or elevation for the drainage system, and the plant palette requiring protection. Minimum depth (minimum cover) in mineral soil should be 2 ft (0.6 m). The depth may vary from 30 to 60 inches (76 to 152 cm). Increasing the depth of the drain will widen the drain spacing.

Drain Depth and Spacing. Drain spacing is greatly influenced by the presence and depth to an impervious barrier and whether the soil is homogeneous. Because this condition is so prevalent in landscape design, either as a compacted horizon in an urban soil or the presence of a structure surface (rooftop), two approaches to spacing are required. The most extensivley used formula, called the ellipse equation, is

$$S = \frac{\sqrt{4K(m^2 + 2am)}}{q}$$

where S = drain spacing, ft
K = average hydraulic conductivity, in./hr
m = vertical distance, after drawdown, of the watertable above the drain at the midpoint between lines, ft ($m = d - c$)
a = depth of barrier below drain, ft
q = drainage coefficient, in./hr
d = depth of drain, ft (see Fig. 7.10)
c = depth to watertable desired, ft (see Fig. 7.10)

The equation makes some assumptions that require its use to be restricted to the following conditions:

1. Where groundwater flow is known to be mostly in the horizontal direction. Examples are stratified soils with relatively permeable horizons.
2. Where soil and subsoil materials are underlain by a barrier at relatively shallow depth (twice the depth of drain or less) that restricts vertical flow and forces the groundwater to flow horizontally toward the drain pipe.
3. Where open ditches are used or when drains with sand and gravel filters or porous trench backfill materials are used.

The modified ellipse equation is used for the following conditions:

1. Where soils and subsoils are deep homogeneous (isotrophic) materials without horizontal stratification.

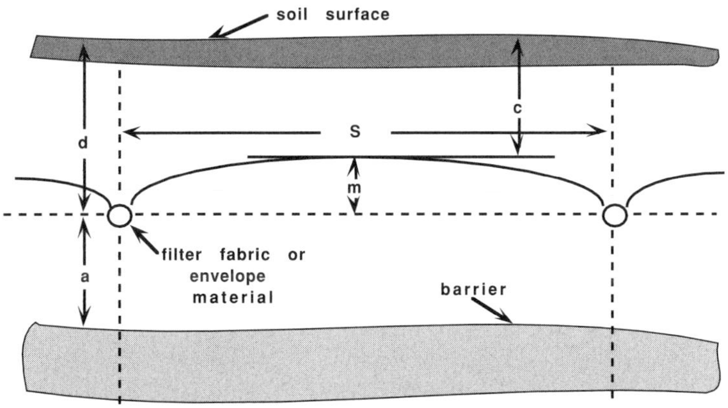

Fig. 7.10 The cross-section of buried drains and the flow net.

2. Where barriers, if present, are at depths greater than twice the drain depth.
3. Where drains are placed without porous filters and when the trench backfill materials have low permeability.

The determination is made through the use of a series of graphic solutions that are not reproduced here since most landscape architecture project soils do not fall in the preceding set of conditions. The graphs are given in *Drainage of Agriculutral Land* (Soil Conservation Service, 1973) and the reader is referred to them for application.

Sand and Gravel Envelopes. Gravel envelopes are employed as a means of preserving the flow characteristics of the buried tile drain. A properly designed gravel or sand envelope should meet two requirements to prevent piping: (1) it should be more permeable to water than the soil or base material and hence allow water to flow freely into the drain, and (2) it should prevent the base material from moving into the drain or onto the filter. Piping is also discussed in Chapter 4.

Materials most susceptible to piping are very fine sand and coarse silt. These materials are entirely lacking in cohesion and the particles are small enough to be carried by very low velocities of water with a velocity of about 10^{-4} cm/s as critical. Proposed ratios for filter stability are (Luthin, 1957; Soil Conservation Service, 1973)

$$\frac{D_{50 \text{ for filter}}}{D_{50 \text{ for base}}} = 12 \text{ to } 58 \quad \text{or} \quad \frac{D_{15 \text{ fine filter}}}{D_{15 \text{ fine base}}}$$

$$= 12 \text{ to } 40 \quad \text{or} \quad \frac{D_{\text{fine filter}}}{D_{85 \text{ fine base}}} \leq 5$$

The third ratio is useful for soils that are more uniformily graded. The first two ratios are applied to nonuniform sands.

Many successful tile drains have been installed without gravel envelopes, but these are mainly in humid regions with stable soils. Where soils are stable there is no problem with sediment movement into the drain. In the western United States, the soils tend to be more unstable, and gravel envelopes are necessary to preserve buried drain flow. Successful gravel envelopes in the Imperial and Coachella Valleys of California (Luthin, 1957) have a particle size distribution including mostly fine gravel, coarse and medium sand, and very little fine sand. It contains nothing finer than fine sand (Fig. 7.11). The particle size distribution curve of the filter should nearly parallel that of the base material.

Where perforated drains are used, the D_{85} size of filter envelope should be no less than one-half the diameter of the perforations and no more than 10%

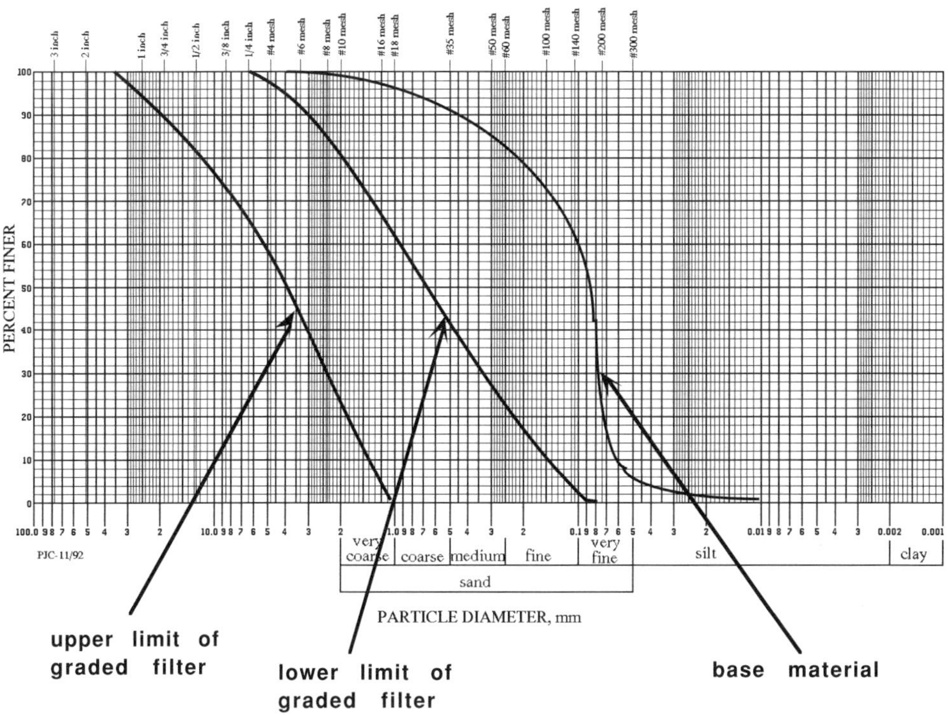

Fig. 7.11 The particle size distribution of a gravel envelope for a buried drain.

of the filter material should pass the No. 60 sieve. The thickness of the envelope should be 3 inches or more around the drain pipe.

Geotextile Filters. It is a common practice to use geotextile fabric as a filter material. The fabric is manufactured with a range of pore size openings. Here, the pore size must be matched with the base material; otherwise, piping occurs. At present, there is some controversy over the placement of the filter fabric. Some engineers recommend that a gravel envelope be placed around the drain pipe and the envelope is encased in the filter fabric. Others recommend the encasement of the drain pipe itself with the filter fabric, with or without the gravel envelope. Because the gravel envelope has been used successfully so extensively, the author prefers that method. Table 7.1 provides a guide for the need for filters.

Organic Filters. Sawdust, wood chips and other organic materials have been used in agricultural applications (Soil Conservation Service, 1973). However, the results have been mixed and the materials so variable that it is probably wise not to recommend use of these materials for drain filters.

TABLE 7.1 Need for Drain Filters Based on Soil Classification

Unified Soil Classification	Filter Recommendation
SP (fine) SM (fine) ML, MH	Filter needed
GP SC GM SM (coarse)	Subject to local on-site determination
GC, CL, SP, GP (coarse) GW, SW, CH, OL, OH, Pt	None

Source: Soil Conservation Service, 1973.

Pumping Systems

A system for groundwater removal by pumping involves the installation of deep wells into the soil in a configuration to effectively lower the watertable. Obviously, pumps must be employed in the system. The use of pumping systems is restricted to high-value areas where the technique is the most effective. Two landscape examples are the pumping system for draining excess water in oak tree pits at the White House, Washington, DC (Craul, 1992), and excess water at the apex of the wall of the Vietnam Memorial, Washington, DC.

Mole Drains

These are drains created in the soil by a moling plow. This is an egg-shaped implement that is pulled through the soil to form a channel or cavity at a set depth. It is formed parallel to the ground surface. The resultant cracks and fissures in the soil permit water to flow into the cavity, which then carries the water to a disposal point. Mole drains are most effective in medium to fine textured soils.

The mole drains are unlined and subject to collapse and crumbling. The life expectancy is about 3 to 5 years. Cultivation of moled soils with heavy equipment reduces their effective life.

Vertical Drains

The use of vertical drains is applied to the drainage of water from a perched watertable to a better-drained horizon below the perched watertable. The technique is successful only when a horizon with higher permeability and no watertable exists at lower depth. Figure 7.1 illustrates a successful example.

Detailed logging of the soil profile and percolation testing is required to detect the proper conditions. The situation is commonly encountered in urban soils.

Interception Drains

Interception of subsurface lateral flow is necessary for protection of areas where subsurface structures exist, or a specific source such as a spring contributes execess water to an area. The drain may be buried pipe or a trench filled with gravel or rip-rap placed perpendicular to the source flow (Fig. 7.7). The drain pipe or the trench dimension is designed to be adequate to carry away the excess water to a disposal system. The area downstream from the interception drain is then subjected to moisture only from precipitation that reaches its surface. The main difference between interception ditches and subsurface drains is the type of flow intercepted: surface versus subsurface.

Outlets for Subsurface Drains

Outlets must be designed carefully. They must carry the drainage load, which may be continuous flow, and they must withstand potential intrusion by high water from the disposal water body or stream or tides in the case of tidal applications. Because of the need for substantial design, most outlets must be permanent and "hardened." The use of concrete or steel culverts, bank rip-rap or other stabilization measures around the outlet pipe, and hardened spillways is always necessary. Standards are provided in many hydrology, wetland, and agricultural planning texts.

DRAINAGE RESTORATION SYSTEMS

These are cases in which the established vegetation requires improved drainage for survival or for more optimum growth. Difficulty arises because any construction to place drainage facilities potentially seriously disturbs the soil and established root systems. The construction must be carefully planned using appropriate materials and equipment, and the installation must be carefully carried out under very close supervision. Any drainage system installed must have an outlet of some type. The author has seen some crude installations that did not make provision for an adequate outlet, and the attempt to improve drainage failed. The same principles of drainage must be applied to simple or crude systems just as they are to the more sophisticated ones.

Because of the high degree of variability from one restoration situation to another, it is not possible to propose any standard design method. The creativity and ingenuity of the designer are called into service.

8

IRRIGATION DESIGN

INTRODUCTION

Irrigation is the deliberate application of water to soil and plants so that the plants may complete their life cycle or provide a high level of productivity when harvested for food or fiber. Irrigation was historically applied only in arid areas where limited precipitation caused poor crop production or limited the kind and variety of crops grown. In more recent times, irrigation has been used in humid areas to nearly eliminate the effects of moisture stress on productivity. The lack of stress also reduces the susceptability of the crops to some insect and disease infestations. However, it should be pointed out that irrigation may increase the threat of other insects and diseases. In landscape design, irrigation provides a means for maintaining plants in an otherwise arid urban situation. It also provides a means to broaden the plant palette for a project that would not be possible under ordinary conditions.

There are three ways in which the water utilization efficiency of a site may be improved:

1. Design a soil that will provide a sufficient supply of water from one precipitation event to another in humid regions.
2. Design an irrigation system that provides the minimum amount of water for satisfactory plant growth.
3. Design a plant palette comprised of plants that are efficient water utilizers and adapted to the microclimatic conditions of the site; thus, a xeriscape.

The purpose of this chapter is to explain the basic principles of irrigation and to connect irrigation to the biophysical analysis of Chapter 3. Various systems are discussed, covering the advantages and disadvantages of each. More particularly, the significance to differences in potential evapotranspiration is covered with illustrations of some urban situations. The design of the soil and the volume of the rooting medium are important factors.

Though this chapter deals with irrigation, it is the author's belief, knowing that many people will sharply disagree, that active irrigation of landscape projects should not be encouraged, and if used, every effort should be made to create a very water-conserving design (Smith and Rakow, 1992). Only when the project requires irrigation for sustainability, such as during a prolonged drought or temporarily for a major maintenance or restoration project, should irrigation be used. It has been the author's observation that the first component of a project design to fail or become damaged is the irrigation system; the lighting system is second (Fig. 8.1).

The concept of sustainability embraces the conservation of our limited water resources even in those climatic zones with a precipitation surplus. As population and technology increases, greater awareness of water utilization and its quality must also increase (Fazio, 1993). Recycling of water from other uses to plant irrigation must be and is being examined (Sopper and Kardos, 1973). Some work of this nature is already being undertaken, pri-

Fig. 8.1 A failed irrigation system, Salt Lake City, UT.

marily in California (Lindsey and Waters, 1995; Waters and Lindsey, 1995). However, there are some costs and risks involved that must be resolved before the widespread application of some of the techniques can occur. Something as fundamental as the redesign of our metropolitan storm sewer systems is required. This activity may seem beyond that expected of the landscape architect, but the redesign planning would involve landscape projects such as constructed wetlands and wet retention basins and ponds. The water from these bodies could be used for irrigation much like the golf course designers have already instituted (Stuller, 1997). There are some lessons to be learned from golf course technology. However, it is always appalling to see attempts of creating an oasis in a desert. When most people go to the desert, they should want to see the desert and its natural attributes, not the greenery of the humid regions. A natural oasis is something different.

A failure of much advice and methodology on irrigation system design is that the plant water requirements receive primary attention and the soil is sometimes nearly neglected or given a secondary role. If more attention is paid to the design of the soil, providing sufficient waterholding capacity and total rooting volume, the need for irrigation could be greatly reduced and used only in those exceptional cases stated previously.

Much of the following discussion on irrigation theory and technique depends on the subject material of heat loading and evapotranspiration covered in Chapter 3, which should be read before continuing with this chapter.

BASIC CONCEPTS OF IRRIGATION

The Hydrologic Cycle Applied to Irrigation

The concept basic to irrigation is the hydrologic system of inputs, storage, and outputs of water within a watershed or an ecosystem. Inputs include precipitation, lateral flow through the soil, upward movement by capillarity from the watertable, and the formation of dew on the soil or vegetative surfaces. Storage is within the soil, plants, or the watertable, which may include permeable bedrock. Outputs include evaporation from the soil and plant surfaces, transpiration from growing plants, surface runoff during storms, subsurface lateral flow to springs, and percolation downward to the watertable. The net difference between inputs and outputs determines whether there is storage of water or created deficit.

Here, we are concerned primarily by the loss of water through two pathways. The soil loses moisture by evaporation at the surface. Conduction upward through the soil profile supplies the water until the water column in the soil pores no longer can be maintained and is broken (Craul, 1992). The soil may continue to lose moisture in vapor form, but this is an extremely slow diffusional process. The soil is already too dry at the surface for most plants.

Plants lose water through the stomata of the leaves by the process of transpiration. The loss continues until the plant becomes under stress and the stomata close as a result, or the soil can no longer supply sufficient moisture to maintain the leaf turgor. The plant now experiences wilting and the soil moisture content has reached the wilting point (Craul, 1992). Because it is difficult to measure separately, the evaporation from the soil and the transpiration from plants, the two are combined in the term *evapotranspiration.*

Thus, there are periods of time during the year in various climates that exhibit a lack of soil water. These periods occur when there is a deficit between the amount of soil water and that utilized by plants. The deficit is greatest and most prolonged in arid regions. Recharge of soil water is by precipitation or subsurface water flow. If recharge or soil water storage is inadequate, the deficit occurs. At this point irrigation has its greatest benefit. In some cases where stress is caused by transplanting into compacted and poorly aerated soils, irrigation may worsen the situation.

The purpose of irrigation is to reduce or eliminate the plant stress effects of the soil moisture deficit (shown as D in Fig. 8.2). For crops, irrigation ensures a high level of productivity. For landscape plants, the irrigation reduces moisture stress so that the plants may thrive for aesthetic purposes, or to maintain plants not adapted to soil moisture deficits. The potential evapotranspiration is a measure of the potential water utilization rate of plants under a given climate or in a particular environmental situation.

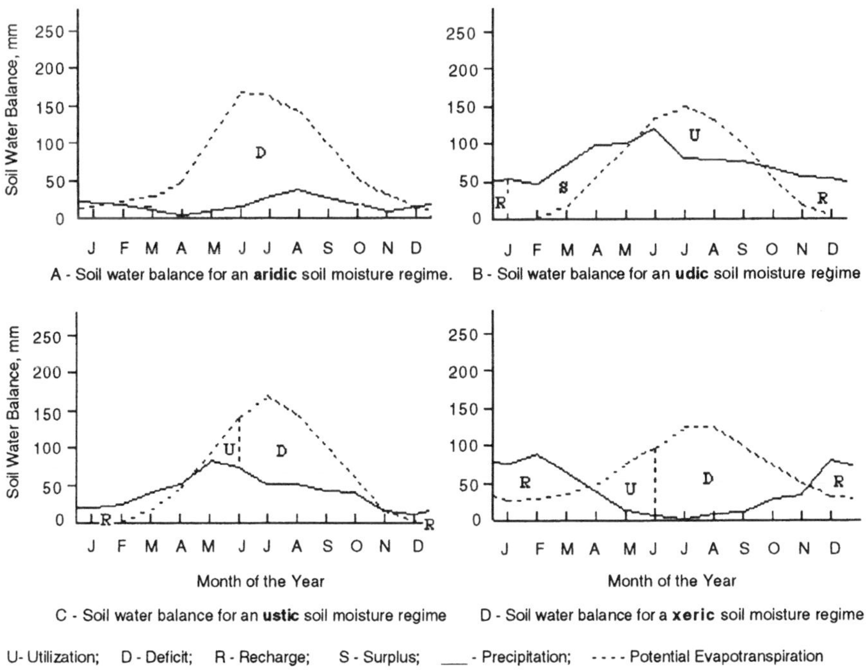

Fig. 8.2. Water budgets for four soil moisture regimes. (Adapted from Soil Survey Staff, 1975, in Craul, 1992.)

Determination of Soil Moisture Deficit

The easiest and most direct way to evaluate the soil water balance is to check the soil moisture content, and then irrigate as necessary to replenish to the desired moisture content. But, in urban areas, the soil may be covered and not accessible for observation and inspection unless remote measurement is installed.

Evaluation of soil moisture deficit must be based on the waterholding capacity of the soil. Knowledge is necessary of the moisture relationships such as field capacity, wilting point and the resulting available water (Craul, 1992). These parameters are determined most specifically from laboratory analysis, which is not always available to the landscape architect. Tabular data of some common textures and their moisture parameters are available in the literature and are reproduced in Table 8.1. Unfortunately, these data do not account for the effects of specified organic matter content and bulk density on the moisture relationships. Some attempts have been made to overcome this shortcoming, but not always with complete success (Rawls, 1983). The problem is discussed further in Chapter 6 with respect to rooftop soil design.

TABLE 8.1 Soil Storage Capacity and Availability

	Total Storage Capacity		Macropore Space	Plant-Available Capacity	
Texture Class	%	in./ft	%	%	in./ft
Coarse sand	24.4	2.93	17.7	6.7	0.80
Coarse sandy loam	24.5	2.94	15.8	8.7	1.04
Sand	32.3	3.87	29.0	13.3	1.59
Loamy sand	37.0	4.44	26.9	10.1	1.21
Loamy fine sand	32.6	3.91	27.2	5.4	0.65
Sandy loam	30.9	3.71	18.6	12.3	1.48
Fine sandy loam	36.6	4.39	23.5	13.1	1.57
Very fine sandy loam	32.7	3.92	21.0	11.7	1.40
Loam	30.0	3.60	14.4	15.6	1.87
Silt loam	31.3	3.75	11.4	19.9	2.39
Sandy clay loam	25.3	3.04	13.4	11.9	1.43
Clay loam	25.7	3.08	13.0	12.7	1.52
Silty clay loam	23.3	2.80	8.4	14.9	1.79
Sandy clay	19.4	2.33	11.6	7.8	0.94
Silty clay	21.4	2.57	9.1	12.3	1.48
Clay	18.8	2.25	7.3	11.5	1.38
25% peat to 75% sand	51.0	6.12	6.0	—	—
50% peat to 50% sand	60.0	7.20	10.0	—	—
75% peat to 25% sand	71.0	8.52	11.0	—	—
100% peat	80.0	9.6	15.0	—	—

Source: Craul, 1992.

Methods of Determining Soil Moisture Content. Normally the landscape architect would not make soil moisture determinations. The project irrigation specialist or soil scientist would do the work. The material is provided here to give the landscape architect a better understanding of the analyses involved and to be able to assess the work of the specialists.

Several facets of soil moisture relationships must be kept in mind. Plants wilt at a soil moisture tension of 15 bars (1.5 MPa: the new SI units are kilopascals, kPa, or megapascals, MPa, where 1 bar = 0.1 MPa = 100 kPa). A soil, when saturated and drained by gravity after 24 hours, will have a moisture tension value of 0.10 to 0.33 bar (0.01 to 0.033 MPa). This point on the moisture retention curve called "field capacity." The water held by the soil between field capacity and wilting point is considered the plant-available water. Most of the drained water existed in the macropore space before draining. The greater the range, the greater the plant available water. Sands have low waterholding capacity and low plant-available water, whereas silts and clays have moderate to high waterholding capacity and variable plant-available water, depending on structure, orgainc matter content, and bulk density.

Gravimetric Sampling. A soil sample of known volume is obtained. The moist soil is weighed, then placed in an oven, and dried at 105°F for 16 hours. The dry soil is weighed with the difference in the two weights being the water content. The value is expressed as a percent or as a ratio on a volume basis. The method is destructive and not appropriate where multiple samples are required and disturbance to the soil is not desirable. The method is used to calibrate the other moisture determination methods as it is the most accurate.

Tensiometer. An instrument that measures the soil water tension and is related to content using the soil moisture tension curve (the *moisture characteristic* as used by soil physicists). It is the most common method of soil moisture determination used by irrigation specialists. The instrument consists of a porous cermic cup attached to a plexiglas column at the below-ground end. A vacuum meter is attached to the above-ground end and measures the tension of the water column within the tube in centibars. The tension of the instrument water column is theoretically in equilibrium with the soil moisture tension. As the soil dries, the tension becomes greater and the meter reading increases, and conversely as the soil is wetted. The technique is most useful for soil moisture measurement in the wet range up to a tension of about 0.5 bar (50 kPa). Unfortunately, the water column in the tensiometer "breaks" at about 0.8 bar (80 kPa) and becomes inoperative. If the plant palette consists of plants with high water requirement and the soil must be kept at high moisture content, the tensiometer method is appropriate. Tensiometers require frequent maintenance and freeze in winter temperatures.

Electrical Resistance Block. Two mesh electrodes are placed parallel to each other within an asbestos or nylon block. The electrical resistance is measured between the two electrodes by a wheatstone bridge. As the block dries, the resistance increases, and conversely as the block is wetted. At saturation, the resistance approaches zero, and as the soil nears a moisture content just below wilting point, the resistance becomes infinite. The moisture block technique is primarily useful for the midrange of available water; it is not accurate in soil with the moisture content at either of the two extremes. The resistance is greatly affected by soluble salts in the soil and may not give accurate results in saline or alkali soils or those subjected to heavy loads of deicing salts. A combination of tensiometers and electrical resistance blocks covers most of the plant-available moisture range if long-term monitoring is required. Continual maintenance is required.

Neutron Attenuation. Soil moisture content is measured by the number of neutrons attenuated by the hydrogen contained in water in this method. A radioactive source and a Geiger tube coupled to a counter on the soil surface by cable is lowered into an aluminum tube inserted in the soil. A count may be taken at any level in the soil. An increase in moisture content shows an increase in the number of attenuated neutrons. This method is accurate when calibrated properly. It is excellent for continuous monitoring over long periods, but has the disadavantage of being very expensive equipment and requires the installation of the semipermanent access tubes. Installation can be very difficult in stony or rubble soils.

The reader should refer to Hillel (1980) for more details on soil moisture relationships and their determination.

Determination of Potential Evapotranspiration

Direct Observation. Potential evapotranspiration, as defined in Chapter 3, is the amount of water lost from a continuously watered soil in a pan and planted to fescue clipped to 12 cm (approximately 5 in.). Obviously, the use of a standard Type A pan is not appropriate for most landscape applications and other methods of determination are utilized. Very few Weather Service stations report ET and these are usually located at major airports. Airport conditions do not resemble the common urban conditions, and the project location may be a considerable distance from the airport.

Climatic Data. The ET equation given in Chapter 3 contains observable climatic factors, including net radiation, air temperature, wind speed, and air vapor pressure. Air temperature, wind, and vapor-pressure data are commonly available from climatological tables issued by the Weather Service. Some states have a special computerized weather network, such as the California Irrigation Management Information System (Hartin and Pittenger, 1992; Perry, 1992). The difference between potential evapotranspiration and precipitation is a simple way to calculate the approximate amount of water to apply. However, it does not provide for the distinction of irrigation differences required within a particular project.

Net radiation is not usually measured and thus not compiled. Estimates are required. These can be inserted into the ET equation to obtain an estimated potential evapotranspiration. It should be kept in mind that this method really does not approximate the conditions of the project site unless a biophysical analysis of the site is made.

Generic estimates based on the definition of climatic zones are provided by some irrigation publications (Table 8.2; Clark, 1994). As indicated in the table title, these are generic values and cannot be validly applied to a specific point on a project. Adjustments are required for plant type (tree, shrub, etc.), planting density (low, medium, high), and microclimatic effect (low, medium, high).

From Biophysical Analysis The analysis decribed in Chapter 3 is the best approach to obtaining evapotranspiration estimates because it reproduces the physical conditions of the site and includes the estimates of net radiation from solar radiation estimates for that site.

THE WATER UTILIZATION OF PLANTS

Unfortunately, there is little documentation on the actual water utilization by individual plants or street trees (Craul, 1992; Hartin and Pittenger, 1992; Kourick, 1992). As a result, we are not able to calculate accurately how much

TABLE 8.2 Generic Potential Evapotranspiration Data

Climatic Zone	Definition for Midsummer	Greatest Daily Potential Evapotranspiration, in./day
Cool humid	$< 70°F$ $> 50\%$ humidity	0.10 to 0.15
Cool dry	$< 70°F$ $< 50\%$ humidity	0.15 to 0.20
Warm humid	70° to 90°F $> 50\%$ humidity	0.15 to 0.20
Warm dry	70° to 90°F $< 50\%$ humidity	0.20 to 0.25
Hot humid	$> 90°F$ $> 50\%$ humidity	0.20 to 0.30
Hot dry	$> 90°F$ $< 50\%$ humidity	0.30 to 0.45

Source: After Clark, 1994.

water should be applied to each and every individual plant. Most nurserymen and irrigation farmers look at the total crop cover for application rate estimation. The irrigation application rates and timing vary as weather conditions change over the growing season and as the plants reach the various stages of physiological maturity. Since the landscape architect is not producing a harvested crop, the approach to irrigation is different; it is a case of survival and optimum growth for aesthetic purposes. Further, the environmental conditions of an open crop field or a series of nursery beds are drastically different from an urban planting surrounded by structures.

Several authors provide some data that approximate plant water utilization, which are helpful in irrigation design. Perry (1992) provides an extensive western plant list showing estimated relative water requirement for the several physiographic provinces of California. The estimate classes are adapted to the region—little or low water needs; low supplemental water needs; low to moderate supplemental water needs; moderate supplemental water needs; moderate to high supplemental water needs; and not well suited to the region. Quantitative values for irrigation rates are recommended to be derived from the California climatic observation network. Kourik (1992) gives several tables providing plant water use.

ACTIVE IRRIGATION

Active irrigation is defined here as an irrigation system requiring active involvement of humans or mechanical equipment to apply the water to the soil and plants. Even the simplist flooding system usually requires at least the periodic opening and closing of gates or valves.

Flooding

Flooding is the method of active irrigation in which the entire field or crop area is flooded. Crops irrigated by this method include rice and sugar cane. On a smaller scale, entire household lawns have been flooded with several inches of water in Phoenix. The infiltration of the water into the soil creates a wetting front that moves down through the profile until equilibrium tension is reached.

The water is contained by series of levees and dikes with the water delivered through ditches and special canals. Some irrigation ditches may extend for miles on western U.S. ranches and farms, many of them originally dug by hand.

The flooding will have the beneficial effect of transporting downward through the profile salts that have accumulated on the soil surface by evaporation, reducing the soluble salt content and, to some extent, the pH of the surface soil horizon. The disadvantages of the method are that it is not an efficient use of water because of the high evaporative potential from the large water surface area, and the plant palette must withstand the flooding effects (periodic lack of oxygen to the root systems). The method is hardly ever used in landscape projects.

Trench or Furrow

Rather than flooding the entire field, the water is allowed to flow down the intertilled rows (the furrows) of the crop from one end to the other. A distribution ditch is located on the higher side of the field and water gates or pipes feed water to each furrow. This method is very common in irrigated farmland row crops. It is less wasteful of water than flooding or the sprinkler system. However, there is the problem of "salting out" on the upper surface of the mounded rows in saline or alkali soils. The method has been widely used on a small scale for gardens in the western United States.

Sprinkler

Sprinkler irrigation has been used for at least 60 years. It is based on two types of heads: the spray head and the rotor head. The spray head is designed simply to form a circular or semicircular spray pattern over a small area of 7 to 15 feet (2 to 4 m) radius and less than 45 feet (14 m) in width (Smith, 1997). Spray heads are good for areas with irregular shape. They can apply about 1 to 2.5 inches (2.5 to 6 cm) per hour (high). The rotor head is able to throw water to a greater distance and can water a larger area of 30 to 90 feet (9 to 27 m) and more than 35 feet (10 m) in width. The rotor head can apply about 0.30 to 0.75 inch per hour (0.7 to 2 cm) (low). It is the preferred equipment in most sprinkler irrigation systems that must cover large areas such as field crops and orchards areas with regular shape and especially turf areas. Large spray rigs consisting of a axis or boom carried on large diameter wheels may cover 160 acres (65 ha) or more. The boom has mulitple spray heads attached to cover the large area.

In fixed systems, the spray heads may pop up when water pressure is applied. When turned off, the head drops down below the soil surface and only the surface of the head or rotor is visible. The design prevents accidental damage or vandalism.

For most landscape applications, a mix of rotor heads and spray heads are recommended. The mix allows large areas to be covered and yet prevent overspray or underspray in irregular-shaped areas and overspray onto hard surfaces.

Sprinkler systems are not water-efficient because much of the water evaporates to the atmosphere as spray or water droplets land on the vegetative foliage and evaporates from there. Even with the small spray heads adapted for flower beds and small gardens, there is considerable water loss. Despite this disadvantage, the sprinkler system remains very popular and will probably continue to be the system of choice for turf areas unless water costs rise sharply (Smith, 1997).

Drip Irrigation and the Low-Volume Concept

Drip or trickle irrigation involves the direct application of water to the individual plant through calibrated emitters. In the usual scheme, lateral lines carry the water underground as in a fixed sprinkler system. Sublateral tubes carry the water to the above-ground emitters. The water is applied close to the soil or mulch surface in close proximity to the plant stem. Thus, each plant has its own emitter. If the water utilization of the plant is high, additional emitters are installed by that plant. This is the common procedure for large shrubs and trees. Some irrigation specialists prefer not to use drip irrigation for trees, using rotor-head sprinklers instead because of the large rooting area to be irrigated.

An example of drip irrigation is shown in Fig. 8.3. Here, each tube delivers 1 gallon (4 L) per hour and the number of tubes is determined by the water utilization of the individual plant or tree. Larger tubes are installed for trees and large shrubs. The irrigation systems are usually installed after rough grading is complete and before the stock is planted. In the case of the drip irrigation system, final grading may be completed before its installation.

Generally, the drip irrigation method is more conservative of water than any of the previous methods discussed. It requires somewhat greater care in planning and installation but is simpler. Maintenance is much easier than other systems. The features of the method are as follows (Clark, 1994):

1. The system is based more on the planting scheme and the use of the area rather than the size of the area being irrigated.
2. Plant water requirements are determined first.
3. Plan for irrigation of the base plants (ground cover or shrubs) first.
4. Calculate the system run time.
5. Then irrigate the "nonbase" plants.
6. Layout the system.
7. Calculate the system hydraulics.

Some advantages of the method include the following (Smith, 1997):

1. There is precise placement of water in the root zone for individual plants.
2. Minimal evaporative losses; almost negligible.
3. Evaporation and overland flow minimized, reducing erosion hazard.
4. High application efficiency.

Fig. 8.3 Drip irrigation of a shrub, Scottsdale, AZ.

5. Weed growth is reduced.
6. Almost invisible, reducing damage and vandalism.

The disadvantages include the following:

1. Filtration required to prevent emitter clogging.
2. Proper management is more complex than other methods.
3. First indication of malfunction may appear only after the plants are stressed.

Generally, drip irrigation is recommended to be frequent and to shallow depth. This procedure is somewhat in contradiction to the opposite advice of less frequent but deep (long duration) irrigation. The difference arises in the type of planting that is irrigated. Gardens and small shrubbery that are drip irrigated usually have shallow root systems and the frequent watering is beneficial. Irrigated trees or other large, deep-rooted vegetation should have the deep, thorough watering.

System Layout

Irrigation systems for large areas such as turf areas tend to be laid out in a more mechanical design. The reason is that the efficiency of water coverage by sprinkler is the major consideration. The design takes on the patterns illustrated in Fig. 8.4.

The layout for drip irrigation is more complex but flexible since the plants become the focus of attention (Fig. 8.5). The lines are irregularly placed to enable the efficient watering of the base plants coupled with the watering of each individual tree.

In a complex landscape design comprised of trees wtih a shrub or ground cover understory surrounded by turf, separate stations should be installed for each component (Fortier, 1994). Trees should be on one circuit, the base plants or understory on a second circuit, and the turf on its own circuit. Trees will benefit most with slow water application over a longer application period to ensure that the entire soil profile is moisted. Shrubs require less watering and may be accomplished over a shorter application period. Irrigation and maintenance is greatly simplified though the system layout may be more complex.

Spacing between emitters for drip irrigation is dependent on the texture of the soil as it affects the wetting pattern (actually, the *diffusivity*—see Hillel, 1980). Generally, sandy or coarse soil exhibits a deep wetting pattern with the horizontal and vertical dimensions in a ratio of about 1:2. For medium soils, such as loams and silt loams, the pattern is wider than it is deep by a ratio of about 1:1. Fine-textured soils like clay loams, silty clays, and clays have a ratio of about 2:1. The differences here reflect the difference in infiltration rates for the three major groups of soils.

SOIL DESIGN

Some irrigation methods require only a rough differentiation of soil. One recognizes just three broad types based on texture alone: coarse, medium, and

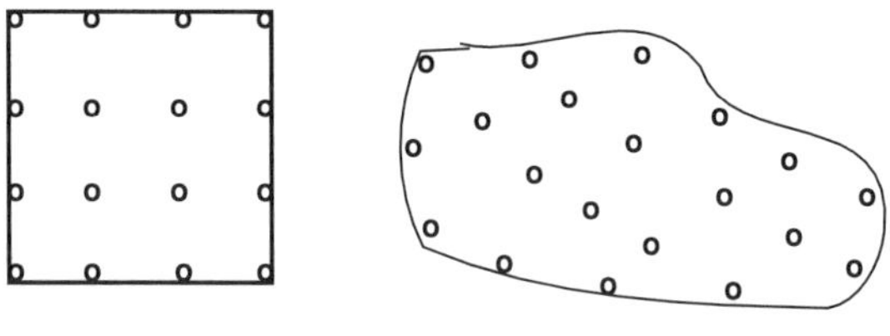

Fig. 8.4 Two examples of sprinkler irrigation layout.

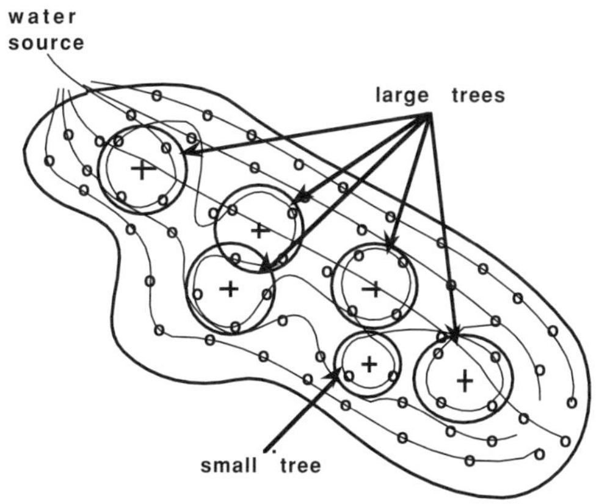

Fig. 8.5 Layout of a drip irrigation system.

fine. It is likely that the coarse includes coarse sand, loamy sand, and sandy loam (USDA classification). The medium class probably includes the fine sandy loam, loam, silt loam, and perhaps sandy clay loam. The fine class includes the sandy clay, silty clay, clay loam, and clay. Maximum infiltration rate, maximum wetted diameter, and available water data are given for each of the three classes. Another method requires the specific USDA soil texture class and the data provided in Table 8.1. These data are used in the Simple Water Budget Method.

In the case of the Biophysical Model Method, the soil is specifically designed for application to the various portions of the project site, following the guidelines given in Chapter 5 and further elucidated in Chapter 12. The design is based on texture and estimated density, waterholding capacity, and required soil volume in profile form, differentiating between topsoil and subsoil, and so on. Examples are provided in the next section.

DETERMINATION OF APPLICATION RATES

There are several methods followed to determine and rate the amount of irrigation. The methods vary on the basis of the data available to make estimates of irrigation needs and the basic design of the system. Generally, plants will require more water in the following situations (Hartin and Pittenger, 1992):

1. The denser the planting, the more transpirational surface present, and more water is required.
2. Trees in median strips and those surrounded by pavement or structures will require more water than those surrounded only by vegetation.

3. Trees in shade, such as the north side of buildings (in the Northern Hemisphere; south side in the Southern Hemisphere) require less water than those in full sun.

The Simple Water Budget

A simple method employed for turf, and sometimes applied to trees, when the potential evapotranspiration and soil waterholding capacity are known is given by Harris (1991).

1. The monthly evapotranspiration values are obtained from observations or compiled historical climatic data. The data are than converted to daily value (Table 8.3).
2. Soil waterholding capacity is obtained from Table 8.1 and inserted in Table 8.4.
3. Determining irrigation rate (after Harris, 1983). For Table 8.3:
   (a) Duration of available water: (available water/ET).
   (b) Suggested interval between irrigations: (75 to 80% available water/ET).
   (c) Amount of water to apply: (75 to 80% available water).
   (d) Time required to replenish water; calculated at a typical infiltration rate for each soil (Table 8.4).

Obviously, this method does not account for possible signficantly different ET rates across a project, as discussed in Chapter 3. The method is most appropriate for a project area that is broad and open and planted primarily to turf.

The Checkbook Method

The method is based on the input and output of water to the system; hence, the "checkbook" concept. A daily record is kept on soil water content at the beginning of the day, effective precipitation or irrigation, water loss by evapotranspiration or by deep soil percolation, and soil water content at the end

TABLE 8.3 Monthly and Daily Evapotranspiration Values for a Rooftop Garden Example

ET	Mar.	April	May	June	July	Aug.	Sept.	Oct.	Nov.	Dec.
Monthly	2.0	5.0	6.0	7.0	9.0	6.3	4.5	3.6	—	—
Daily	0.07	0.17	0.20	0.23	0.3	0.21	0.15	0.12	0.07[1]	0.07[1]

[1]Estimated from the March value.

Source: Data Derived from U.S. Weather Bureau Data for the locality (Boston), in inches.

TABLE 8.4 Duration, Interval, Amount to Apply, and Time to Replenish Water for the Boston Site

Available Water (in.) (AW)	Duration (days) AW/ET[1]	Suggested Interval (days) 75–80%AW/ET	Amount of Water to Apply (in.) 75–80% AW	Time to Replenish (hr)
1.00	3.3	2.5	0.66	2/3[2]
4.78	19.9	12.7	3.82	7.6[3]

[1]ET = 0.30 in./day.
[2]Based on the infiltration rate of 1 in./hr.
[3]Based on the infiltration rate of 0.50 in./hr.
Source: Format after Harris, 1983.

of the day. An example of the record sheet is given in Table 8.5. Daily estimates of evapotranspiration are required, which is a disadvantage in light of the relative unavailability of valid data for many locations and specific sites. Soil moisture content must be monitored daily or at least for the observation period. Deep percolation is difficult to estimate. The record keeping is also time-consuming. The method has the advantage of simplicity in concept and is appropriate for field crops. It is not applicable to most landscape projects unless daily or frequent attention to irrgation management is possible.

The Tabular Adjustment of Evapotranspiration

This system of determination makes adjustment to the evapotranspiration values as given in Table 8.2 according to soil group (texture), plant species, planting density, and microclimatic factors (Clark, 1994). In addition, "hydrozones" of different plants are differentiated.

TABLE 8.5 The Balance Sheet for the Checkbook Irrigation Method

Day or Date	Irrigation Depth, inches (+)	Effective Rainfall, inches (+)	ET, inches (−)	Deep Percolation, inches (−)	Soil-water Content Beginning (inches)	Soil-water Content End (inches)
1	—	—	0.25	—	3.00	2.75
2	—	—	0.28	—	2.75	2.47
3	—	0.40	0.20	—	2.47	2.37
4	—	—	0.32	—	2.37	2.05
5	0.75	—	0.32	—	2.05	2.48
6	—	3.55	0.08	2.50	2.48	3.00 (3.45)[1]

[1]Soil water above the field capacity drains in response to gravity and contributes to deep percolation. The soil-water content is adjusted back to field capacity of the soil.
Source: Adapted from Smith, 1997.

The first differentiation is individual plants versus dense plantings. Individual plants have crowns more than 3 feet apart. Dense plantings are those with plant crowns less than 2 feet apart. Individual plants are irrigated by a determined number of emitters that supply the appropriate amount of water to the plant root zone for that plant. Plants irrigated individually are trees and shrubs. Dense plantings are irrigated by emitters that supply the appropriate amount of required water over the entire planted area. Dense plantings include ground cover only or shrubs only or in combination with other plant types. Trees and shrubs with a ground cover would be a dense planting. The ground cover would be the base plant.

The next series of estimates for adjustment to potential evapotranspiration are species, density, and microclimatic factors. Those species with with low requirement have a factor value of about 0.2; those with medium needs, 0.5; and those with high water needs, 0.7 to 0.9, with the maximum applied to trees alone or a mixed planting of trees, shrubs, and ground cover. The density factor values follow the same trend as for species. Low density has a value of 0.5; average, about 1.0; and high density, 1.1 to 1.3, with trees and mixed plantings with the maximum value. Examples of the microclimatic factors include open-grass park areas at 0.5; partially shaded areas next to driveways or sidewalks, 1.0; street locations in full sun have the highest values as 1.2 to 1.4. The three factors are then multiplied together (K_c), which is then multiplied by the potential evapotranspiration (PET) value.

The water requirement for individual plants is then based on the area of the root system and the area of the crown canopy. The formula used by Clark (1994) is

$$\text{Gallons/day/plant} = (\text{root zone area} \times \text{canopy area} \times K_c \times \text{PET})/(\text{application efficiency})$$

The base plants or dense plantings are watered-based on the soil group, rooting depth, and minimum watering depth. The spacing of the emitters and their required capacity are then determined.

The method is fairly comprehensive in considering various factors that affect the water needs of various types of plants in several situations. The major weakness is that in urban situations, it is difficult to estimate root area under pavement. Roots are extremely plastic and opportunistic and will extend themselves wherever the conditions are favorable such as along pipe traces and building foundations and walls (Craul, 1992). If the root system is confined to a below-ground vault, the procedure is invalid. The real danger is either overwatering or underwatering the plants. The old saying goes "house plants are overwatered and urban trees are underwatered."

IRRIGATION FREQUENCY AND TIMING

Though several irrigation methods include determination of frequency and inference on timing of irrigation, Smith (1997) suggests an examination of the slope of the historic evapotranspiration curve as an indicator for timing and intervals of irrigation (Fig. 8.6). The steeper the curve, the more frequent the irrigation, and the shorter the interval between irrigation events. Soil moisture deficit would show a similar trend. This monitoring method is expensive, but would be more site-specific unless evapotranspiration data are available for the specific site.

PASSIVE IRRIGATION

Introduction

There are some techniques in the design of soil and drainage systems that permit the passive irrigation of plants. These methods are most desirable from the viewpoint of sustainability because they usually involve water from precipitation or recycled water. Unfortunately, little specific empirical research has been accomplished to develop the technology to the degree that landscape architects can accept the techniques without some doubts on their workability. However, it is most appropriate to cover these techniques here as an opportunity to encourage some creative designs on the part of the reader. The author is aware of several successful applications of passive irrigation but does not have access to the details for elucidation here. The basic designs discussed in the following section are based on general soil physics and agricultural land drainage knowledge. If the fundamental principles are followed, there is no reason why the systems should not work.

The Cistern-Overflow System

Underground cisterns were used in old castles and forts. The water is collected from the hard surfaces of the fortress, piped to large stone cisterns located below the floors of the castle or the ramparts of the fort. The water is then available to the inhabitants without being exposed to enemy fire. Old forts in the United States such as Fort Totten in New York City harbor still have operational cisterns. The same idea can be applied to a passive soil irrigation

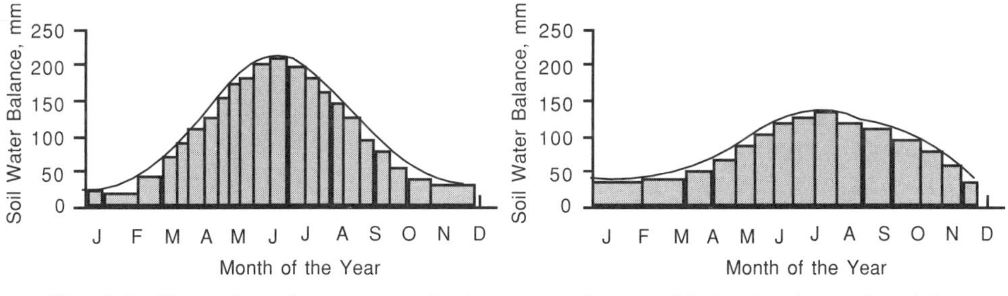

Fig. 8.6 Examples of evapotranspiration curve slope and irrigation interval and frequency. (Adapted from Smith, 1997.)

system except that the system is designed to apply water to the soil as needed. No moving parts are required in the ideal system as gravity is used throughout. The original idea successfully applied to a park is derived from that of Robert Wulkowski (personal communication, 1993), formerly with the Chicago Department of Parks and Recreation.

9

TREE PLANTING TECHNIQUES

It's better to put a $100 tree in a $200 hole than a $200 tree in a $100 hole.
—Anon. (*Arborists News,* February 1995)

INTRODUCTION

Tree planting is usually considered a simple and straightforward routine carried out by foresters, homeowners, landscape contractors, nurserymen, and arborists. The technique is to provide a hole sufficiently large to accommodate the root ball or root system, backfill with an amended soil, while thoroughly watering the soil and root ball as backfilling progresses. Postplanting methods may include mulching, pruning the top, fertilizing, and watering. Few problems are encountered and the tree usually survives the transplanting shock and may reach a reasonable life expectancy if favorable soil conditions and sufficient space for root extension exist. However, experience by many practitioners and researchers indicates that these favorable soil conditions and sufficient space for rooting do not exist for most trees planted in the urban environment. Thus, a major reason for the short-lived expectancy of urban trees. This fact suggests that modified or entirely new techniques and designs must be developed to improve the potential life expectancy of urban trees as well as their relative vigor (Urban, 1989).

Tree planting techniques have existed for as long as trees have been planted. Certain techniques have been developed through folk lore, and, later, specifications were drawn up by various arborists, foresters, landscape architects, and some engineers. In the United States, the present specifications usually followed are derived from practices developed during the latter part of the nineteenth and the early part of the twentieth century when large efforts were made to plant trees in newly developed metropolitan areas of the country or to replace older plantings that had deteriorated. Another press for replanting occurred during the 1930s as part of the old WPA (Works Progress Administration) employment activities during the Great Depression, and in the 1960s and 1970s in efforts to replace the widely planted American elm, which was killed by Dutch elm disease. Other insect and disease outbreaks have caused serious mortality in the urban tree population such as the Gypsy moth on oaks, the oak wilt disease, sugar maple decline, syrcamore and dogwood anthracnose, and many others.

The tree planting techniques followed in most of these more recent tree planting efforts have supposedly been the old "tried-and-true" techniques. But there have been many failures. Many are unexplained because no post-planting analyses have been made. But observations by some practitioners suggest several causes. One is that present urban conditions have changed or been greatly modified from the "old days" and the old techniques are no longer applicable. Others claim that the failures are due to the wrong species or cultivars being planted in the wrong situations or just plain poor planting practices. Failure to appreciate the problems presented by the modern urban soil is another overwhelming reason. More and more plantings are being made as part of redevelopment and so-called restoration projects. The soils on these sites are highly disturbed and modified compared to the conditions existing at the initial plantings, if there were preexisting ones.

Modification of some older techniques and the use of new techniques are required for application to the modern urban condition (Patterson, 1976, 1985). The development of new technology on root morphology and tree physiology and better knowledge of the importance of the root-shoot ratio should be used in the search for modified or alternate tree planting techniques.

The techniques discussed here attempt to take into consideration all of the factors previously mentioned, using the present state of our knowledge and, to some degree, in anticipation of future research developments. It is hoped that future basic and applied research will provide more specific information. Until the results are available, we must employ our best professional judgment coupled with reasonable application of our knowledge, and then make certain that the practitioners are made aware of the information.

Cox (1916) discussed the five criteria that tree planting, or more properly, the urban soil, must provide to the tree. These are

(1) a sufficient amount of good soil
(2) sufficient moisture
(3) proper drainage
(4) proper aeration of the soil
(5) a supply of plant food

It will be seen that the techniques given here attempt to meet these five criteria. Obviously, one technique may do a better job of meeting one or several of the criteria than another. The key to the choice of the most successful planting technique is to match the proper technique to the condition presented by the urban situation. In some situations, a special, unique design may be required that depends on the creativity and knowledge of the practitioner. The totally engineered design is the final one discussed as it is the most expensive and is applied in those situations that require unqualified success.

Special techniques required for semiarid and arid region applications are discussed with the primary concept of water conservation in mind. There are too many instances in the southwestern United States where irrigation remains a consideration in tree planting. With the increased water demands on a shrinking resource, irrigation should be looked on as a "none" alternative. There will come a time when the same approach should be used for the humid parts of the country.

Finally, there needs to be a reliable process for assessing urban sites prior to planting and for postplanting evaluation to determine the reasons for failure. Unfortunately, practitioners cannot now usually tell why trees thrive in one location but can usually determine why trees die in a another location. Several processes are given in the final section of the chapter.

TREE PLANTING CRITERIA

Volume

In a forest stand, there is great intermingling of roots from adjacent trees with ropelike laterals extending outward from the framework roots. It is not uncommon to find the roots of four or more trees intermingled through a square foot of forest floor surface. Even in a closed forest, the roots extend well beyond the dripline. This widespread extension of roots from individual trees suggest that a large volume of soil is occupied by an individual root system. Lyford (1980) observed individual roots of red maple and red oak extending to a length of 40 to 60 feet in a mixed hardwood forest. Since most lateral and fine feeder roots occur within the surface 8 inches of the soil (Craul, 1964; Lyford, 1975), the maximum volume of a horizontal disk 80 feet in diameter and 8 inches thick would be 3317 cubic feet (94 m^3). With competition only from turfgrass roots in a park setting, the rooting disk of an open grown tree may be even greater. Perry (1982) reports roots of open-grown oaks reaching five to six times the height of the tree.

These observations suggest that the optimum rooting volume for a "traditional" mature tree is quite large. In the urban environment, these volumes are attained only in open areas such as parks, large medians, and greenbelts. Urban (1992) has examined the apparent rooting volumes of 1500 trees in five northeastern cities of the United States and evaluated their relative vigor and survival. He found that in the urban environment of sidewalks, curbing, and less than optimum soil that trees exhibiting near optimum or healthy vigor had rooting volumes in excess of 600 cubic feet (17 m^3) and a maximum of

1200 cubic feet (34 m^3) for trees of greater than 25-inch caliper (63.5 cm) at maturity.

The question of adequate volume dwells in the realm of soil moisture relationships and the evapotranspiration at the planting site. Thus, the water-holding capacity and the supplying power of the soil are considered together with the water utilization of the plants.

Moisture

One of the major problems associated with the evaluation of moisture for a planting site lies with the lack of data on the actual moisture utilization of trees for optimum growth in the presence of unlimited available water. It is known that the amount required will vary with species and with evapotranspirational conditions (Kramer, 1987; Kramer and Kozlowski, 1979). Therefore, some general rules of thumb combining species water utilization coupled to evapotranspirational stress must be adopted. A principal rule of thumb is the moisture requirement of most plants to be about 1 inch (2.5 cm) per week in a humid temperate climate. This amount will vary with the season of the year, the regional climate, and the heat-loading situation in the urban setting. The amount supplied by the soil is determined by the available water in the soil profile. The available water is determined by the pore size distribution as influenced by the texture, structure and bulk density, the total soil depth, and moisture content.

Kramer (1987) estimates that a 35-foot (10.6-m) high tree with an actual leaf surface area of 2000 square feet (186 m^2) loses about 35 gallons (132 L) of water per day, whereas Vrecenak and Herrington (1984) estimate about 250 gallons (945 L) of water are used by a tree with a 64-foot (19.5-m) diameter canopy.

Lindsey and Bassuk (1991) have developed a scheme for estimating the required soil volume under various climatic conditions. A preliminary study by the authors (together with reports by others) showed that pan evaporation could be validly correlated to tree water use. A three-step process leads to the adequate soil volume:

1. Determine daily tree water use.
   (a) Calculate crown projection.
   (b) Select the approximate leaf area index.
   (c) Determine the evaporation rate.
   (d) Use the evaporation ratio as a constant applied to local water utilization estimates.
2. Determine an adequate soil volume.
   (a) Select available waterholding capacity of the soil.
   (b) Determine rainfall frequency.
3. Calculate possible bed dimensions.

The method is better applied to confined or limited rooting conditions. Open-grown trees seldom have limited rooting unless the soil is compacted. The method may also be used as a guide for potential required rooting volume under arid or semiarid climate. Two major drawbacks of the method are that pan evaporation data may not be available for a local area or it may not applicable to the planting site, and the selection of the leaf area index is critical and is not well established for many tree species. Another method is to develop the heat budget of the site as suggested by Craul (1992) and Chapter 3, following an initial study by Stark and Miller (1977) using sky-view. Again, the process requires further research and elaboration.

Drainage

Soil drainage primarily deals with the depth in the soil profile to a permanent or fluctuating watertable as well as the rate of water movement from the soil surface. Permeability deals with the rate of water flow through the soil itself. As defined by the National Resources Survey (Soil Survey Staff, 1993), the seven drainage classes are as follows.

Excessively Drained. Water is removed very rapidly. The occurrence of internal free water commonly is very rare or very deep. The soils are commonly coarse-textured and have *very high hydraulic conductivity* or are very shallow.

Somewhat Excessively Drained. Water is removed from the soil rapidly. Internal free water occurrence commonly is very rare or very deep. The soils are commonly coarse-textured and have a *high saturated hydraulic conductivity* or are very shallow.

Well Drained. Water is removed from the soil readily but not rapidly. Internal free water occurrence commonly is deep or very deep; annual duration is not specified. Water is available to plants throughout most of the growing season in humid regions. Wetness does not inhibit growth of roots for significant periods during most growing seasons. The soils are mainly free of the deep to moderately deep redoximorphic features (formerly called *mottling*) that are related to wetness.

Moderately Well Drained. Water is removed from the soil somewhat slowly during some periods of the year. Internal free water occurrence commonly is moderately deep and transitory through permanent. The soils are wet for only a short time within the rooting depth during the growing season, but long enough that most mesophytic crops are affected. They commonly have a *moderately low* or *lower saturated hydraulic conductivity* in a layer within the upper 1 m (39 in.), periodically receive high rainfall, or both. Redoximorphic features may occur below 50 to 75 cm (19 to 30 in.).

Somewhat Poorly Drained. Water is removed slowly so that the soil is wet at a shallow depth for significant periods during the growing season. The occurrence of internal free water commonly is shallow to *moderately deep* and transitory to permanent. Wetness markedly restricts the growth of mesophytic crops, unless artificial drainage is provided. The soils commonly have one or more of the following characteristics: *low* or *very low saturated hydraulic conductivity,* a high watertable, additional water from seepage, or nearly continuous rainfall. Redoximorphic features occur below 20 to 50 cm (8 to 20 in.).

Poorly Drained. Water is removed so slowly that the soil is wet at shallow depths periodically during the growing season or remains wet for long periods. The occurrence of internal free water is shallow or very shallow and common or persistent. Free water is commonly at or near the surface long enough during the growing season so that most mesophytic crops cannot be grown, unless the soil is artificially drained. The soil, however, is not continuously wet directly below plow depth. Free water at shallow depth is usually present. This watertable is commonly the result of *low* or *very low saturated hydraulic conductivity,* of nearly continuous rainfall, or of a combination of these. Redoximorphic features occur below 10 to 20 cm (4 to 8 in.) and not to the surface.

Very Poorly Drained. Water is removed from the soil so slowly that free water remains at or very near the ground surface during much of the growing season. The occurrence of internal free water is very shallow and persistent or permanent. Unless the soil is artificially drained, most mesophytic crops cannot be grown. The soils are commonly level or depressed and frequently ponded. If rainfall is high or nearly continuous, slope gradients may be greater. Redoximorphic features occur to the surface.

The preceding drainage classes apply to natural, relatively undisturbed soils. Soils in the urban environment are usually disturbed and exhibit varying degrees of compaction (Craul, 1985). As a result, many urban soils typically have restrictive drainage that are characterized by moderately well drained to very poorly drained. This condition is one of several that causes the high degree of mortality of planted urban trees (Berrang et al., 1985). No matter what attempts are made to create a well-drained backfill material, a dug root pit will drain only at the rate of the surrounding soil, unless underdrainage to the pit is provided. Excessively drained to well-drained urban soils are uncommon unless they remain in an undisturbed state from initial well-drained conditions. Existing urban soils initially designed to be well drained are an exception to this general observation.

The predominance of restricted drainage in many urban soils places constraints on the range of plant species that may be safely planted in these soils. Planted species adapted to periodically wet soils appear to be more successful in urban areas than those adapted to well-drained soils (Kozlowski, 1986; Kramer and Kozlowski, 1979).

Aeration

Soil aeration status involves mainly the proportion of air-filled pore space in the soil and the concentration of various gases in the soil atmosphere, especially oxygen and carbon dioxide.

The relationship between air-filled pore space and water-filled pore space is determined most directly by soil moisture content. As moisture content increases, the amount of air-filled pore space decreases until the saturated soil contains no air-filled pore space. Large pores drain first in response to gravity and the remaining pores generally contain available water, except for the smallest, or micropores, where the water is unavailable due to low potential forces. The dry soil contains little water and most of the pores are air-filled. The soil properties involved include texture, structure, and bulk density as they affect the pore size distribution. Coarse-textured or sandy soils are well drained and contain a large proportion of air-filled pore space at field capacity. Gaseous diffusion into and out of the soil readily occurs, sometimes to great depth. Silty and clayey soils without a well-developed structure will exhibit poor or restricted drainage. These soils will contain high concentrations of methane, ethane, ammonia, and hydrogen sulfide in place of the oxygen. Chemical reducing conditions prevail and anaerobic biological processes are dominant. Well-developed structure will improve the aeration. Increased bulk density due to compaction, finer texture, or lack of well-developed structure, or all three, leads to smaller average pore size and a decreased total pore space. A greater proportion of water is held in the small pores and less pore space is air-filled. The water-filled pores act as a barrier to ready gaseous diffusion, and the oxygen concentration is reduced with a conconmitant increase in carbon dioxide concentration.

Nutrients

Plant nutrients supplied by the soil must be evaluated on the basis of the native nutrients supplied by the soil together with their rate of cycling in the system. The latter involves the organic matter cycle between the soil surface horizons and the associated vegetation, and the mineral cycling within the soil profile and the parent material. In the case of native nutrients, the amounts present are dependent on the texture, organic matter content, and mineralogy of the soil. Open-park and greenspace areas may have little limitations for nutrients and their cycling. Planting sites in built areas with an open-soil surface may have adequate cycling through the application of fertilizers and mulch. On the other hand, hard-covered sites may require fertilization through irrigation water.

TRANSPLANTING TECHNIQUES

There are several methods of transplanting stock from the nursery to the planting site. These range from bare-root stock to container-grown stock. Each has its own advantages and disadvantages that must be evaluated on the basis of the type of stock to be planted, its size, the desired aesthetic effect in the design, and the situation in which it is to be planted, for example, at curbside, in a covered plaza, or in an open yard.

Bare-Root Planting

This planting technique is actually very old, having been used for years in planting young orchard stock and roses as well as small forest tree seedlings. Obviously, the method is restricted to small-size tree stock, usually smaller than 2-inch (5 cm) caliper (diameter at the root collar). Because the stock is smaller, a larger proportion of the inital root system may be retained on the transplanted stock. The shipping weight is much less, but the entire plant must be wrapped to prevent drying and freezing. Dug stock must be stored under cold conditions. It is usually dug after dormancy in the fall, stored during winter and shipped and planted the following spring. Planting of small stock in urban areas is risky unless staked and well protected. Small specimens can be easily damaged by vehicles, vandalism, or careless maintenance operations.

The dug pit should be slightly less than the depth of the root system. The width should at least twice the diameter of the root system. This provides loosened soil around the root system. The backfill soil is carefully tamped to stabilize the soil around the roots and to remove any air pockets. Staking is frequently required for bare-root hardwood stock since the tree lacks the stability provided by a root ball. Conifers usually do not require staking. Staking should follow the various styles suggested by Harris (1991). He recommends staking only when necessary for protection or in wet soils. The best method is to have two stakes on opposite sides of the trunk and 400 mm from it. The stakes and ties should be removed as soon as the trunk has attained adequate strength to support the crown. A preferred alternate method to staking hardwood saplings is the use of plastic treeshelters or large tubes (Evans and Potter, 1985). Originally developed by the English, these are translucent plastic tubes about 6 cm (2 in.) in diameter and about a meter in length. The tube supports the trunk, prevents damage or vandalism, and enhances early growth (Burger et al. 1996; Frearson and Weiss, 1987; Jones et al. 1995; Potter, 1988). The latter effect results from higher trunk temperatures, elevated relative humidity, and carbon dioxide levels inside (Svihra et al. 1994). The treeshelters are very effective in preventing injury from wildlife. However, redwood sapling growth is inhibited by plastic treeshelters and should not be used for this species (Svihra et al., 1996).

Ball and Burlap Planting

This is the traditional method of tree planting. It has the advantage of keeping the root system contained within the root ball relatively intact. Very large trees may be transplanted by this method. Standards for root ball size and size of the tree have been established (Himelick, 1981). See Table 6.1 for the data. There are several disadvantages to the method. A major one is that most of the initial root system remains in the nursery when dug unless a vigorous program of root pruning is carried out before transplanting. A second disadvantage is the sheer weight of the soil in the root ball causing difficulty in handling without the proper equipment and also contributes to high shipping costs. The root balls must be protected against drying or freezing during storage.

The depth of the tree pit should be slightly less than the height of the root ball. The base of the pit needs to be only slightly larger than the root ball diameter. The tree pit flare will provide sufficient space for removing the burlap and for the backfilling operation. The burlap, if not biodegradable, should be cut and removed or rolled down to expose at least half of the root ball. No twine should be left uncut unless it is biodegradable. Even biodegradable twine should be cut in several places to prevent any girdling. If the stock was shipped in a wire basket, the basket should be cut at numerous places and the pieces removed. Research has found that root girdling does not affect overall growth and extension of the roots, but does have some significant effects on the distribution of carbohydrates above and below the girdle as well as the presence of a water potential gradient (Goodwin and Lumis, 1992). The claim that root balls will fall apart if the burlap is removed is unfounded if the root ball is handled properly and has not dried out. Obviously, care must be taken not to walk on or crush the root ball while removing the burlap and backfilling the pit, even if the burlap is retained in place.

Containerized Planting

This technique for tree planting has gained in popularity during the last few years. The seedling tree is usually started in a seedbed and transplanted to a quart or half-gallon (1 or 2 L) plastic or metal pot after 2 to 3 years. It is then transplanted again to a 5- or 10-gallon (19 or 38 L) plastic pot. The pots may be placed on the soil or a gravel surface or partially buried in the soil (Appleton, 1994). An advantage of the method is that after the initial tranplant from the seedbed, the root system is intact. However, there is great potential for pot bounding of the root system at each repotting stage. If initially present, the pot-bounding persists through the various transplanting stages unless deliberate effort is made to cut and straighten the roots at each stage. Appleton (1994) has found that coating the inside of the pots with paint containing copper sulfate enhances the multiplication of roots and reduces the potential for pot-bounding. The stock is transported in the containers and they are partially removed after insertion in the planting hole or some landscape contractors will slip the root balls out of the containers before placement in the pit. The root ball must be coherent and handled carefully for this operation. Dry, loose root balls will tend to fall apart, partially defeating the purpose of the container technique. The treatment of the planting pit is the same as for ball and burlap stock. Whitcomb (1987) and others have used containers of various fabrics with mixed success. The technique eliminates the burlapping process and facilitates the removal of the container in the planting pit. It is poor practice to leave plastic pots in place. Svihra and colleagues (1994) found that treeshelters are successfully used on container-grown stock.

Backfill Material

There has been a continuous difference of opinion about the composition of the backfill material (Corley, 1984a, 1984b; Pellet, 1971; Watson et al., 1992; Whitcomb, 1975) for tree planting pits. The oldest (questioned by some) recommendation known (oral tradition) to the author, besides adding nothing, is to simply add an amount of animal manure to the bottom of the pit and backfill with the excavated soil. Informal observation of this technique shows that the manure decomposes very slowly, especially in moderately well-drained or slower soils. Even when the manure releases nutrients on decomposition in well-drained soils, they are well below the fine feeder roots that primarily absorb water and nutrients close to the soil surface. Further, the manure has no beneficial effect of the surface soil physical condition. There is some benefit if mixed with the soil before backfilling. Manure is now scarce and there is reluctance to handle it. Composting of manure and sewage sludge is a viable alternative for an organic amendment to the backfill. Peat moss has been frequently used as well. Vermiculite, Perlite, and other inorganic amendments have also been used for the same purpose.

Enhancement of porosity (primarily the granulation) and nutrients is the main purpose of backfill amendment. Root extension and proliferation in the backfill is faciltiated, aiding the recovery of the stock from transplant shock. Research (Spomer, 1983) and some practical observations show conflicting results. The conflict most likely arises as the result of planting soil differences. The various viewpoints may be summarized as follows. Amendment of the backfill is usually not necessary in well-structured (granular), medium- to fine-textured soils. The water-stable structural aggregates generally persist through the excavation and backfill process. If the soil is already well granulated and well drained, the amendment has some, but limited, beneficial effect. In small proportions, the amendment can offset the loss of aggregation during planting. Whitcomb (1987) and others have found that amending backfill had no effect as compared to unamended backfill. Thus, some authorities recommend no amendment whatsoever. Others (Watson et al., 1992) have found varying degrees of effectiveness in root-growth enhancement depending on the kind of amending material. Organic matter amendment of backfill for a compacted soil definitely enhances the porosity of the backfill material, which is already loosened by the excavation. However, it creates two other problems. The greater porosity (reduced density) of the backfill increases its infiltration rate as compared to the surrounding soil. Surface water and precipitation can enter the backfill more readily than the surface of the compacted soil. The root ball pit, without drainage, fills up like a teacup saturating the backfill and root ball soil. The roots suffocate. The second problem arises as the soil dries. Because the pores and channels are smaller in the surrounding compacted soil than the large ones of the amended backfill, an outward water-tension gradient is created so that water is drawn from the amended backfill to the surrounding soil, eventually drying the former to the wilting point before the surrounding soil (B. L. Kays, personal communication, 1982). The same phenomenon may occur within the root ball itself. If a specified soil mix is used in an isotropically homogeneous soil profile, these problems do not arise; organic matter should be a routine component in this instance.

The recommendations of the author are that amendment is unnecessary in well-drained, well-granulated, medium-textured soils. Smalley and Wood (1995) found no differences in red maple (*Acer rubrum,* L.) shoot and root growth between unamended native backfill and backfills with aged pine bark or a proprietary soil amendment or the latter alone. Amendment with organic matter is desirable in compacted soils where underdrainage is provided for each tree pit, or in deep sandy soils where the waterholding capacity and fertility need to be enhanced. The backfill should remain unamended if no drainage is provided in a compacted soil or if the soil is poorly drained. In these cases, the tree pit should be made shallower so that the root ball is raised several inches (cm).

Mulching

This practice has had a resurgence in popularity during the last few years as the result of interest in and sensitivity to organic waste recycling. Most organic wastes were burned or landfilled as cheap methods of disposal. Originally, partially composted manure, hay, straw, and especially salt hay where readily available near coastal areas, were used (Solotaroff, 1911). Leaves and grass clippings have been used in home applications. These have a tendency to pack when used alone or in mixture. They are best applied after composting with other organic wastes and soil. Sawdust alone is useful as a sterile mulch for forest seedling seedbeds, but not widely used for other applications because of its very high carbon:nitrogen ratio and tendency to blow away when dry. If it is mixed with other organic wastes, the disadvantages are reduced. Commercial organic mulching was expensive with the nearly exclusive use of peat moss since other sources were not widely available. Later, bark was used as a substitute in areas where it was readily available around large pulp mills or sawmills. Wastepaper pulp has also been used locally. However, some of these materials have been found to have high dioxin contents from the chlorine bleaching process and must be avoided as a human hazard.

As the value of organic wastes has risen, it is now profitable to ship the wastes long distances, so that they are available in localities that had no source. Also, localities are producing their own sources of organic wastes. At present, the most popular materials used for mulch include wood chips, bark chips or shredded bark, composted sewage sludge mixed with chips, composted brewers' waste, and composted mixes of leaves, bark, and wood chips. Most have been found to be satisfactory as mulches for plants.

There are several purposes in mulching newly planted trees or established plants (Craul, 1992):

1. The soil surface is protected from compaction and crusting.
2. Soil moisture loss from evaporation is greatly reduced.
3. Soil temperatures are reduced during daytime and increased at night.
4. It is a means of adding organic matter and its nutrients to the soil.
5. Root competition from grass and other vegetation (weeds) is controlled (Watson, 1988).

Mulch may also provide nutrients not otherwise available. Himelick and Watson (1990) found that oak chlorosis was reduced with woodchip mulch treatment. The type of mulch used is generally determined by its availability and the plants to be mulched. Most of the organic mulches can be safely used for tree planting. The layer of mulch should be only 3 or 4 inches (7 or 10 cm) thick and additional amounts should not be added until the layer thins somewhat (Greenly and Rakow, 1995; Watson and Kupkowski, 1991). There are many instances in which mulch applied too thickly—exceeding 6 inches (15 cm) or more—has actually caused tree mortality. The mulch layer should taper toward the trunk. The root collar should be essentially free of mulch. Mulch piled up around the base of the trunk tends to cause rot at this point, as moisture is preserved (Smiley, in Stone, 1993). The mulch layer should extend beyond the root ball and cover the backfill soil in the flared portion of the tree pit. The irrigation rim is constructed just outside of the circle of mulch. It is generally unnecessary to underlay a plastic or woven fabric liner below organic mulches. The practice arises as a method to prevent rooting of weeds and to facilitate their removal. It seriously defeats the purpose of eventual incorporation of the organic matter into the underlying soil. The liner has been primarily black or clear plastic and is now being supplanted by spunwoven polymer fabric. The plastic has several major disadvanatges. Water and air cannot penetrate through it and, thus, the soil below either remains saturated and anaerobic or totally dry, or both conditions may persist for long periods during the year. Severe stunting of plant growth or mortality almost always occurs unless the plastic liner fails by being torn, lifted by frost or wind (when the mulch layer becomes too thin being blown or washed away), or otherwise deliberately punctured. The spun-woven polymer fabric is porous to air and water and does not have the disadvanatge of the black plastic. However, the fabric must remain covered for good water transmission. Certain brands of fabric tend to become hydrophobic if exposed to sun and wind.

In some landscape designs, it is appropriate to use inorganic mulches. These include fine or coarse gravel, brick or marble chips, and so on. These tend to encompass the same benefits as organic mulches except that organic nutrients are absent and no organic matter is added to the soil. The inorganic mulches may be spread in a thicker layer than the organic mulches. Their relative coarseness and good stability maintain large continuous channels for water infiltration into the soil while reducing evaporation. Air exchange between the atmosphere and the underlying soil is not restricted, as might happen with compacted or moist organic matter. A highly successful example is shown in Fig. 9.1. To prevent intrusion of soil into the mulch layer and to reduce weed establishment, a liner is usually placed over the soil before the inorganic mulch is spread. The same conditions described before for plastic versus woven fabric liners hold for inorganic mulches. Generally, the liner is nearly always necessary for use under the inorganic mulches. An exception may be made when the mulch layer is at least 6 to 8 inches (15 to 20 cm) thick.

Fig. 9.1 Gravel as a mulch under a landscape planting, Kamloops, BC, Canada.

Amending with Hydrogels or Super Absorbents

Recently, several natural and synthetic water-absorbent polymers have been developed and marketed for use as soil amendments to increase waterholding capacity and reduce irrigation requirements. According to manufacturer's claims, the material holds 300 to 500% of its weight in water, which is released over time. The purpose is to increase the waterholding capacity when mixed with soil. It was first used as a potted medium mix and is now used as an amendment to backfill soil in ornamental plantings, including trees. The early natural material, a form of starch, was quickly broken down in the soil and the favorable effects were lost within several months. The newer synthetic material, cross linked polyacrylamide copolymer, is more resistant to breakdown and may last several growing seasons—4 to 5-years or longer according to one manufacturer. Several superabsorbent trade names are Terra-Sorb, Super Slurper, and Viterra Planta-Gel.

Superabsorbents are manufactured in several particle sizes, depending on their use. Fine material, about 0.5 to 1.0 mm, is used in turf areas and hanging baskets. The size recommended for landscaping planting or container growing is 1.0 to 3.0 mm and has an absorption capacity 325 to 400 times its weight, with 90% or more plant available. One manufacturer claims that for a 3-inch (7.6-cm) caliper tree, 4 ounces (113 g) mixed into the backfill soil will increase the waterholding capacity by 5 to 7 gallons (19 to 26 L).

Its use as a potting medium component in greenhouse and container applications has shown some benefit (Whitcomb, 1987). However, there are reports of mixed results when used in the field. In several studies reported by Whitcomb, there were no differences in growth and survival of containerized or field plantings between superabsorbent-treated plants and those without superabsorbent amendment. A major disadvantage for field use in large bulk volume is that it absorbs moisture very quickly under humid conditions, giving it a sticky characteristic. It readily adheres to equipment surfaces and clothing. Several landscape contractors do not like the material for this reason alone. In many instances, packets of superabsorbent are simply dumped in a mass on top of each of the rootball burlap of recently planted trees with no attempt to mix the material in the backfill soil. Under these circumstances, the superabsorbent is essentially ineffective.

The superabsorbent is also used as a root dip by mixing with water to form a gel. It increases the moisture retention on the roots and tends to protect them from dessication. Hensley and Fackler (1984) found no benefit from root dipping. Ingram and Burbage (1985) showed a slight benefit on live oak (*Quercus virginia,* L.) transplanted on Florida sandy soils.

Some brands of superabsorbents are not stable. These may be repackaged health care superabsorbents. They should be avoided for horticultural/arboricultural use.

A properly specified and prepared landscape backfill material does not require superabsorbent amendment. It does no harm, but cannot correct or compensate for serious moisture deficiency in a poorly prepared backfill material.

Irrigation

Use of irrigation in urban areas is irregular, depending on the location and aesthetic value of the planting. Streetside trees are seldom planted with an installed irrigation system, being watered after establishment only for a specified period by the contractor. Susceptibility to accidental damage and vandalism, and high maintenance costs are the major reasons for its omission. Other more valuable plantings may have a complex system designed with zones to provide different application methods and rates for a variety of planting types. Simple irrigation may vary from manually watering with a hose from a fire hydrant or a tank truck to the installation of watering bags about the tree trunk (Fig. 9.2).

Harris (1991) estimates that a large solitary tree may use two to three times as much water as turf occupying the same area. The recommended amounts of watering are based on indirect factors such as precipitation, rough estimates of potential evapotranspiration, and soil profile moisture storage. The usual rule of thumb is that trees and turf should have at least 1 inch (2.5 cm) per week. When applied by hose or bucket to the watering basin around a street tree, this amount is difficult to estimate other than to simply fill the basin once a week. To make certain that the backfill, root ball, and immediate surrounding soil are wetted, the dike of the watering basin must be placed over the surrounding soil and outside the rootball diameter, as illustrated in Fig. 9.3. Otherwise, the root ball may be the only portion watered. The best, most direct method to determine moisture distribution is the use of a soil probe. Probing the soil during irrigation will determine if sufficient water has been applied to be distributed throughout the root ball, backfill, and the edge of the surrounding soil. Overwatering is also detected; however, most watering operations apply insufficient water to thoroughly moisten the desired portion of the root zone. Constant vigilance is required. Moisture monitoring instruments such as tensiometers and electrical resistance blocks may be installed, but are susceptible to damage and vandalism. See Chapter 8 for more details on irrigation.

THE PLANTING SITUATION ARRAY

Ideally, every planting occasion includes the optimum condition for all the rooting criteria needed for the long-term success of the plants. The only two situations that may be considered to meet all of these conditions are the planting in an open, unrestricted site in a deep, well-drained, medium-textured fertile soil, or in a highly "engineered" design in which all limiting factors are compensated to near optimum levels by the design (Patterson, 1985). For

Fig. 9.2 Watering bag installed on a tree trunk, Harvard Yard, Cambridge, MA.

example, some trees may be adapted to imperfectly drained conditions, but they almost always exhibit good growth in a well-drained soil; few show poorer growth. In the engineered situation on the other hand, irrigation provides moisture in a soil designed to be well draining that provides sufficient aeration when field capacity is attained on gravitational drainage. The alternatives have conditions and degrees of design intermediate between the two

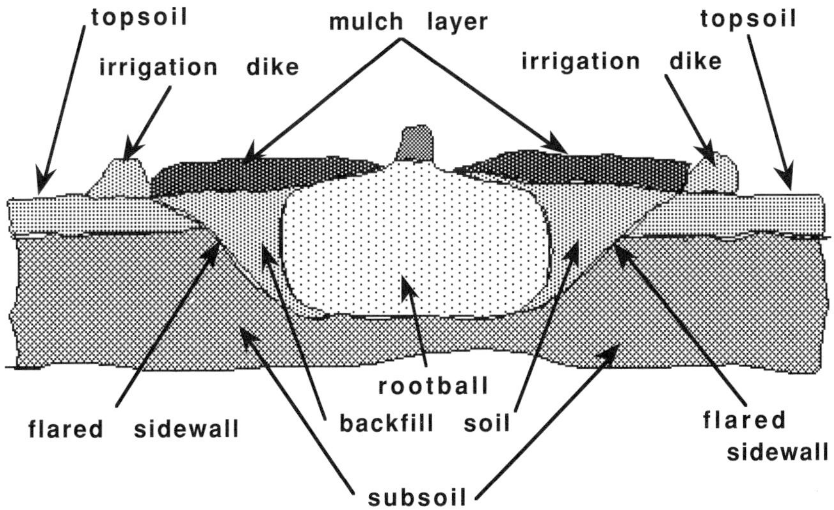

Fig. 9.3 The ideal tree pit.

extremes and must employ compromises between the optimum or desired levels of the various criteria.

The planting situation array may be divided into eight classes of limitations: open, with few, if any, constraints; limited space; limited moisture; limited drainage; limited aeration; limited nutrients or the presence of toxics; highly engineered to overcome a combination of several; or all of the listed limitations. These are placed into a situation array in Table 9.1. It is clear that each of these situations may have more than one limitation. For example, a site with limited volume will probably have limited moisture and nutrients and perhaps limited aeration. These interactions are discussed with modifications to show potential applications. Examples are provided where available.

Situation A: Individual Sites with No Limitations of Space and Soil (Open)

This situation exists where trees are to be planted in an open-grown condition. The trees may be planted individually or in a group (copse). There are no constraints on crown and root space, and the soil surface is or will be open or planted to some type of vegetation and perhaps covered with mulch. The soil in which the tree is to be planted has the properties of texture, structure, and density that lead to the desired characteristics of favorable waterholding capacity, good infiltration and drainage, a pore size distribution and arrangement that provide adequate air-filled pore space and ease of root penetration, moderate cation-exchange capacity, and at least a minimum amount of organic

TABLE 9.1 The Array of Planting Situations and Their Limitations

	Planting Limitations[1]					
Planting Situation	Volume	Moisture	Drainage	Aeration	Nutrients	Toxics
A—open	Limited only by the backfill material and surrounding soil conditions. Probably requires minimum input of energy and maintenance.					X?
B—curbside and sidewalk.	XXX	X?	X?		X?	X?
C—covered by impervious material.	X?	XXX		X?	X?	X?
D—compacted soil; mechanical barrier below; high watertable; rooftop	X?	XX	XXX	XX?	X?	X?
E—infertile soil; nutrient imbalance; presence of contaminants.	X?	X?	X?	X?	XXX	X?
F—presence of toxics.						XXX
G—highly engineered	Limited only to the extent of the engineered design, but requires high energy inputs and a high level of maintenance.					

[1]The number of X's indicates the probable severity of the limitation(s). The (?) indicates that the limitation may or may not be present.

matter. The soil has no extremes in pH and is relatively free of contaminants. No rooting barrier exists between the upper and lower portions of the profile, the latter being the subsoil or the parent material, or both; not a rooftop application. The characteristics of a desirable soil were given earlier in Chapter 5, Table 5.1.

The soil must be considered as limiting if soil conditions do not meet those given in Table 5.1. The planting situation would change to another having that limitation as contained in Table 9.1 and falls under the appropriate technique to be covered later.

The tree pit design appropriate for this situation is simple and straightforward (Fig. 9.3). Several modifications from traditional technique are required, depending on the type of planting stock used, such as bare-root, ball-and-burlap, or containerized. The pit may be dug by backhoe, tree spade, or by hand. Each of the methods have advantages and disadvantages. The backhoe and tree spade excavation methods are the fastest methods. The backhoe operator needs to tilt the boom left and right from the vertical to obtain the desired flare in two directions. Some touchup by hand is required; otherwise, the entire flare is formed by hand. Excavation by tree spade creates steep sides unless the spades are tilted to a shallow angle with the horizontal soil surface. This shallow angle is limited on most tree spade machines; flaring is then done by hand. The latter operation also removes the disadvantage of slick side walls in heavy soils. The smooth, slick sides need to be roughened to prevent "pot-bounding" of the roots within the pit (Bridel et al., 1983; Cool, 1976). Excavation by hand gives the best control on the quality of the tree pit, but is the most expensive of the methods. Apparently, the shape of the pit has no effect on the transplant success of some species (Arnold and Welsh, 1995). In the case of valuable trees being planted in a high-visibility site (aesthetically valuable), hand digging may be the most appropriate method.

Situation B: Curbside and Sidewalk Sites

Most urban trees are planted curbside or along the sidewalk, perhaps between the sidewalk and the curb (Fig. 9.4). In residential areas, the soil is not cov-

Fig. 9.4 Optimum curbside plantings, State College, PA.

ered, so that a strip of open soil exists parallel to the curb. In downtown business districts and commercial areas, the soil is almost always covered with paving. The latter situation is discussed under Situation C.

The amount of open soil is dependent on the width of the linear planting strip. Strips with trees as narrow as 18 to 24 inches (45 to 60 cm) have been observed. Many of the examples have trees obviously in poor health and others have healthy appearing trees. This contrast appears to be related to the area of paving adjacent to the trees. If the roots are able to extend under the sidewalk and into a front yard or an area with open soil, as well as linearly along the curb, the trees appear to survive (Fig. 9.5). There is copious practical evidence to show that roots very seldom elongate under the curb foundation and under the street paving; thus, the curb acts as a root barrier. On the other hand, if the only open soil available to the roots is the linear curb strip, the trees do not thrive. In cases when the strip is wider than 3.3 feet (1 m) tree roots apparently have sufficient open soil for adequate water infiltration and gaseous diffusion. Cox (1916) recommended a width of at least about 6 feet (2 m). In a recent study based on a questionnaire by Carlson (1997), 24% of 116 cities responded that their avearge curbside width is equal to or less than 3 feet, 31% have an average of 4 feet, 22% have 5 feet, 14.3% have 6 feet, and 7% have an average of 7 feet. In response to the question of how wide the curbside planting should be, 18% said 6 feet, 37% said 8 feet, and 23% said 10 feet. Large open-soil areas in pavement provide the same amenities (see Situation C).

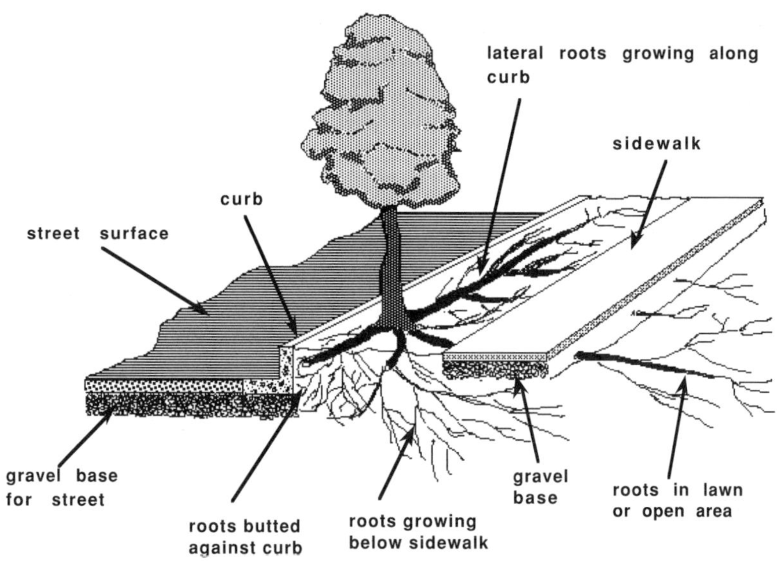

Fig. 9.5 Section of the curbside planting.

The preparation of the tree pit and backfill material for curbside planting is similar to the open planting. Some modifications are required when dimensions are restricted by planting strip width.

When the curbside soil is compacted or otherwise unsatisfactory, consideration should be given to replacing the entire soil volume in the curbside strip and replace it with specified soil (Patterson, 1985). This design becomes what is now commonly called a "linear planting pit." This technique has been employed separately by the landscape architects Jim Urban in Baltimore and Heidi Schustermann in Philadelphia (Craul, 1992). The general plan of the applications is illustrated in Fig. 9.6. Bernatzky (1978) also suggests this design.

There are several advantages to this design. The undesirable soil is removed and underdrainage is provided to ensure removal of excess water during the dormant season. The surrounding soil is likely to remain compacted and exhibit restricted drainage and will create a perched watertable unless the drainage system is provided. The design also provides "shared rooting space" in that the roots are able to elongate into the specified soil along the entire length of the pit rather than being comfined to the immediate tree pit backfill in a conventional tree planting design of separate pits in a compacted soil. In this case, underdrainage is required for each individual tree pit. Note that in the general plan, the soil surface remains open, being covered by mulch, gravel, or a groundcover. Situation C becomes the case in which the pit is covered by paving.

A linear application to overcome compacted soil and the attendant restricted drainage was installed on the campus quadrangle of the State University of New York College of Environmental Science and Forestry, Syracuse, in 1970. The design has successfully supported sweet hackberry trees (*Celtis laevigata,* L.). The orignial soil was a compacted glacial till further impacted by construction siting.

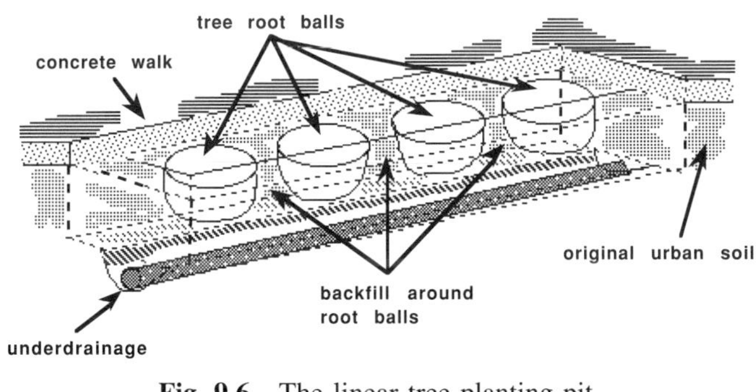

Fig. 9.6 The linear tree planting pit.

Situation C: Sites Covered by Impervious Material

This situation occurs when trees are planted within a paved sidewalk or plaza. The root ball is placed below the paving in many cases. Therefore, the roots must elongate into the soil below the impervious surface because the paving is almost always brought up close to the tree trunk at about the root ball diameter. The open soil is only the area of the root ball. In most cases, this opening is inadeqaute for tree survival unless the opportunistic roots are able to find their way into a storm sewer or some other favorable rooting environment such as the gravel base of the sidewalk and pavement cracks. This may cause failure of the infrastructure. As Jim Urban, a landscape architect in Annapolis, Maryland, states (personal communication, 1993), "The tree survives if the infrastructure fails; the tree fails if the infrastructure succeeds!"

Successful planting below an impervious layer must include some provision for root elongation space and gaseous exchange with the atmosphere (Fig. 9.7). Yet, the soil material must support the weight of the paving and the traffic. There are several ways to accomplish this objective. One method (Grabosky and Bassuk, 1995) is to employ stone as the support matrix. A volume of specified soil just slightly less than the stone pore space volume is the actual roofing medium. A pilot study on a Cornell University campus walking plaza has been constructed. The authors caution that further pilot studies are required to assure success for a variety of tree species.

The use of spaced Belgian blocks, brick, concrete pavers, or other jointed paving reduces the problems normally caused by impervious surfaces (Kopinga, 1985). Pervious (or porous) paving is an alternative surfacing material (Field et al., 1982). It is either low-density asphalt or various grades of gravel held together by a special binder or mastic (McGavin, 1974). Its strength is adequate to support pedestrian or automobile traffic, but is too weak for heavy traffic such as trucks, buses, and recreational vehicles. Precipitation passes

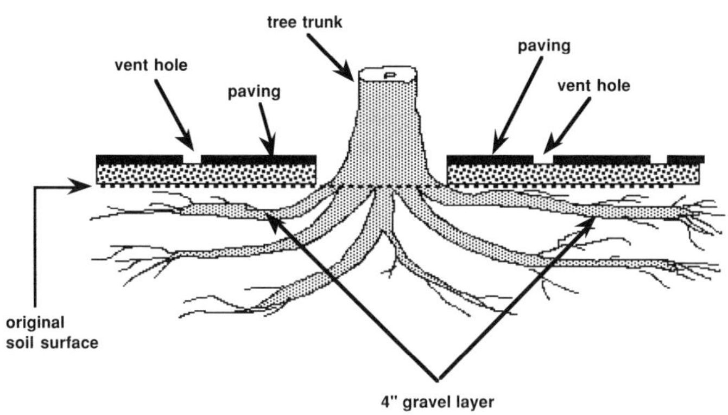

Fig. 9.7 Planting below an impervious surface. (Adapted from Harris, 1991, in Craul, 1992.)

through to the soil below, eliminating or reducing water runoff. The parking lot system for the University of Pennsylvania Morris Arboretum, Germantown, is a successful application of pervious surfacing (Carol Franklin, landscape architect, Philadelphia, personal communication, 1993).

A slightly different situation arises when tree roots are paved over with an impervious surface. However, the potential problem remains essentially the same: lack of adequate rooting space and gaseous exchange (Yelonosky, 1964). Experience shows that a sidewalk placed over the framework roots of an established tree, without severing the roots, has little affect on the root system since most of the absorbing roots may be beyond the sidewalk and are in the yard. However, if the paving covers most of the root system, provision must be made for aeration below the paving. Harris (1991) suggests the placement of a thick gravel layer over the original soil surface together with the creation of grated openings in the paving arranged in a grid pattern. A spacing of about 6 feet (2 m) is recommended. Sometimes, piping is laid horizontally from one opening to another to enhance the aeration of the covered intervals.

Situation D: Sites in Compacted Soil or Presence of a Subsoil Barrier or High Watertable

Though the application of the linear planting pit to compacted soil is discussed in Situation C, there are some conditions when this design is not suitable or practical. The situation arises when it is not practical to remove all of the undesirable soil, or the presence of a high watertable or a shallow physical barrier (such as massive infrastructures) precludes excavation. Rooftop gardens have special features that must be addressed and are discussed later in this section. Rooftop gardens were also discussed earlier in Chapter 6.

In the case of a high watertable, the installation of drainage tiles and interception of surface water is one solution. The technique has been employed for many years in agricultural fields. Thus, additional drainage structures to those recommended for the linear planting pit are required.

If installation of a drainage system is not feasible, or if the compacted soil cannot be removed, the construction of a berm above the existing soil level is indicated. The berm is a mound of specified soil in which the root balls are subsequently planted (Fig. 9.8). The specified soil should be self-draining. The areal extent of the berm may range from mounding of individual trees to a linear large berm encompassing the root balls of a group of trees. The soil is mounded to a depth equal to at least the height of the tree root ball to be planted or higher. The sides are sloped gently, so that soil erosion potential and washing of any mulch is minimized. If the berm is large, grass cover is established and must be maintained. The effectiveness of the berm is greatly enhanced if drainage tiles are laid on the original soil surface. This procedure reduces the potential for rising of the watertable by capillarity up into the

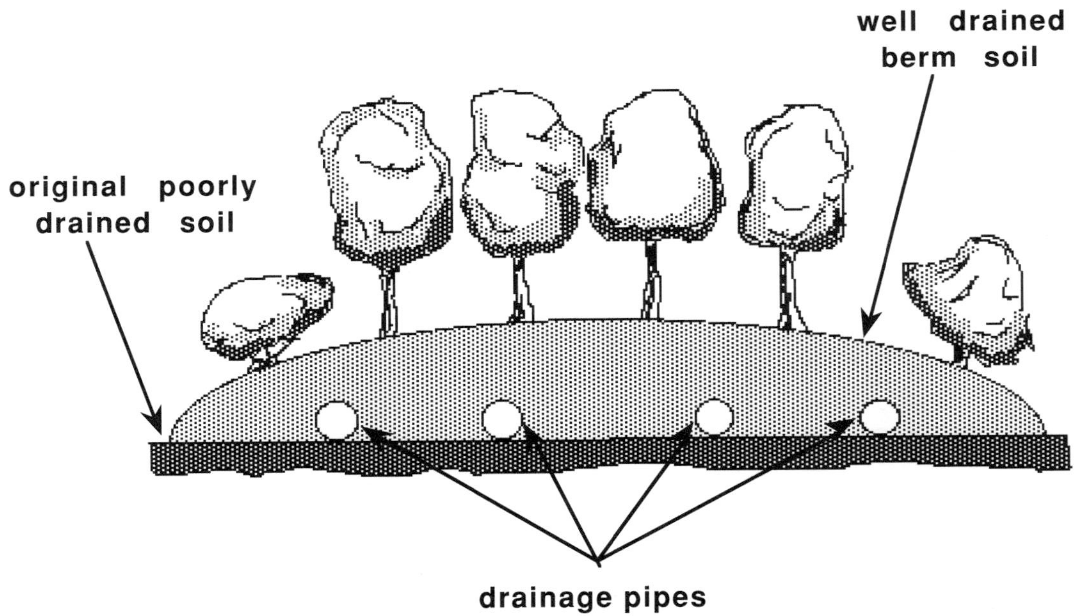

Fig. 9.8 The planting berm.

berm and aids drainage from the bottom of the berm when placed over compacted soil.

Situations E and F: Sites with Limited Fertility or with Toxic Levels of Contaminants

Solutions to problems created by these types of sites deal with the treatment of the problems or selection of species rather than the planting techniques themselves. The planting techniques discussed for other situations apply to this group as determined by the physical site conditions.

Fertilization of Low-Fertility Soils

The question arises as to what causes the low fertility. Is the soil inherently of low fertility? What nutrients are lacking? Is it a pH problem that influences nutrient availability? Is the proposed planting stock too nutrient-demanding for the planting soil? These questions need to be answered before a solution can be adopted. Another question is the physical condition of the soil. Plants on a compacted, poorly aerated soil with impeded drainage and low organic matter content will not respond to fertilization. The soil physical condition must be improved before fertilization is contemplated.

Soil testing for unknown soils is the best approach in planting situations. Though the soil may contain most nutrients, there could be an imbalance of one nutrient or several, or the pH may be in the wrong range for the planting stock. A designed soil should always be tested.

If the test results show that addition of nutrients is required, fertilization should be done prior to planting by working into the soil, or, better, after the first growing season. Most authorities do not recommend fertilization at the time of planting if a complete fertilizer is used (Harris, 1991; Whitcomb, 1987). The reasons offered are that the roots will not be able to take advantage of the fertilizer since they must extend out of the root ball into the backfill before the nutrients can be absorbed. Fertilizer salts of quick-release materials can be injurious to roots on direct contact, especially with bare-root stock, and nitrogen fertilizer applied at planting may even suppress root growth (Warren, 1993). Sometimes fertilizer packets placed in the backfill are not highly effective and are too concentrated in location (personal observation on a tree pit planted after 1 year in Syracuse, New York, June 1994). Best fertilization results in most cases are obtained by fertilizing in the fall following planting (Appleton, 1988; Whitcomb, 1987), although Hensley and colleagues (1988) found that fertilizing magnolia at planting in the spring was better than at other times as did Harris (1991). Hensley and colleagues (1988) also found that fertilizer placed at the bottom of the pit was just as effective as surface application or mixing with the backfill. Differences in results are to be expected because of differences in site conditions, species preferences, and procedural techniques.

The fertilizer material used should include at least some slow-release nitrogen. The various forms include sulfur-coated urea, polymer-coated material, and slowly available insoluble forms such as isobutylidene diurea and urea formaldehyde. These materials vary in their nutrient availability period, ranging from approximately 8 weeks to 1 year. The use of these materials prevents the excessive leaching loss of nitrogen and yet supplies it to the plant over a long period. Also, because of their slow-release feature, they can be applied alone (without phosphorus and potassium) up to moderate amounts at the time of planting, although this recommendation remains controversial. A complete fertilizer containing nitrogen, phosphorus, and potassium is necessary only when all three elements are deficient. The best method is to apply only the nutrient that is seriously deficient, which is usually nitrogen. Coarse sandy soils may be deficient in phosphorus or potassium or both.

Broadcasting of the fertilizer over the soil surface for open-grown trees is by far the best method of application (Good, 1989). A ring around a small tree beginning 6 inches (15 cm) from the trunk and extending outward to about 6 inches (15 cm) beyond the edge of the root ball pit is the area broadcasted. The width of the ring is increased as the root system extends farther outward in subsequent years. Most of the fine feeder roots will be within the surface 6 to 8 inches (15 to 20 cm) of the soil profile and will be able to absorb most of the surface-applied fertilizer. Broadcasting fertilizer is also recommended for large established trees. Drilling holes or injection need not exceed 12 inches (30 cm) are used only under special circumstances of limited space or when reduction of soil compaction is necessary. The fertilizer may be mixed with compost as a filler.

Fertilization of trees planted in vaults or below impervious surfaces is best accomplished by the addition of liquid fertilizer to the irrigation water or by manually watering with liquid fertilizer on a regular schedule. This technique partially compensates for the limited soil volume available to the roots for nutrient absorption. However, there must be some provision for access to the roots extended under the impervious surface. Watering only the tree pit under the grate or pit opening may not be adequate to supply the feeder roots with the necessary amount of nutrients.

Injection of fertilizer directly into the tree trunk is recommended only for established large trees with specific nutrient, disease, or insect problems. Young trees should not be injected. Organic mulching is often overlooked as a form of fertilization.

There are some plant species that have low nutrient requirements (efficient nutrient utilization), but they are not necessarily the most attractive or desirable for urban planting. Generally, these are the so-called "pioneer" species that withstand relatively harsh, infertile sites. The best procedure is to improve the soil with amendments.

Choice of Plant Species Resistant to or Tolerant of Toxic Contaminants

Some woody plant species are better able to tolerate certain contaminants than others. One of the contaminants present on landfill plantings is the gases produced by the landfill itself. Craul (1992) and Harris (1991) provide lists of species that are appropriate for landfill plantings.

Many plants can tolerate relatively high heavy-metal concentrations provided that the heavy-metal content does not create an ion antagonism problem or the site is not used to provide foodstuffs or a space for human activity. The planner must consider these implications before producing the final planting design.

Elimination or Reduction of Soil Contaminants

The complete elimination of soil contaminants can seldom be attained without total replacement with uncontaminated soil—a very expensive process. The usual feasible process is to reduce the concentration levels below the threshold values of tolerance for the plants and animals. If the contaminated soil is extensive and human activity is to be present, such as playgrounds and athletic fields, more care in elimination or reduction is required. However, these concerns are generally beyond those of tree planting and are not considered further here.

The common contaminants in urban soils include the heavy metals, deicing salts, and residual pesticides. The solubility of heavy metals can be significantly reduced by maintaining a high pH (above about pH = 8.0). Unfortunately, this solution greatly restricts the potential composition of the plant palette. Most woody plants in the eastern United States grow best in a soil with a pH within the range of 5.6 to 6.5. Western native species are adapted to higher pH levels and may grow satisfactorily in soils that are purposely maintained at high pH. However, these species may not be appropriate for planting in the east.

As mentioned earlier in this chapter, deicing salts can be removed by flushing with freshwater. The reduction of pesticide effects can be accomplished by the addition of charcoal to the soil and thoroughly mixing it (Harris, 1991; Whitcomb, 1987).

Complete replacement of the soil before planting may be the best alternative in highly contaminated soils. Replacement before planting as indicated by soil analysis results can solve potential problems. The soil has been successfully replaced even around existing trees in many instances in Europe, and now some cases have been reported in the United States, particularly in California and the midwest (Watson et al., 1996).

Situation G: The Totally Engineered Sites Where All Criteria Are Limited

There are instances in which the funding is available and objectives of the client permit a highly "engineered" design. The tree plantings in the sidewalks along Pennsylvania Avenue, Washington, DC, are an example of this application. Figure 9.9 illustrates the the below-ground design. The sidewalk needs to support the weight of fire equipment as required by the District of

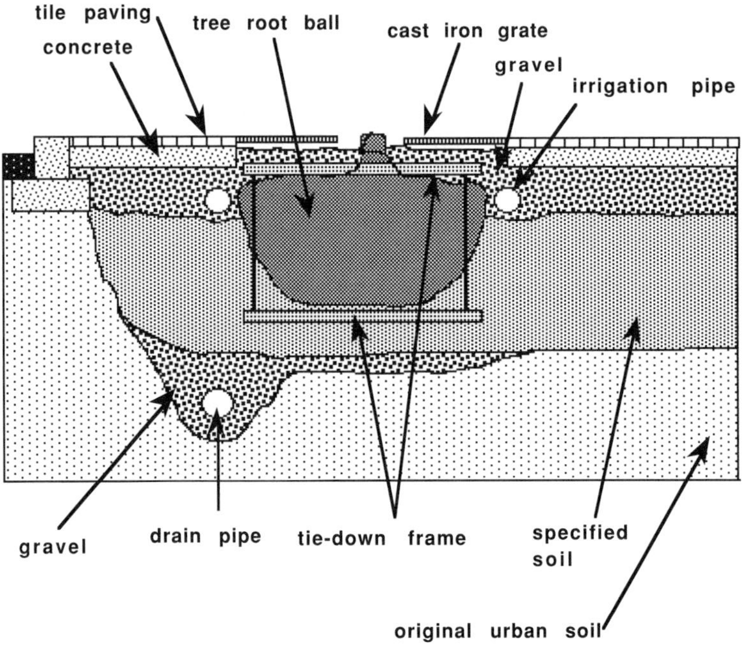

Fig. 9.9 The section for the Pennsylvania Avenue, Washington, DC, street tree installation. (Adapted from Jewell, 1981, and in Craul, 1992.)

Columbia. Thus, it is reinforced concrete and quite impervious. Aeration piping is installed and radiates like wheel spokes outward from each tree pit and connects one pit with another. The pipes are also used for irrigation and fertilization, being connected to a common valve box. The box is filled for a prescribed time period by maintenance personnel to apply the appropriate amount of irrigation water. Liquid fertilizer is applied in the box. The willow oaks (*Quercus phellos,* L.) are surviving and growing very well with little mortality.

Another example is the Denver Transit Mall (Fig. 9.10). The sidewalk is suspended over the soil. Aeration is through the trunk opening. If the tree requires replacement, the concrete cover is removed in two half sections by crane. Slots in the sidewalls provide exits for root extension into the surrounding soil. The type of soil surrounding the box is unknown to the author. Personal observation of the plantings in 1991 shows some trees exhibiting poor growth but the cause is unknown.

The Use of Root Barriers in Planting

There is much controversy concerning the employment of root barriers in tree plantings or for application to existing trees (Anon., 1990, 1995). Gilman

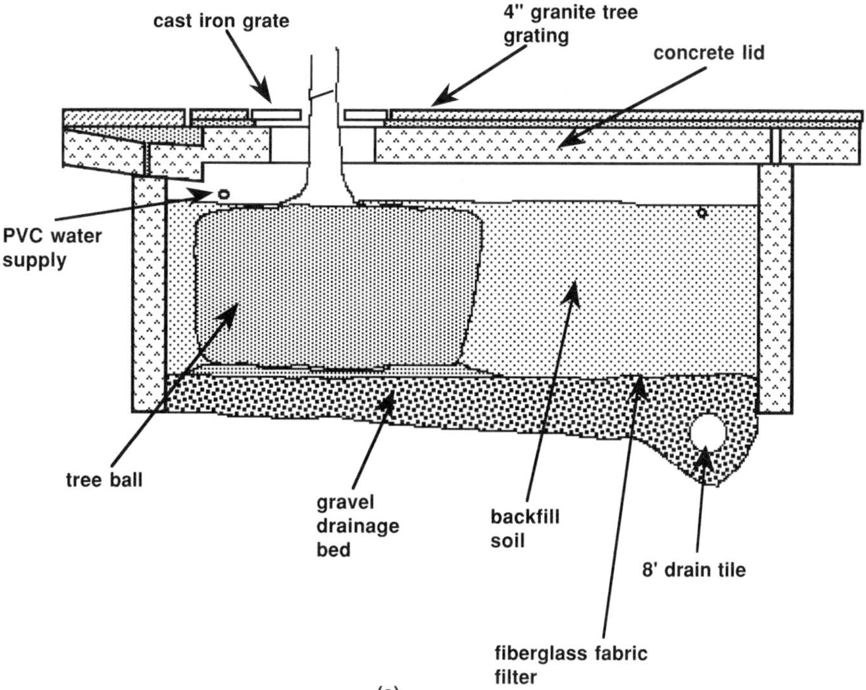

(a)

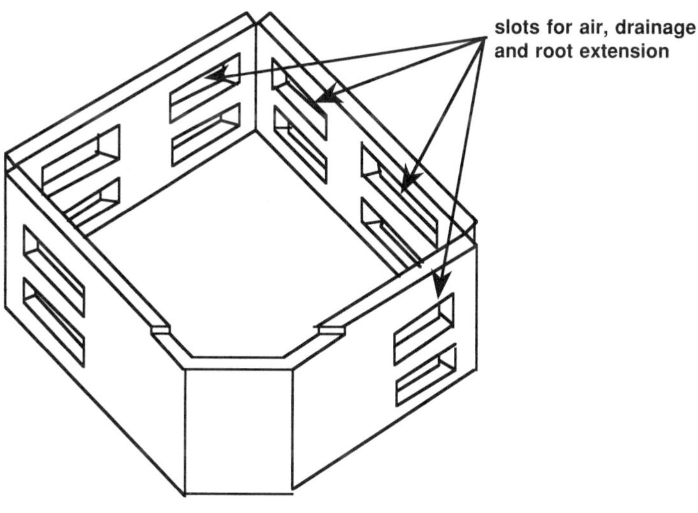

(b)

Fig. 9.10 The (a) section and (b) plan of the Denver transit mall tree installation. (Adapted from Jewell, 1981, and in Craul, 1992.)

(1996) has shown that root barriers do force roots to greater depth in the soil at the barrier, but that they return to the original depth closer to the surface within 4 feet (122 cm) or less outside the barrier. This phenomenon is not unexpected, as Lyford (1980) shows that roots tend to extend in a straight line from their origin at the root collar. If diverted by obstructions such as rocks or other roots, they will curve and return to the initial axis. Roots also return to the surface because the environment for them is more favorable unless the lower soil is loosened (Costello et al., 1997). This effect is observed most strongly in compacted urban soils (Fig. 9.11).

Fraedrich (1995) gives recommendations on a combination of root barriers and root pruning as a solution to sidewalk lifting by mature tree roots. Placement of root barriers at planting will aid in forcing the roots to greater depth or remain within the area bounded by the barrier. However, the use of the barrier increases the potential for girdling roots to form, which can have a detrimental effect on the trees (Watson et al., 1990).

If sufficient rooting space is provided in the planting design, root barriers should not be necessary. Their utilization becomes necessary only when the roots are to be confined into an insufficient space for elongation and expansion, which must be considered a design or planning failure. Their use is probably most effective only in very well-drained soils or specially designed rooting media.

SPECIAL APPLICATIONS

Utility Corridors

Utility corridors include pipelines, power and telephone lines, and irrigation channels. Transportation rights-of-way should also be included, as they are usually corridors with special soil conditions that influence planting techniques. Highway interchanges contain large areas of open spaces subsequent to construction that are amenable to planting of trees and ornamental shrubs. The soil has been disturbed or the profile truncated to some degree. Strips of open land that become isolated between utility corridors and/or rights-of-way, sometimes considered greenspace or greenbelts, have soil conditions similar to the general urban situation with some natural soils interspersed with disturbed soil. Therefore, relatively intense soil sampling is required to establish the intricate soil mosaics that result from construction disturbance interlaced with undisturbed soils. Linear patterns are anticipated on these sites. The appropriate planting technique for most cases would probably be Situation A since most areas are open and unpaved. Construction activity and traffic may have compacted the soil. Amelioration measures are required.

A serious impact detected along many major regional and interstate highway rights-of-way is salt-spray contamination of adjacent tree foliage and both wet and dry deposition of salt on soil. The effects may carry tens of meters into the surrounding landscape (Townsend and Kwolek, 1987). Conifers such as eastern white pine (*Pinus strobus,* L.), red pine (*Pinus resinosa,*

Fig. 9.11 Roots forced to the surface by compacted soil, Syracuse, NY.

Ait.), and the cedars are especially susceptible. Mortality is common. Many hardwoods including sugar (*Acer saccharum,* Marsh.) and red maple (*Acer rubrum,* L.), London plane (*Platanus* x *acerifolia*), littleleaf linden (*Tilia cordata,* Mill.), and others also suffer, usually showing water sprouts and witches' brooms. Obviously, the dead and dying trees should be replaced by salt-resistant species. Lists are provided by Harris (1991), Craul (1992), and others. Flushing the soil with large quantities of freshwater effectively reduces the soil-salt concentration (Brady, 1990). In areas of high spring rainfall (more than 30 in., 750 mm, annually), the precipitation may be sufficient to flush the soil by the middle of the growing season (unreported research on deicing salts by the author in Syracuse, New York).

Various state agencies are becoming aware of the large acreage of vegetated land along corridors that require some type of management. Limited budgets restrict the amount of care and attention that is presently applied to these areas.

THE CONCEPT OF *le Preverdissement* (PLANT TODAY, BUILD TOMORROW)

This is a concept of preplanting by Guinaudeau (1987) in France that has strong potential application to those portions of the United States undergoing development or restoration. It is comprised of a carefully developed planting plan for an area that fits or conforms to a anticipated potential use on the site. The design arranges the plants so that they do not interfere with planned structures or the process of construction. In this way, they are able to become well established before construction occurs. However, care must be taken to prevent construction damage to the plants. Thus, a combination of design and protection provides a planting with advanced establishment and growth, reducing the amount of time required for the vegetation to reach desired size. Even if intended use is not apparent, the planting plan can be devised in such a way that is amenable to a wide range of potential uses. In the meantime, the aesthetic quality of the open space is greatly enhanced as compared to an ordinary vacant piece of land with fugitive vegetation.

If properly conceived, the plantings can be handled as in an open-soil surface with future structures, including paving, constrained not to intrude over the plant rooting space. The site assessment should determine the need for underdrainage or other subsurface infrastructure that can be installed before planting. A distinct disadvantage of the concept is the capital outlay required prior to development. Normally, a developer would wish to capitalize the costs over a period of 5 years or less. With some intelligent forethought, the concept leads to early establishment of a landscape plan in and around which structures are built, rather than the landscape plan fitting around the structures. In turn, the architect may take advantage of the aesthetics of the landscaped site.

TRANSPLANTATION OF LARGE TREES

The existence of very heavy specialized transplanting equipment makes it possible to transplant mature or very large trees. The operation involves the preservation of a very heavy and large root ball to ensure transplant success. There are various methods in carrying out the operation and certain arborist firms specialize in this work. It is a specialized art and is beyond the scope of this chapter. However, many of the principles discussed herein apply to transplanting large trees as well as to small trees.

ASSESSMENT OF THE TREE PLANTING

It is important to fully assess soil and site conditions before planting trees and other plants. This procedure ensures the success of the planting not only to the length of the contractor's guarantee, but also to the mature stage of the planting design. There are several simple steps in the assessment process that produce valuable information with little effort and expense. The process is far superior to making no assessment to save time and money and just taking the chance that "everything will work." The cost of replacement of unsuccessful plantings easily exceeds any costs of prior assessment.

Tree Health Evaluation

Various visible physical tree features can be observed to indicate the survival and general well-being of a new planted individual. These include the examination of the foliage, twigs, bark, and roots.

The color of the foliage is an excellent diagnostic tool. A uniform bright green color over the entire crown with leaves fully formed is a reliable indicator of good health on hardwood species. Yellowing indicates water stress from too little or too much soil moisture or may indicate various degrees of nutrient deficiencies. The latter take on rather specific features and are listed in Table 9.2. Care must be taken to prevent confusion among the various symptoms. Yellowing and incomplete maturity of leaves early in the spring of newly planted trees usually indicate stress due to soil waterlogging during the dormant season. This condition commonly occurs on urban trees planted in compacted soils. Improving the soil drainage will aid survival. Pruning and fertilization are not recommended and may do more harm than good (Whitcomb, 1987). Diagnosis for moisture stress of conifer foliage is somewhat more difficult. Many times the foliage will become brown long after the damage has reached mortal proportions. Yellowing of the new or current needles gives some clue to moisture stress but can be easily confused with several nutrient deficiencies. Examination of the soil conditions and roots may eliminate moisture stress.

Length of mature current twig elongation as compared to previous growth is another indicator of plant health (Fig. 9.12). An abrupt shortening of elongation as compared to previous growth in the nursery indicates a problem. Some twig shortening should be expected due to transplant shock, although some species and individuals may show very positive response due to removal of nursery competition and perhaps much improved growing conditions in the new planting.

The bark is inspected for mechanical injury from handling, shipping, and planting. Any wounds should be treated by good arboricultural practice. Severe injury should be grounds for stock rejection and replacement. Search for insect and disease evidence should be made on the foliage, the bark of twigs, and the trunk. The root collar should be inspected to make certain that the collar is planted at the proper depth. Planting the root collar too deep can be fatal to many species, cause rotting and eventual failure, or unwanted sprouting. Those species sensitive to deep planting include sugar maple, California live oaks, dogwoods, black pines, and white pines (E. T. Smiley in Stone, 1993). Planting the root collar too shallow has much fewer detrimental consequences and may be a suitable practice where a shallow watertable or barrier exists.

Though root examination is difficult, it is recommended when serious growth symptoms appear. The color of healthy, vigorous tree root tips is white to light yellow, with smooth surfaces. When cut in cross-section, the same color appears. Roots affected by poor aeration as in waterlogged soils are

TABLE 9.2 Nutrient Deficiency Symptoms of Trees

Nutrients	Deficiency Symptoms
Nitrogen	Small needles or leaves, general yellowish-green color, thin crowns, poor growth, reduced chlorophyll content
Phosphorus	In *conifers,* needles turn gray or bluish-green or sometimes (in pines) turning to purple, brownish-purple, or red. Symptoms most pronounced in late summer on needle tips, especially older needles which are retained and contrast with less-affected younger ones. In *broad-leaved trees,* leaves (especially oaks) become dark green or reddish, foliage is less dense, and growth is poor
Potassium	*Conifer* needles are small and turn yellow-green and yellow with symptoms most pronounced on tips of older needles in autumn, winter, and spring. With severe deficiency tip-browning, needle-shedding, and branch and twig death occur. Younger twigs often remain green. *Broad-leaved* trees exhibit dark bronze leaf colors with bright yellow margins with necrotic spots; finally leaves shrivel and die
Magnesium	*Conifer* needles show golden or yellow-tip "halo" effects in late summer or autumn; a sharp transition to the green portion. The color extends farther back the next year and tip necrosis may occur. Symptoms are more severe in moist years. Careless observation may cause confusion with potassium deficiency. *Broad-leaved* trees show irregular yellow spots between the leaf veins and are more severe on older leaves. In severe cases, entire leaves yellow and fall prematurely. Symptoms disappear slowly after fertilization if leaves do not fall
Sulfur	Symptoms are similar to those of nitrogen deficiency. Soil and foliage tests will show which is deficient
Iron	Symptoms are chlorosis of young *leaves* or *needles,* especially in wet or cold years. In oaks, leaves may be yellow on emergence or develop interveinal necrotic spots and light color, finally curling and dying. Youngest leaves are most severely affected. Spraying with iron salts results in quick recovery if effects are not too severe. Best solution is to correct high soil pH, the usual causal agent
Manganese	*Conifer* symptoms are essentially the same as for iron deficiency. *Broad-leaved* species show marginal leaf chlorosis, gradually extending between the major veins, with bands of green along the main veins and the midrib. Necrotic spots may develop in the chlorotic areas. No gradient of decreasing symptoms from basal to terminal leaves occurs as in magnesium deficiency
Boron	Generally characterized by death or distortion of meristematic tissues. Symptoms on various species include tip wilting, bending, or dying; distorted, small or closely spaced terminal leaves; leaf or bud dieback at the end of the season

TABLE 9.2 (*Continued*)

Nutrients	Deficiency Symptoms
Copper	For *conifers,* generally exhibit weak, drooping, wavy or pendulous shoots; crooked stems; recurved branches; terminal bud death; terminal needle coloration; and needle fall late in the season. In *broad-leaved* trees, new foliage may be dark or bluish-green and be distorted by cupping or stunting. Necrosis or irregularity of leaf margins and interveinal chlorosis may occur. There may be a "rosette" effect from shortened terminal growth. On poplars, large dark-brown spots or blackish leaves are symptoms
Zinc	Symptoms are marked chlorosis of younger leaves, particularly in the upper crown and on the sunny side of the tree. "Rosette" effect occurs. There may be chlorotic mottling of interveinal areas and shoot dieback on severe cases. Extreme shortening of branches, needles, and needle spacing in the upper crown may occur, plus general yellowing and loss of older needles in *conifers*

Source: Craul, 1992.

dark brown, purple, or even black, and have rough-looking surfaces. The cross-section color is usually darkly stained. Deciduous tree roots may also be examined for stored carbohydrates (Wargo, 1975, 1988). Those trees with high stored starch have a better chance of surviving transplant shock. Roots with optimum stored carbohydrates will give a dark blue test color throughout the cross-section when an iodine solution is applied. Medium stored starch is indicated by only the rays being colored. Those with little stored carbohydrates and probable growth deficiency will give no color in any tissues. Tested roots should be at least 3/8 inch (1 cm) in diameter and 1 foot (30 cm) from the root collar. Branches may be tested in the same way, but roots are the ultimate starch storage location. If branches show medium or high amounts of starch, the roots need not be tested. The season of the year will influence the results.

Soil Evaluation

Soil evaluation may be carried out in conjunction with root examination. Infiltration rate, drainage and aeration, moisture-holding capacity, bulk density, and fertility all require scrutiny (Craul, 1988; Kays, 1982).

Infiltration may be observed during irrigation or rainfall events. Generally, the rate should exceed 1 inch (250 mm) per hour. If the soil is ponded in heavy rain for only a short time (less than 1 hr), the infiltration rate is acceptable provided that subsoil permeability is adequate to carry away the excess water. Water should not accumulate anywhere within the profile and especially in the bottom of the tree pit. This condition can be detected by probing the soil. If accumulation occurs, underdrainage should be provided.

(*a*)

(*b*)

Fig. 9.12 (a) General view and comparision of shoot lengths for (b) a *stressed* tree and (c) a *normal* tree, Salt Lake City, UT.

(*c*)

Fig. (*Continued*)

The soil moisture should be distributed throughout the profile with no dry areas. The backfill and surrounding soil needs to be moist as well as the root ball or bare-root area. Soil density may have changed from a predicted value due to placement and backfill operations (Patterson and Mader, 1982). Bulk density of the backfill and surrounding soil should not exceed threshold values for root penetration: 1.65 Mg/m^3 for clays and 1.70 to 1.80 Mg/m^3 (Veihmeyer and Hendrickson, 1948; Vepraskas, 1988). Fertility levels and pH should be those as recommended elsewhere.

10

PROJECT EROSION AND SEDIMENTATION CONTROL

INTRODUCTION

Erosion may be defined as the wearing down of the earth's surface by the forces or agents of degradation: water, ice, wind, gravity, and humankind's activities. *Geologic erosion* is the process of degradation that occurs naturally and is part of the geomorphic cycle that all land surfaces undergo. Human activities tend to contribute to this wearing-down process and is called *accelerated erosion*. This phenomenon is most prominent on land that has been cleared of native vegetation and used for agriculture or for urban development.

Soil erosion is *the major environmental problem* faced by most countries of the world (pollution problems not withstanding) because it has direct bearing on food production—when the major sociological problem of the world is hunger. In some instances, erosion is spectacular and easily observed, whereas in others, it is subtle and insidious.

Erosion generally occurs at the soil surface and thus removes the most valuable portion of the profile—those horizons containing incorporated organic matter and high levels of nutrients, and having optimal physical conditions for root penetration and growth, and organism activity. Severe erosion causes rills and gullies, so that not only is the valuable soil mantle lost, but obstacles are created that hinder agricultural operations and make reclamation difficult. These soils may be incapable of supporting agricultural production, reducing the amount of arable land—a limited resource in today's world. Soil losses may range from 1 to 25 tons (2.2 to 56 metric tons per ha) or more per acre per year. And, it has been estimated that the United States has lost about 50 million acres (20 million ha) to erosion and that an equal amount is approaching that condition. An additional 100 million acres (40 million ha) have lost one-half or more of the surface soil. Besides the loss of soil on the land, the streams, lakes, and rivers are filled with the sediments. Loss of vegetation and the storage capacity of soil for water reduces baseflow of streams and has a drying and warming effect on the local and regional climate such as the present condition in the Sahal of Africa (although this phenomenon is also partially due to a macroscale climatic shift).

Soil erosion has had tremendous impact on civilization. The most obvious examples are the decline of the early civilizations that occupied the area between the Tigris and Euphrates rivers—the site of the original Garden of Eden. It was a lush, subtropical land in the time of King Solomon. In fact, the timbers for his palace and temples came from the forest of what is now Lebanon. It is thought that overzealous land clearing and misuse of water (*this should sound familiar!*) caused changes in the regional climate, further exacerbating the problem. The same situation appears to have contributed to the decline of the North African civilizations as well, although the destruction of the invading Roman legions was also a factor. A further example is the disappearance of the Mayan civilizations of Yucatan and Guatemala. Apparently, northern Europe never experienced widespread serious erosion, perhaps due to a higher level of husbandry ethics and more favorable climate.

Soil erosion associated with land clearing and plowing of the American midwest prairie reached its peak in the 1920s and caused the Dust Bowl of the early 1930s. In 1935, the forerunner of the present-day Natural Resources Conservation Service was formed as part of the Department of Agriculture, to initiate plans and carry out assistance to farmers and other landowners on controlling soil erosion by the agents of wind and water. A considerable degree of success has been achieved during the subsequent 55 years, but much yet needs to be done.

Because of the conditions cited earlier, and the recognition of the problem by regulatory agencies, more requirements for erosion and sediment control on construction and restoration projects are now part of the permitting process in many states and localities. Therefore, every landscape architect needs to be sensitive to the problem, to take necessary preventive action during the construction process, and to consider permanent erosion and sediment control as part of every design.

THE PROCESS OF EROSION

The erosional process consists of three phases: *detachment* of the soil particles, *transport* of the detached particles, and *deposition* of the particles as sediments.

Soil particles are detached by the kinetic energy of the falling raindrop striking the bare soil surface. The kinetic energy of a natural storm may amount to 6.2×10^6 joules per hectare-centimeter. An additional factor causing detachment occurs after some water becomes ponded on the soil surface—the splashing of raindrops into the ponded water detaches soil particles. The detached particles then become suspended in the water.

Once the soil particles are detached, they are carried along suspended in moving water that is channelized over the soil surface, or as bedload, being bounced along the channel bottom when the volume and velocity of the flowing water is above a critical level. This suspended material acts as an abrasive agent and the flowing water contributes to the detachment of soil particles along the shoulders or banks of the gullies. The greater the velocity and/or the greater volume of the moving water, the larger the suspended load that water can carry, and the larger the diameter of the particles carried. Slow or sluggish water movement would carry only suspended clays; all larger-diameter material would settle out. Torrent flow occurring during flooding or spring snowmelt can carry large loads of gravel- or stone-sized material. In fact, many mountain torrents can carry boulders several feet in diameter as bedload during periods of heavy, high-velocity flow.

Deposition of sediments occurs when there is sufficient reduction in the velocity or volume of flowing water, or the kinetic energy is dissipated in some way. Larger material settles out first and generally closest to its source, whereas fine particles, such as clay or fine silt, may be carried the entire length of the drainage system and deposited off-shore in the ocean. Other deposition may be more local—within the same field or at the bottom of the slope. The sediment is deposited in stratified form because of changing patterns of deposition among erosional events, or by changing velocity or volume of flowing water within one event.

Interflues and slopes are worn away through this process, and valleys and depressions are widened. The subsoil is exposed with its finer texture and lower infiltration rate: Erosion potential is increased. Soil compaction by mechanical equipment during tillage, construction, and so on, also increases erosion potential by infiltration.

It should be recognized from the preceding discussion that surface erosion occurs only when the soil surface is bare. In the case of forest land on rugged terrain, the soil surface is generally protected by the vegetation canopy and forest litter that covers the surface. Yet erosion still occurs. Different mechanisms are involved and are due to the force of gravity, coupled with water, providing the processes of geologic erosion.

In undisturbed vegetation, most of the destructive kinetic energy of rainfall is absorbed by the vegetation and litter, and the plants protect the soil from erosion. Soil compaction from construction or recreation causes decreased infiltration rates and consequent increases in surface runoff and erosion, but can be controlled by some simple conservation practices or people management.

Creep

Creep is defined as the slow downslope movement of soil material in response to gravitational force. This movement may amount to as much as 15.2 millimeters/year with most movement in the upper profile. Most movement occurs during the periods when soil moisture content is high, because water acts as the lubricant for the soil movement. Creation of very steep slopes on cut and fill can cause serious examples of creep. The rather continual and subtle movement is sufficient to contribute a constant supply of material at the streambank for subsequent erosion by streamflow in a natural watershed situation. It has been estimated that under natural conditions, 2 millimeters/ year movement will supply a metric ton of material per 100 meters of streambank length.

Slumping

Earthflows occur where creep deformation of the soil material has exceeded the shear strength of the soil. A slump is characterized by up-slope failure of the soil and the piling up or accumulation of material on the lower slope. It is a faster-acting event contrasted to the subtle, long-term event of soil creep. Slumping occurrence is related to deep, cohesive soils, especially when underlain by hard impermeable rock. These areas can be avoided by careful design and layout of the project.

Debris Avalanche

Debris avalanche is a rapid, shallow mass movement on steep slopes. The mass is a mixture of soil and the vegetational cover, and the action is triggered by a heavy rainfall event. It commonly occurs on shallow, noncohesive soils on steep slopes where subsurface water may be concentrated by topography or an impermeable bedrock. Road building on steep terrain (probably greater than 40% slopes) commonly results in accelerated debris avalanche events. Debris avalanches are very common in rugged or alpine terrain.

Debris Torrent

Debris torrent involves the rapid movement of water-charged soil, rock, and vegetational debris down steep stream channels. They are triggered during extreme catastrophic events such as slides entering the stream from a side slope. Movement of very large amounts of material and streambed scouring is common. Some events may involve as much as 10,000 cubic meters of material. Debris torrents are typical of rugged, steep mountainous terrain. Indiscriminant road building and other construction may contribute to the occurrence of debris torrents.

THE UNIVERSAL SOIL LOSS EQUATION

Through the development of controlled studies at specific locations across the United States since 1930, Walter Wischmeier of Purdue University has derived an equation for the prediction of soil loss from agricultural land. The equation is

$$A = R \times K \times L \times S \times C \times P$$

where A = computed soil loss per unit area (tons per acre per year)
R = rainfall factor, the number of erosion-index units in a normal year's rainfall; the erosion index is a measure of the erosive force of a specific rainfall event
K = the soil-erodibility factor, the erosion rate per unit of erosion index for a specific soil in cultivated continuous fallow (the worst possible condition) on a 9% slope that is 72.6 feet (22m) long
L = slope-length factor, the ratio of soil loss from the field slope length to that from the 72.6-foot length of the same soil type and gradient
S = slope-gradient factor, the ratio of soil loss from the field gradient to that from the 9% slope
C = cropping management factor, the cropping rotation practice determined by the use crop rotation, of winter cover, plowing time, and residue practice,
P = erosion-control practice factor, the ratio of soil loss with contouring, strip-cropping or terracing to that with straight-row farming, up and down the slope.

The rainfall factor, R, is a measure of the erosive force of specific rainfall. As mentioned earlier concerning the kinetic energy of the raindrop, storms have energy available that becomes the erosive force and may amount to 6 million foot-pounds of kinetic energy in a 2-inch rainfall falling at 20 miles per hour. The R factor is the sum of the kinetic energy times the maximum 30-minute intensity for each storm during the year. The values range from 50 in the midwest to 600 along the Gulf Coast. The iso-erodent map in Fig. 10.1 illustrates the distribution of these values for the humid eastern United States. It should be noted that the rainfall factor measures only the erosivity of rainfall and associated runoff. It does not measure or predict soil loss due to thaw of soil, snowmelt, or wind. In areas where these losses are significant, they must be estimated separately and combined with the predicted value.

The soil-erodibility factor, K, is a measure of the soil factors that influence erosion by water. These include infiltration rate, permeability, and total water capacity; and factors that resist dispersion, splashing, abrasion, and transporting forces of the rainfall and runoff. The K factor has been determined experimentally for 23 major soils of the United States, and these values are listed in Table 10.1. These values have been determined for the worst possible condition for erosion—continuous fallow with tillage up and down the slope. The soil loss on the plots is determined and divided by the rainfall factor. For application in the equation, the values are obtained from the table.

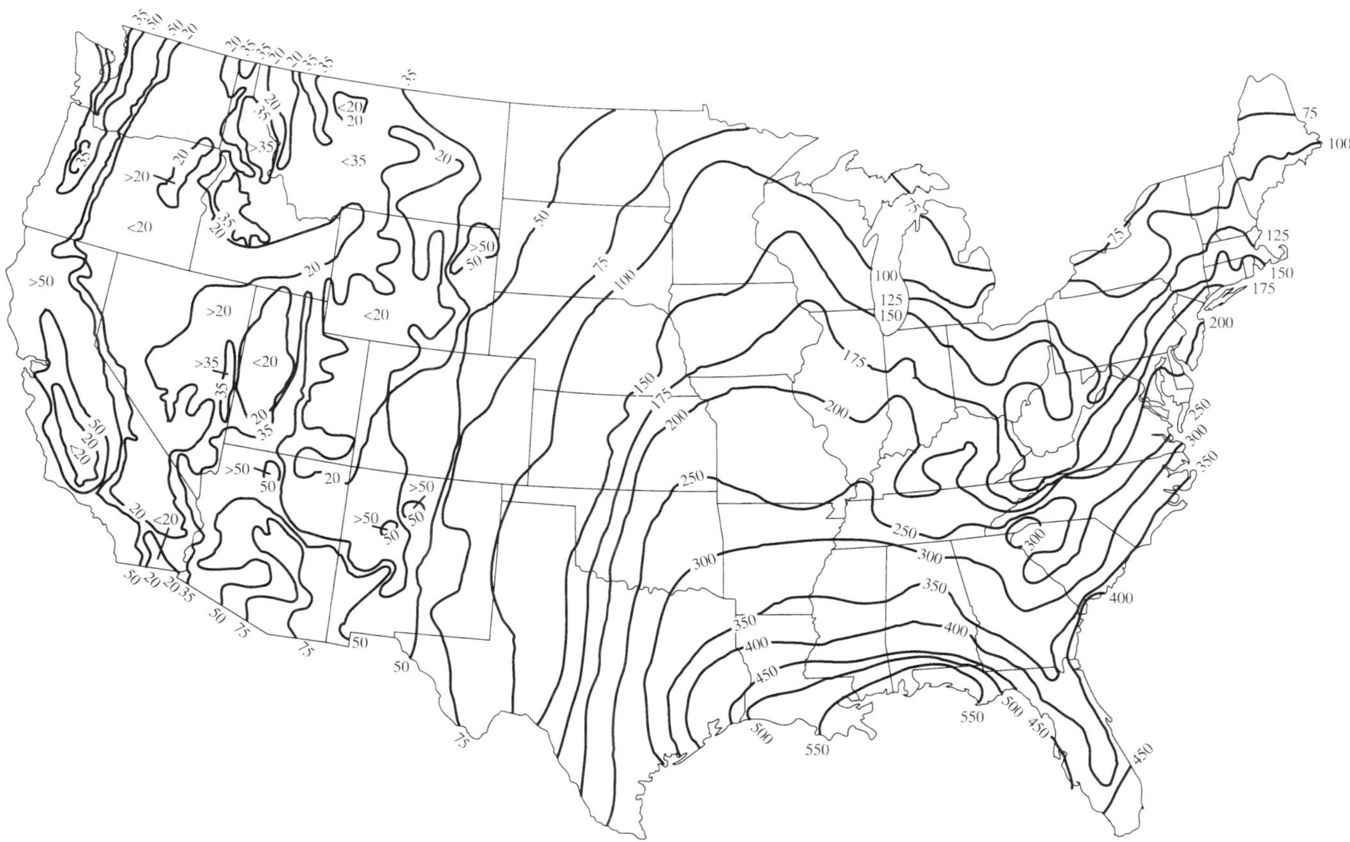

Fig. 10.1 Rainfall factor, *R*, for the eastern United States.

TABLE 10.1 Computed *K* Values for Soils on Erosion Research Plots

Soil	Source of Data	Computed *K*
Dunkirk silt loam	Geneva, NY.	0.69
Keene silt loam	Zanesville, OH	0.48
Shelby loam	Bethany, MO	0.41
Lodi loam	Blacksburg, VA.	0.39
Fayette silt loam	LaCrosse, WI	0.38
Cecil sandy clay loam	Watkinsville, GA.	0.36
Marshall silt loam	Clarinda, IA	0.33
Ida silt loam	Castana, IA	0.33
Mansic clay loam	Hays, KS	0.32
Hagerstown silty clay loam	State College, PA	0.31
Austin clay	Temple, TX	0.29
Mexico silt loam	McCredie, MO	0.28
Honeoye silt loam	Marcellus, NY	0.28
Cecil sandy loam	Clemson, SC	0.28
Ontario loam	Geneva, NY	0.27
Cecil clay loam	Watkinsville, GA	0.26
Boswell fine sandy loam	Tyler, TX	0.25
Cecil sandy loam	Watkinsville, GA	0.23
Zaneis fine sandy loam	Guthrie, OK	0.22
Tifton loamy sand	Tifton, GA	0.10
Bath flaggy silt loam with surface stones 2 in. removed	Arnot, NY	0.08
Albia gravelly loam	Beemerville, NJ	0.03

Source: Wischmeier and Smith, 1978.

Slope length, *L*, is the distance from the point of origin of overland flow to either a point where the slope decreases to the extent that deposition occurs or the point where runoff enters a well-defined channel. Runoff from the upper part of a slope contributes to the runoff produced on the lower portion of the slope. This increases the water running over the lower slope, creating more erosion there than on the upper slope. Erosion increases as the 0.5 power of slope length; hence, the longer the slope, the greater the erosion. This results in about a 1.3 times greater soil loss per acre with a doubling of slope length. Short slopes would have little, if any, erosion because there is little chance for the buildup of running water.

Increased erosion results with an increase in slope gradient. This is due to the increased velocity of water on steeper slopes, increasing its erosive power. A doubling of runoff water velocity increases the kinetic energy four times and causes a 32-time increase in the amount of material of a given particle size that can be carried. Splash erosion causes a net down-slope movement of soil and increases with slope gradient. Because slope length and gradient are interacting (L and S), they are combined for use in the equation. The values are illustrated in Fig. 10.2.

The cropping-management factor, C, takes into account that vegetative cover can absorb the kinetic energy of falling raindrops and greatly reduces rainfall erosive potential. Also, vegetation itself retains a significant amount of the rain and slows the runoff flow of water. Therefore, the presence or absence of a complete vegetative cover essentially determines whether erosion will present a problem or it will be negligible. The C factor measures the combined effect of the interrelated cover and managment variables, including type of tillage, residue management, time of soil protection by vegetation, and so on. Complex tables have been developed for the C factor and a simplified one is given here (Table 10.2). Generally, the C value would decrease as more cover is left on the soil in the cropping. For example, continuous corn, a row crop, would have a higher C value than that for continuous hay.

The erosion-control practice factor, P, refers to the practices that are used to control runoff water and thus reduce the amount of soil loss. These practices include contour tillage, stripcropping on the contour and terrace systems. The maximum protection for most gentle slopes (3 to 7 percent) is provided by contour tillage and stripcropping. Table 10.3 provides some P values for contouring.

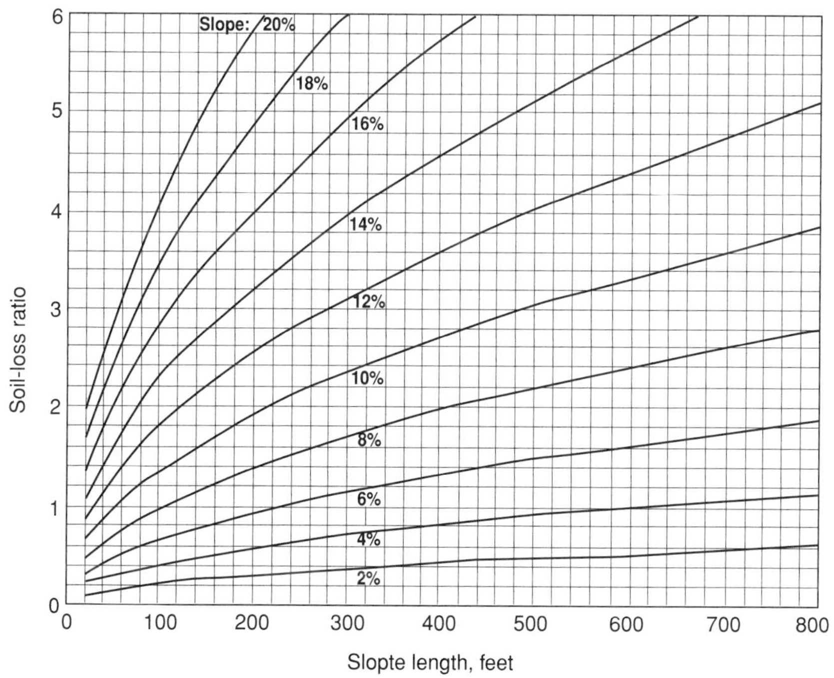

Fig. 10.2 Slope-effect chart for the topographic factor LS. (Wischmeier and Smith, 1978.)

TABLE 10.2 Cropping-management Factor *C*

Crop and Management	Winter Cover	Residue Use	*C* Value for Area
Continuous corn, 3+ years spring plow	None	Removed	0.460
		Left	0.359
	Cover	Removed	0.346
		Left	0.299
Spring plow minimum	None	Removed	0.355
		Left	0.240
	Cover	Removed	0.242
		Left	0.181

Source: Wischmeier and Smith, 1978.

Table 10.4 summarizes the factors in the soil loss equation, and the range in their values and whether the factors are fixed or managed.

Application Example

To illustrate the usefulness of the Universal Soil Loss Equation, an example is presented to illustrate contrast in erosion hazard.

An agricultural field cropped to contiuous corn is 50 acres in area. It has an average slope of 8% and the average slope length is 500 feet. The silt loam soil has a K value of 0.28 (obtained from the county soil survey report). The erosion index R is 100 (from Fig. 10.1). The LS factor is 2.2 from Fig. 10.2). Contouring is used, but not alternate strips. The C factor is 0.359 from Table 10.2, and the P factor is 0.60. By using the Universal Soil Loss Equation:

$$A = R \times K \times L \times S \times C \times P$$

$$= 100 \times 0.28 \times 2.2 \times 0.359 \times 0.60 = 13.27 \text{ tons per acre per year}$$

TABLE 10.3 Practice Factor, *P*, Values for Contouring[1]

		Strip-Contouring	
Land Slope, %	Contouring	Alternate Small Grain	Alternate Grass
1.1 to 2	0.60	0.45	0.30
2.1 to 7	0.50	0.40	0.25
7.1 to 12	0.60	0.45	0.30
12.1 to 18	0.80	0.60	0.40
18.1 to 24	0.90	0.70	0.45

[1] 1 For no practice, *P* 1.0.
Source: Wischmeier and Smith, 1978.

TABLE 10.4 Summary of Soil Loss Factor Values

Fixed	*R*—rainfall factor		50 to 600
	K—soil erodibility		< 1
	S—slope gradient	LS	0 to 6
Managed	*L*—slope length		
	C—cropping mgt. practice		1 and less
	P—conservation practice		1 and less

From Table 10.5 a conversion factor of 0.87 is used to calculate the cubic yards of the silt loam soil, which is $13.27 \times 0.87 = 11.54$ cubic yards per acre per year. Since the area is 50 acres, the annual total loss is 663.50 tons, or 577 cubic yards.

Converting the entire area to hay crop changes the *C* and *P* factors, but the remaining factors are unchanged. For continuous hay (meadow), the *C* factor is 0.006 and the *P* factor becomes 1 since there is no practice involved. The estimated loss now becomes

$$A = 100 \times 0.28 \times 2.2 \times 0.006 \times 1 = 0.369 \text{ ton per acre per year}$$

or

$$0.369 \times 0.87 = 0.32 \text{ cubic yards per acre per year}$$

The annual total loss becomes 18.45 tons or 16 cubic yards. Erosion tolerance for natural soils ranges from 5 to 15 tons per acre per year, depending on its weathering processes. The change in the cropping and practice factors illustrates the effects of different kinds of vegetation on erosion losses. This effect becomes important in erosion and sediment control practices.

TABLE 10.5 Conversion Factors from Weight to Volume for Soil Texture Classes

Soil Texture Class	Conversion Factor
Sands, loamy sands Sandy loam Fine sandy loam	0.70
Loams, sandy clay loams Sandy loam Silt loam Silty clay loam, silty clay	0.87
Clay loam Clay	1.06

Source: Lake, 1989.

Though most of the foregoing data and information were developed for application to agricultural and later forested land, the same principles form the basis for erosion and sediment control in urban areas as discussed in this chapter.

URBAN AREA EROSION

Erosion from urban areas is even more severe than that from agricultural land (Wolman and Shick, 1967; Yorke and Herb, 1978). For example, annual sediment yield amounts to 218 tons per acre (489 metric tons per ha) from construction sites in Washington, DC. Annual sediment yield from agricultural land may range from 5 to 20 tons (11 to 45 metric tons per ha) per acre (U.S. Department of Agriculture, 1982) and less than 1 ton (2.2 metric tons per ha) from forest land (Yorke and Herb, 1978). The reasons include (modified from Lake, 1989)

1. Increase of large soil areas exposed to erosion from water and wind
2. Increased water runoff and peak flows, soil movement, and sediment accumulation caused by
   (a) Removal or drastic thinning of plant cover
   (b) Decrease in the area of soil that can absorb water because of construction of streets, buildings, sidewalks, and parking lots with impervious surfaces
   (c) Changes in drainage configuration or reduction of the floodplain caused by grading operations, diversions, and streets
   (d) Changes in volume and duration of water concentrations caused by altered slope, length, and surface roughness, usually causing high peak flows and short flow duration ("flashy")
   (e) Soil compaction by heavy and light equipment that can reduce the water intake of soils from 1/16 to 1/20 of the original rate, further influencing the effects of (d).
   (f) Prolonged exposure of unprotected sites and service areas to poor weather conditions, enhancing soil compaction
3. Altering the groundwater regime that may adversely affect drainage systems, slope stability, survival of existing vegetation, and establishment of new plants
4. Exposing subsurface materials that are too rocky, too acid, susceptible to compaction, or otherwise unfavorable for establishing plants
5. Obstructing streamflow with new buildings, dikes, and landfills, which in turn can be damaged by high flows
6. Improper timing and sequence of construction and development activities that cause further damage to the soil and drainage system

7. Abandonment of sites before completion of construction, prolonging exposure, and loss of erosion and sediment control structure maintenance.

URBAN AREA EROSION AND SEDIMENT CONTROL PLANNING

Planning for erosion and sediment control involves the elimination or reduction of the seven factors listed in the previous section. Thus, critical points or elements must be identified, not only for the undeveloped site but also for the modifications caused by the project plan. In some cases, the project design may mitigate some of the critical elements. In many cases, however, the number of critical elements may increase with development or the development may intensify the existing natural critical elements. A good project development plan will mitigate as many as the critical elements as possible as part of the design and not create new ones. This fact is the major reason why an erosion and sediment control plan for a large development should be one of the first products of the drawing board.

The Land Development Plan

The assembly of certain basic data and information is required to properly evaluate the site potential for development. Much of the data and information are required for the development process anyway, so no real additional expenses or costs are incurred for the assembly except for the erosion and sediment control planning process and its execution.

Evaluation of the Physical Site. The data needed here include location, present land use, tract size, topography, drainage pattern, geology, hydrology, soils, vegetation, and climate. On-site examinations are valuable and needed, but most of the data will come from various sources in the form of maps, charts, and reports. The topography and drainage patterns are obtained from a USGS topographic sheet, the geology from a USGS geological quadrangle report or from the county soil survey report, which includes some geologic information. Some states publish their own geologic reports. Land use and vegetation surveys are done on-site. Hydrologic and climatic data are obtained from the NOAA (formerly published by the USGS) hydrologic annual or monthly reports. Soil data and maps are derived from the appropriate county soil survey report if available as published by the USDA-Natural Resources Conservation Service. If one is not available, the local field office may be able to provide the materials.

The Soils of the Area. Basic soil information needed for erosion and sediment control planning include soil mapping unit descriptions, erodibility, engineering properties, suitability for topsoil, gravel, sand, and so on. Suitability for foundations, buildings, on-site waste disposal, artificial drainage, and various landuses are also required. Nearly all of this information is available in the county soil survey report.

In some cases, the mapping scale, usually 1:24,000, and the accompanying detail of the soils may not be appropriate for a complex project. A special detailed survey may be necessary when the scale may be enlarged to 1:6000 or even 1:4800 (1 inch = 400 feet or 1 cm = 122 m).

The Erosion and Sediment Control Plan

The erosion and sediment control plan should be a routine component of the project planning process. Segments of the plan may be temporary, in place only during the construction process, whereas other segments may become part of the permanent structures on the project. For this reason, thought must be given to the control plan early in the planning process; otherwise, drastic changes to the plans may be required in the later phases of the project.

Plan the Development to Fit the Site. Though this facet of planning is a significant component of any landscape architecture project, it is important from the standpoint of minimizing erosion hazard.

The physical characteristics of the site should be examined and assessed so that it can be developed with minimum of risk. Utilize the existing topography to the greatest degree possible so that soil disturbance and grading are minimized. Maintain buffer strips between disturbed and adjacent undisturbed areas. Observe all setback requirements from adjacent properties, streams, wetlands, and water bodies.

Determine Limits of Clearing and Grading. Determine a sequence for construction and then plan for clearing and grading in the same sequence. The tendency in the past has been to clear the entire development area (Fig. 10.3) as the first phase, exposing the site to a high degree of erosion hazard and potentially causing significant damage on-site and off-site with the resulting liability for damages. Many contractors feel that this method is the least expensive, but in the long run, it may be quite expensive, and in some areas, it is illegal under environmental protection regulations (e.g., Maryland, New Jersey, Pennsylvania, and Virginia). Critical areas to be protected until construction are steep slopes, highly erodible soils, and borders of surface water. In some instances, it may be necessary to clear the entire area or large portions because of the complexity of the site and/or the development plan. In this case, the cleared areas need to be protected by soil stabilization practices until construction begins in those areas. Again, stabilization in the past has many times been left to nature. This neglect assures problems later.

Divide the Site into Natural Drainage Areas. As part of the on-site inspection, determine the direction and relative amount of runoff from each section

Fig. 10.3 Topsoil stripping of an entire development area. Subsequent storms created a significant erosion and sediment control problem on this project and the adjacent areas, Syracuse, NY.

of the project area. It is easier to control erosion at the source of runoff than it is farther downstream. Perimeter control is more difficult and more expensive. After each section has been examined and preliminary planning achieved, then proceed with piecing together the entire project area for the overall erosion and sediment control plan.

Select Erosion and Sediment Control Practices. The erosion and sediment control practices fall into three broad categories of soil stabilization, sediment control, and runoff control. Each one contains a separate group of practices with some overlap in objectives. The five-step process culminates in specific actions for control of specific conditions or erosion and sedimentation hazard:

1. Identify the control problem: soil stabilization, sediment control, and runoff control.
2. Identify the problem areas: graded, slopes, large (more than 1 acre or 0.4 ha), or small (1 acre or less) disturbed areas.
3. Identify the required strategy: protect the surface, manage surface water, divert runoff, convey runoff, intercept groundwater, and so on.
4. Identify the control measure group: vegetative or nonvegetative soil cover, diversions, waterways, enclosed drainage, spillways, outlets, sediment basins, sediment filters, and mud and dust control.

5. Select the specific control measure: temporary seeding, sodding, and so on.

Soil Stabilization. There are some basic principles that apply to establishing vegetation for soil stabilization (adapted from Lake, 1989).

1. Slopes should be stabilized before being vegetated. Slumping and creep will occur even after slopes are vegetated and the stand cannot become established.
2. Excess surface and subsurface water must be controlled.
3. Stockpile topsoil on-site and apply to areas to be vegetated. This action ensures the successful establishment of the vegetation. If topsoil is stored, the piles should be temporarily seeded to grasses.
4. Retain and protect trees, shrubs, and other plants wherever possible. The protected areas will help to reduce erosion hazard.
5. Prepare an adequate seedbed even if the vegetation is only a temporary cover. This includes the application of lime and fertilizer if needed, and mulch.
6. If the seeding is a permanent element, the area should be protected for at least 1 year. In recreation areas or areas to become subject to traffic, 2 years of protection may be necessary.
7. In concentrated flow areas, sod, rocks, erosion mesh, and so on, should be used in conjunction with seeding to adequately stabilize the soil.
8. Seed bare soils within 15 days of exposure unless construction will begin within 30 days. If construction is suspended or sections are completed, the areas should be seeded and mulched immediately.

It should be remembered that trees and shrubs alone do little to control erosion unless densely planted. The foliage breaks the impact of the raindrops, and soil surface litter acts as a soil mulch. Surface runoff can still occur under trees and shrubs so groundcover, especially grasses, should be planted with the trees and shrubs.

From the flowchart of Fig. 10.4, soil stabilization measures are applied to graded areas and slopes, and control is by protecting the surface and/or managing the surface water. Surface protection is done by vegetative soil cover, particularly grasses, or by using artifical surface protection such as jute mesh, paving, gravel, or mulching. A combination of vegetative and nonvegetative measures may be required on large areas with steep slopes. Management of surface water is accomplished by diversion of the water away from the exposed soil areas or by the creation of waterways through the areas. These may be unstabilized or stabilized, depending on the amount of water to be diverted. The channels can be grassed waterwaterways or lined with rock, and so on (Fig. 10.5). Diversion structures may be a simple ridge or a dug

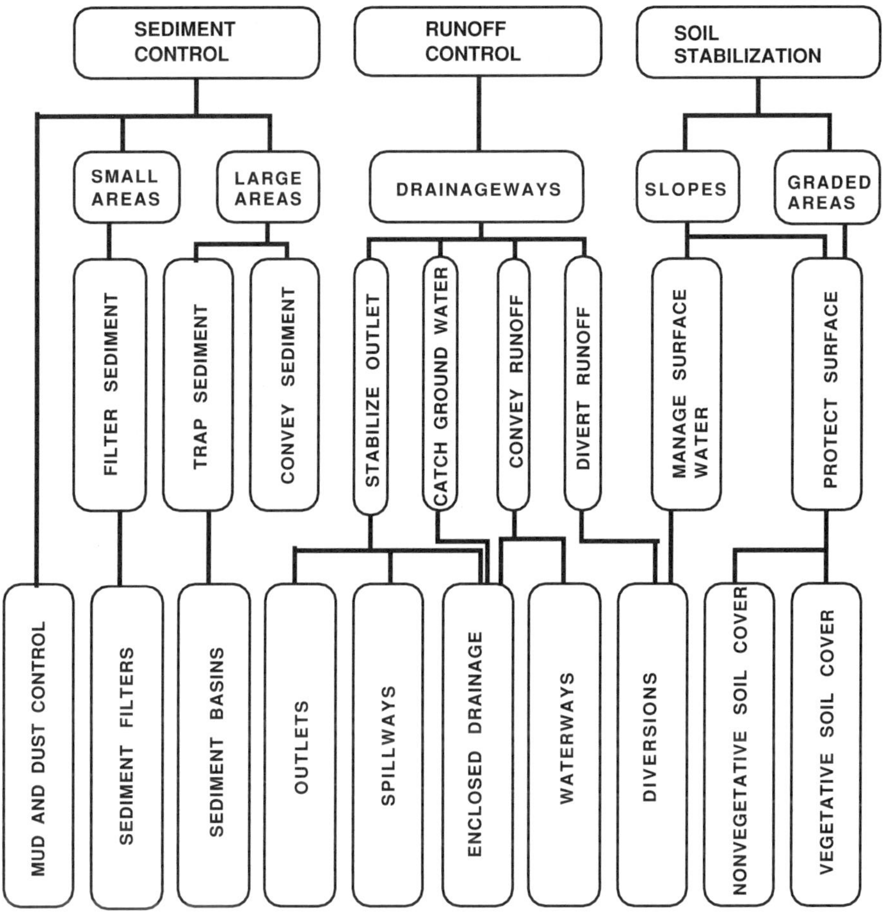

Fig. 10.4 The erosion and sediment planning control flow chart. (Adapted from Lake, 1989; also in Craul, 1992.)

channel or swale to intercept and direct the water. Curb and gutter construction is a more elaborate method of diversion and is a part of the permanent infrasturcture of the project such as housing developments.

Sediment Control. The degree of structure construction and the alternatives are dependent on the size of the watershed. Obviously, flow from a small watershed is easier to control than the flow from a large watershed. The somewhat arbitrary division between small and large watersheds is taken as 1 acre (0.4 ha).

The major goal in sediment control is to retain all sediment on the site. On small areas, the sediment can be contained by installation of filter strips. However, one of the key factors overlooked on the success of the strips is

Fig. 10.5 Waterway lined with rock in an urban development, Manlius, NY.

that they must be correctly installed and maintained. Poorly installed and maintained filter strips are illustrated in Fig. 10.6. The bottom of the strips must be buried at least 6 inches into the soil. Excess sediment accumulated behind the filter strip must be removed to prevent failure, and if failure occurs, the sediment must be removed and the filter strip correctly repaired. A correctly installed filter strip is illustrated in Fig. 10.7. A vegetated buffer strip as a filter is also effective. However, to be effective, the buffer strip must have stabilized soil and dense groundcover. In large areas, sediment traps such as sediment basins may be required for direct interception, or the sediment conveyed by a drainageway from the area to a sediment basin located downstream. The size of the sediment basin must be designed to handle the volume of runoff expected from the watershed or one of its segments. The concept of the sediment basin is to reduce the velocity of the water so that the sediment settles from the water to the basin bottom. The basin must be large enough so that the water from several storms may be retained, as clay and silt particles may settle very slowly. Turbulence by subsequent storms is able to stir the bottom and cause settled sediment to become suspended again.

Mud and dust control measures include the stabilization of the construction entrance and the interior access roads. Periodic application of water or calcium chloride to maintain a moist surface may be all that is required for dust control. Stabilization for mud control may require the surfacing of the entrance and roadways with gravel. Stabilization of the entrance is critical because many localities and states are very strict about the carryout of mud by

Fig. 10.6 Poorly installed and maintained filter strips.

Fig. 10.7 Correctly installed filter strips.

construction vehicles onto paved road and highway surfaces. The construction general contractor becomes responsible for the requirement, but the landscape architect and project engineer are responsible for design of the control measures. Further, there should be no flow of water from the entrance onto the road. The water should be filtered and then diverted to the lateral roadway drainage ditches.

Runoff Control. This control deals mainly with the installation of drainageways (Fig. 10.8). The strategies include the diversion of runoff as covered under soil stabilization, conveyance of runoff through waterways or enclosed drainage, interception of groundwater by enclosed drainage, culminating in the stabilization of the outlet for all of the previous listed methods. Enclosed drainage is defined as the use of storm sewers or underdrainage systems. Some engineering princples are involved in the design of these structures. The stabilization of the outlet usually consists of some kind of spillway or lined apron. Drainage outlets must be carefully designed to reduce off-site impact and to conform to various regulations.

Temporary Structure Practices. These are structures installed for only the period of construction and are removed on project completion and the disturbed areas are permanently stabilized. They include the following structures.

Erosion Blankets or Nettings. These are used to stabilize disturbed soil surfaces where vegetation is not yet established or is too steep for grass estab-

Fig. 10.8 A permanent dry retention basin, Manlius, NY.

lishment alone or for standard hydromulching operations. The material may be a geotextile fabric, woven paper mesh, hemp mesh or netting, or other materials. The blanket or mesh must be anchored to the soil surface by staking or by the use of tackifiers.

Earth Dike. It is a temporary berm of compacted soil to channel water to a desired location such as a sediment trapping structure or device. It may also be used to divert clean water away from disturbed areas.

Straw Bale Dike. It is a very temporary structure, to last no more than 3 months, used to reduce runoff velocity to cause deposition of suspended sediment. It is usually used to prevent sheet erosion from a disturbed site, or placed across channels as a series of sediment dams. The bales must be anchored in place with stakes or re-bars.

Silt Fence. It is a barrier of geotextile cloth (filter fabric) used to intercept runoff from small areas of disturbed soil. The sediment is deposited behind the fence. It is imperitive to have the lower 6 inches (15 cm) of the fabric buried in the soil and tamped.

Temporary Swale. It is a shallow excavation used to intercept runoff and divert it to a desired location. It is no deeper than 1 foot (30 cm) and about 4 to 6 feet (60 to 90 cm) wide. It is usually filled in after project completion.

Perimeter Dike/Swale. It is a temporary berm of soil excavated from an adjacent swale, with the swale upslope of the berm. The berm need not be compacted. It is used to encircle large disturbed areas to divert runoff away. It may also be used along tops of slopes or along the base of slopes. The perimeter structure remains in place until the construction site is stabilized.

Sediment Basin. It is a temporary excavation or dam used to intercept sediment laden runoff. The purpose is to retain sediment on the site yet allow clean water flow from the outlet. It must be designed to retain storm runoff for an area less than 100 acres (40 ha) and be no higher than 15 feet (2.2 m); otherwise, a permanent structure is required. It must have seep control and a hardened outlet.

Permanent Structure Practices. These are structures that are installed before or during the construction operation. Some may be used as control structures during construction and others may used only after the project is completed: the latter may be in the way of construction. Obviously, the most efficient process is to design a control plan so that the control structures serve both purposes wherever possible. They include the following structures.

Diversion. It is a swale or depression cut into the soil. The cross-sectional area, wedge-shaped or trapezoidal, is determined by the volume of water to be interecepted and diverted. It may be hard-lined if large amounts of water are to be diverted; if small amounts, a swale only 5 or 6 inches (13 or 15 cm) deep, broad, and grass-lined may be adequate. It is usually installed above sidehill cuts or other disturbed areas.

Grassed Waterway. This structure is much like the grass-lined diversion swale, but rather than diverting water from a surface, it is used to carry water without erosion and sediment load over a soil surface. It is commonly used in agricultural pastures and strip contour crops.

Lined Waterway or Outlet. Again, it is used to carry water over a soil surface, but because it must carry larger amounts of water, it is hard-lined. The structure is commonly used at the lower portion of a waterway and/or at the outlet itself. A *grade stabilization structure* is much the same type of structure but may be larger due to a greater concentration of water. It may involve grade changes to reduce the flow velocity.

Rock Outlet Protection. Waterflow at the outlet to a drainageway usually exhibits turbulent flow. To reduce the erosive effects of this type of flow, the bed and sides of the outlet, and the plunge basin, if present, are lined with large rock or rip-rap. The size of the outlet is determined by the predicted storm flow.

All of these structures are effective only if they are properly maintained. Maintenance involves removal of excess amounts of sediment from the structures and repairs to the structures themselves. The maintenance factor is one of the most frequent omissions on construction sites and leads to more penalties and fines than any other phase of project construction.

For more details on the preceding structures, their specifications and installation, or for additional structures, refer to manuals by Lake (1989) and Leopold (1968) or a text like Beasley (1978).

Predicting Soil Losses

Prediction of soil losses from urban areas is accomplished in much the same way as illutrated with the Universal Soil Loss Equation earlier in the chapter. However, some of the factor values are different. The process is illustrated in the example at the end of the chapter.

Estimating Sediment Yield

Several authors (e.g., Lake, 1989) provide an equation for estimating the amount of sediment yield from urban land. It does not include bedload sediment, which is assumed to be equal to the amount of sediment carried through the structure outlet. An adjustment must be made because of the variation in rainfall and erosion potential of soils in different regions.

The equation is derived from limited data and must be used with caution. It relates sediment yield to the area under construction:

$$V = [(C \times A)/1320] \times (222{,}000/A)^{0.715}$$

where V is the required sediment storage per year, in acre-feet, A is the uncontrolled drainage area under construction, in square miles, C is an iso-erodent correction factor. Though this equation was developed for the New York State general area, Lake (1989) suggests corrections to the iso-erodent values that may be applied to other regions. Their validity is not assured, but using them to gain guideline information coupled with development of local equations is encouraged.

The value 1320 represents the tons per acre-foot of sediment having an average dry weight of 60 pounds per cubic foot. Lake (1989) provides a graph that gives sediment acre-foot weights for other dry weights.

An Example. Drainage of an area under construction is 0.25 square mile (160 acres), and occurs in the region of the 100 iso-erodent value, which requires a correction of 0.76 (Table 10.6). Then:

$$\begin{aligned} V &= [(C \times A)/1320] \times (222{,}000/A)^{0.715} \\ &= [(0.76 \times 0.25)/1320] \times (222{,}000/0.25)^{0.715} \\ &= 32.08 \text{ acre-feet} \end{aligned}$$

TABLE 10.6 Iso-erodent Values and Their Correction Factors

Iso-erodent Value	C Factor
600	1.78
550	1.76
500	1.74
450	1.71
400	1.67
350	1.62
300	1.56
250	1.47
200	1.34
150	1.12
125	0.94
100	0.76
75	0.57

Source: Lake, 1989.

AN EROSION AND SEDIMENT CONTROL EXAMPLE

The Site

The project is to be constructed on a terrace and lower slope in a junction between Pine Creek and No Name Brook (Fig. 10.9). Most of the project area is covered with forest, with a small opening of grasses on the main terrace. The soils have moderate erosion hazard based on texture, drainage, and general slope. No Name Brook is a perennial stream, having a spring as its source.

The Development Plan

It will consist of a single building cut into the hillside, a parking area on the wider portion of the terrace, with a long access road parallel to Pine Creek and crossing No Name Brook. A portion of the access road must be cut into

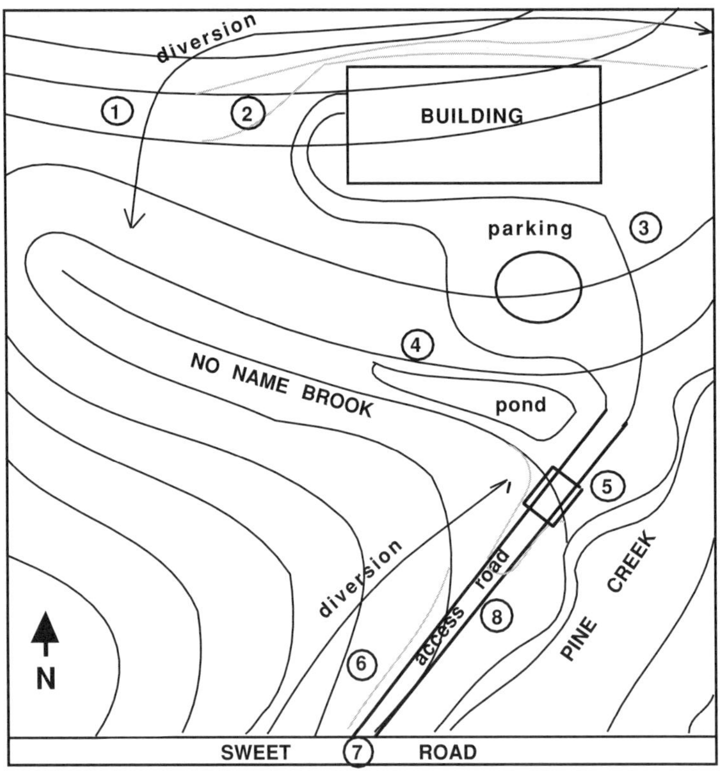

Fig. 10.9 The development superimposed on the physical site. The critical elements are numbered as discussed in the text. The gray lines represent the reformed contour lines.

the hillside forming a cut-fill side slope. A bridge needs to be built over the brook.

The Critical Elements

To provide proper erosion and sediment control as protection to Pine Creek and No Name Brook, eight critical elements are proposed:

1. A grassed diversion swale is constructed above the hillside cut for the building. It will drain in opposite directions, with the east end draining into No Name Brook and the west end draining upstream into Pine Creek.
2. The cut made for the building is stabilized with grasses and other slope erosion measures.
3. Clearing of the forest and grass vegetation is done only where construction is necessary. The remainder vegetation is maintained.
4. A pond is constructed on the lower terrace next to No Name Brook. It is constructed before the land clearing begins for the cut slope, building pad, and parking area. It is intended to serve as the permanent sediment basin during and after construction. Measures are taken to control overland flow into this basin during construction. After project completion, it becomes a permanent pond with water partially diverted from No Name Brook.
5. A bridge is built over No Name Brook. Filter strips are placed around the abutment areas to prevent siltation of the fill. The abutments are stabilized after construction with grasses and erosion control blankets.
6. A grassed diversion swale is constructed above the proposed cut for the access road. It drains into No Name Brook well above the bridge abutement so as to not cause erosion to the bridge abutment.
7. The access road entrance needs to be stabilized so that sediment does not enter the lateral ditches for Sweet Road and eventually Pine Creek.
8. A filter fence is constructed along the access road and parallel to Pine Creek to prevent sediment from entering Pine Creek.

The plan may seem elaborate and perhaps an unnecessary expense, but in many states, the plan is required and is part of the permitting process. Lack of an erosion and sediment control plan is grounds for rejection of the building permit application. It should be looked on as a contribution to sustainability.

11

THE SOIL DESIGN DECISION PROCESS

The soil design decision process is presented to aid the user in choosing design alternatives based on the the project physical requirements, called the element situation, the selected plant palette, the type of proposed use such as traffic or no traffic, other modifying environmental factors, leading finally to the *soil design* itself. The most commonly useful designs are presented. All alternative design possibilities cannot be anticipated here nor is there space to cover them all. In addition, the groups of designs provided are limited to humid, temperate regions. Soil designs for other climatic zones would require an entirely different set of suites. Reference to Chapter 5 should be made to discover modifications that would be required. Future publication efforts by the author may include extension of the decision process to other climatic zones.

Many dimensions are given. The reader may be disappointed by the fact that minimum dimensions are stressed and maximum dimensions may not even be provided. It is amazing to most field forest soil scientists of how well trees and shrubs survive on shallow soil as long as it has the fertility and physical properties amenable to root penetration, the strength to support the trees in wind, and the lateral dimensions for shared rooting space. As in all specifications, dimensions or thicknesses greater than certain minimums are a waste of space, material, and money. As a rule of thumb, soil thickness or depth greater than 4 feet (1.2 m) is unnecessary; for many species and applications, a depth of 3 feet (0.9 m) is plenty and a depth of 2 feet (0.6 m) is adequate for medium-sized trees. A depth of 18 inches (45 cm) is absolute minimum, provided there are compensating factors present or planned.

The user is encouraged to use good professional knowledge and creativity, with some common sense, to develop designs individually, keeping in mind all of the foregoing material in this book and elsewhere. All possible ramifications of climate, soil, and plant species (cultivars, etc.) differences cannot be anticipated here. The experience of local landscape contractors, arborists, and horticulturists may differ with information provided here and should be considered as modifications. Their information can prove to be invaluable.

The soil design includes the various mixes for each layer (horizon) and the sequence of horizons in the profile. Alternative mixes and profiles will be offered where needed or desirable. Rototilling between horizons during installation is required to prevent the development of interfaces between them. The techniques described in Chapter 4 should be followed. Not all of these designs have been employed by the author in actual projects, but they are based on the experience of colleagues and on natural field experience in combination with sound soil science knowledge over a period of 40 years. Some of the material covered in Chapters 3 and 4 should be reviewed before reading the various soil design sections.

The profile descriptions roughly follow the format of the NRCS (see Soil Survey Staff, 1993). Organic matter contents are always given on a weight basis. Approximate volume values may be obtained by multiplying the percent weight by 2.2. To conform with irrigation useage, waterholding capacity is given in inches of water per foot (mm/cm) of soil. In cases where moisture content percent is given, it is always on a volume basis.

Commerical names of products are provided as references only and do not constitute endorsement of the products whatsoever.

Before entering into the specific soil designs, it is necessary to discuss the situation of the plants in a project design to place the soil design process in proper context.

The Element Situation

The element situation is the existing or proposed design physical characteristics governing the planting design element. Examples would include a curbside row of trees, a row of trees in a sidewalk with the majority of rooting area covered by pavement, a copse of trees with an understory and open soil, a single tree on a rooftop, a parterre in an open area, and so on. These element situations form one axis of the decision matrix in Table 11.1.

Open. Open applies to the cases in which the soil and the plant palette rooting volume are not covered by an impervious surface. Open as applied to turf areas may appear to be redundant; the term is retained to conform to the established terminology for the soil design decision process.

Curb. When the planting area is located parallel to and near the street side, the term *curb* is applied. It usually is covered. The designs applied to the "covered" situation apply here as well as the requirement for compaction resistance due to street vibration. The curb structure inhibits root extension in that direction so that roots must extend in the direction, parallel to and away from the curb (Figure 9.7 in Chapter 9). The roots may extend under the sidewalk into a front yard or other nearby open area. Under a continuous impervious cover, the roots will tend to follow the same pattern. If more

TABLE 11.1 The Initial Decision Matrix in the Soil Design Decision Process

	Element Situation			
Palette Structure	1 Open	2 Curb	3 Covered	4 On Structure
A Turf	Section 1A Open Turf	Section 2A Curb Turf	Section 3A DOES NOT EXIST	Section 4A Roof Turf
B Bedding	Section 1B Open Bedding	Section 2B Curb Bedding	Section 3B DOES NOT EXIST	Section 4B Roof Bedding
C Shrubs	Section 1C Open Shrubs	Section 2C Curb Shrubs	Section 3C DOES NOT EXIST	Section 4C Roof Shrubs
D Trees	Section 1D Open Trees	Section 2D Curb Trees	Section 3D Covered Trees	Section 4D Trees on Structure
E Trees with Understory	Section 1E Open Trees w/Understory	Section 2E Curb Trees w/Understory	Section 3E Covered Trees w/Understory	Section 4E Roof Trees w/Understory

favorable conditions exist as a result of infrastructure failure, the roots will follow the paths of opportunity along pipe traces, and so on. (Craul, 1992).

Covered. Covered is applied to those situations in which most of the rooting soil area is covered with an impervious layer. The layer may be continuous or jointed, depending on the paving material employed in the design. The situation occurs most commonly in plaza and continuous sidewalk areas. The term does not define the soil below, which may be residual urban soil or even a designed soil with a highly engineered *pedosphere* such as the Pennsylvania Avenue, Washington, DC, design (Craul, 1992). The term pedosphere is coined here to define the "landscape below ground." See the Glossary for a more detailed definition.

Rooftop or On Structure. The term refers to the situration in which the lower boundary of the soil is a roof or other structure. Thus, rooting depth is limited by the structure surface and the soil depth is limited by the loading limit and other architectural design considerations. The material covered in Chapter 6 should be reviewed.

The Plant Palette Structure

The plant palette structure consists of turf, bedding plants, shrubs, trees, and trees with understory. These palette structures form the second axis of the decision matrix in Table 11.1. The structure is usually simple with a single layer of vegetation and becomes complex when two or more layers are combined in the same rooting space; thus, a two-, three-, or more tiered layered structure. Root competition increases with each additional layer along with the potential imcompatibility.

Turf. Turf is considered as grass cover alone or may be associated with, but separate from, shrubs or plants in beds. Natural or artificial borders separate the turf from the beds. Turf may also be under open trees, but it is not considered an understory within the soil design decision process. Covered turf does not physically exist and is omitted in the matrix. Turf as a component of an interior landscape is not considered here, though it would probably be designed similar to open turf.

Bedding. This is that portion of the project design containing annuals, biennials, or perennials as annuals, or a garden that is to receive frequent cultivation and vegetation replacement. This term includes the general application of a parterre. The area receives no traffic other than routine maintenance. Bedding by definition is open and not covered; the matrix cell does not exist.

Shrubs. Shrubs occur in beds in the open or as an understory beneath small or large trees. The shrubs are placed permanently, receiving only routine maintenance with little soil disturbance. Mulch is a standard soil cover. Shrubs with covered soil do not physically exist. Generally, their root systems are in open, mulched soil, with their root systems more or less confined to the extent of the open soil.

Trees. Single trees or those arranged in a group, require greater rooting volume and area than most other types of plants. Therefore, the soil design must be different from the design for bedding plants and shrubs. Trees are also planted in covered (pavement, etc.) situations requiring still further modifications to the basic tree planting soil design.

Trees with an Understory. This is a complex situation for rooting competition and space. The soil design must be handled differently than for trees alone.

The species, cultivars, and so on, comprising the plant palette have definite soil requirements of drainage and aeration, rooting space, reaction (pH), and fertility. These requirements must be the determinants in the design of the soil and are discussed in the later sections. In the case of the covered situation, the tree roots are covered since they extend under the cover beyond those of the understory, if the design is properly done. Confining the tree roots to the open-soil area by the use of root barriers is not a design alternative unless the designed soil volume is adequate for both trees and the understory. The roots of the understory would be limited by the area of the open soil in all usual cases.

Traffic

Traffic or *no traffic* on the soil or turf is recognized next in the design process sequence because compaction and shear resistance must be considered in the soil design. Some areas will have no traffic and therefore have no further limitations to the design of the soil mix and its profile. Other areas are paved and traffic is not a factor in soil design, but structural support of the paving may be required. On the other hand, areas to receive traffic must have a designed soil that is compaction- and shear-resistant yet retain the root-growth qualities required for plant survival. Both pedestrian and vehicular traffic are considered in open-turf areas.

ELEMENT 1: OPEN

In all cases, the lower part of the undisturbed existing soil profile must exhibit adequate drainage or underdrainage must be installed. A high watertable, perched or otherwise, is not desirable for most plant palettes.

Section 1A: Open Turf

Open turf is very common in many landscape design projects. It is usually a simple situation, especially if the application is similar to the "front-lawn" concept. It is made more complex if the area is to receive traffic, especially playground or playing field activities in all types of weather. The following dicotomy recognizes these divergent applications.

Subsection 1AA: Turf with No Traffic. This is probably the simplest design of all (Fig. 11.1). Essentially, the design emulates a natural grassland soil with an organic rich topsoil and a subsoil serving as mechanical support and as a reservoir of nutrients and water. Minimum thicknesses of the horizons are stressed to prevent inadequate rooting volume and waterholding capacity. The objective is to create a maximum sustainable design requiring little maintenance with respect to irrigation and fertilization. The need for irrigation varies depending on the climatic zone. Fertilization depends on the turf grasses used in the design. The state agricultural experiment station is the best source for turf grass and fertilization information.

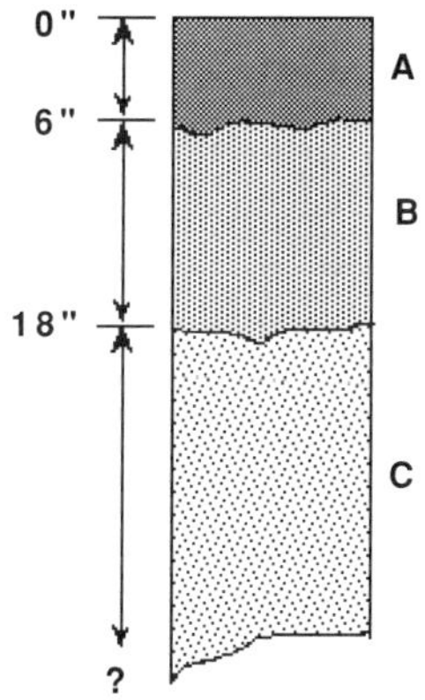

Fig. 11.1 The turf without traffic profile.

The Profile Description

A horizon	0–6 inches (0–15 cm)	Loam to sandy loam with 5% organic matter, preferrably composted biosolids
B horizon	6–18 inches (15–46 cm)	Sandy loam with at least 1% organic matter
C horizon	18–24 inches (46–61 cm)	Sandy loam or coarser, organic matter content not required, but should not contain more than 1%

Notes

Design and installation of the B and C horizons may not be necessary if the existing soil material on the site has suitable physical and chemical properties, such as good drainage, acceptable bulk density, and moderate fertility, and it does not contain toxic contaminants. If poor drainage is the only limitation, and then artificial drainage should be installed. The A horizon may be created by incorporation of organic matter into a satisfactory surface soil material. See Chapter 4 for additional details.

The sandy loam should be a medium sandy loam (USDA).

Subsection 1AB: Turf with Traffic. The soil and the turf must both be resistant to traffic. More specifically, the soil must be resistant to compaction and shear (tear). The turf must withstand damage to the foliage and root system. The situation is very common to golf courses where much research and development has led to soil design and turf cultivar selections that exhibit these attributes.

There are three techniques for designing compaction-resistant soil. One is to employ a loamy sand texture as the matrix with a low content of organic matter. This mix is referred to as a "prepared athletic turf" (PAT) common to many professional playing fields, especially of American football fields. The coarse sandy soil is resistant to compaction, has a high infiltration rate, but has a low waterholding capacity, and does not have a very high shear resistance (somewhat loose). Therefore, the soil requires frequent irrigation and must be covered with a tough, traffic-resistant turf. In most cases, both of these conditions are met.

The second technique is to use a finer texture as the soil matrix, perhaps a loam with a moderate organic matter content, and a structure stabilizer such as expanded slate or diatomaceous earth aggregates. The finer texture reduces the irrigation requirement and the stabilizer increases the compaction resistance. The initial cost is greater than the first alternative, but over time, it may be cheaper because the amount and frequency of irrigation is reduced.

The third technique is to provide an open wearing surface that supports vehicular traffic and is open to permit grass to grow through the openings. Rainfall may infiltrate and moisten the soil. Runoff caused by soil compaction is nearly eliminated. The three specifications follow (Fig. 11.2).

The Profile Descriptions

Profile (*a*)

A horizon	0–12 inches (0–30 cm)	Coarse loamy sand with 5% composted organic matter
B horizon	12–24+ inches (30–60+ cm)	Medium loamy sand with 1% organic matter

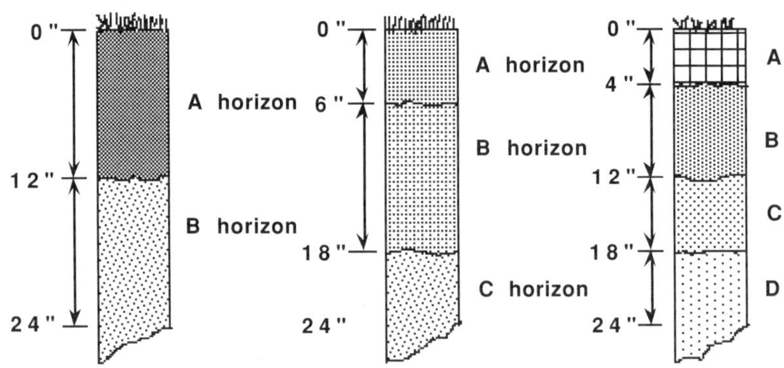

Fig. 11.2 The turf with traffic profiles: (a) the "prepared athletic turf;" (b) the structure-amended profile, and (c) open-paving profile.

Notes
The loamy sand should be medium to coarse and contain only very small amounts (less than 15% total) of fine and very fine sand (USDA). Some authorities recommend no organic matter amendment below 12 inches. The texture is coarse to provide compaction resistance; however, this compromise increases the need for frequent irrigation. The greater thickness (depth) of the B horizon that can be provided, the dceper the potential rooting and concomitant reduction in irrigation. This profile is applicable to pedestrian traffic only.

Profile (b)

A horizon	0–6 inches (0–15 cm)	Loam or sandy loam with 15% structure amendment by volume, 3–5 mm diameter (such as 1/4-in. RIDGELITE®, or ISOLITE® CG-4), and 5% organic matter content
B horizon	6–18 inches (15–46 cm)	Loam or sandy loam with 5% structure amendment by volume, 5–10-mm diameter (such as 1/4–1/2-in RIDGELITE®, AXIS® Coarse, or ISOLITE® GC-6) and 1% organic matter content
C horizon	18–24+ inches (46–60+ cm)	Sandy loam or coarser, unamended

Notes
The inorganic structure amendment is to provide resistance to compression and compaction in both upper horizons. The major differences between the horizons is the organic matter content and the particle size and proportion of structure amendment. The inorganic amendment would be TURFACE® MVP, AXIS®, SOLITE®, RIDGELITE®, ISOLITE® or other similar inorganic materials, choosing the product with the recommended particle size. Uniform particle size provides greater resistance to compaction than a wide range in particle size.

Profile (c)

A horizon	0–4(?) inches (0–10 cm)	Silt loam to sandy loam layer as backfill for pavers. Organic matter content should be 6%

B horizon	4–12 inches (10–30 cm)	If installed soil, it should be loam to sandy loam. If the horizon is of existing soil with moderate structure, it can be silt loam to silty clay loam, but must be well drained. The A horizon should then be silt loam
C horizon	12–18 inches (30–46 cm)	Should be well drained to prevent perched watertable from rising to the surface

Notes
This open-paved profile is designed primarily for vehicle traffic, either as access or as parking areas. The A horizon thickness is dependent on the type of open paving used in the design, being as thick as the paving height. Thus, for open concrete block paving, the thickness needs to be 8 inches. Some paver manufacturers recommend a sandy layer below the pavers as a foundation for the paving to add strength, and then backfill with a medium- to fine-textured soil filling the openings. Unfortunately, the sandy base creates a double interface that reduces water permeability through the layers. The backfill also can become very dry since it is hydraulically separated from the profile below. The base layer is not recommended here. Several types of open pavers have interlocking units. These types do not require a sandy base layer for strength.

The open-paving technique should be applied only to well-drained soils or those without a high watertable. The author has observed a very successful installation of open, nonlocking paving for a parking lot on a well-drained silt loam soil, without a sandy base.

The rough grade below the paving should be slightly compacted if the soil is loose.

Section 1B: Open Bedding

Open bedding is mainly for the planting of annuals, biennials, and herbaceous perennials. The design does not have traffic over the surface other than for occasional access to the plants. The topsoil resembles a horticultural mix but is not identical to that specified for potting mixes. The design is to accommodate annual digging for replanting and necessary cultivation during the growing season, unless mulch is used. The profile is a relatively simple one.

The Profile Description

A horizon	0–6 inches (0–15 cm)	Loam to silt loam with 10% organic matter and 15% fine structure amendment (such as PROFILE® or AXIS® fine)
B horizon	6–12 inches (15–30 cm)	Loam with 5% or less organic matter, unamended
C horizon	12+ inches (30+ cm)	May be an existing soil material. Should not be finer than silt loam nor coarser than sandy loam.

Notes
The A horizon thickness should match the average rooting depth of the plants; 6 to 8 inches is adequate for most annual plants (sometimes 4 inches is adequate for small annuals). The structure amendment is added to enhance the tilth of the topsoil rather than as strictly a preventative for compaction.

Section 1C: Open Shrubs

They are mainly for the planting of succulent or woody perennial shrubs. Traffic over the surface is confined to access. The root systems are generally deeper but are not frequently cultivated or replaced as with annuals and biennials. The planting is more permanent with a mulch surface as the common design.

The Profile Description

A horizon	0–8 inches (0–20 cm)	Silt loam to fine sandy loam with 10% organic matter, mulched with shredded bark or woodchips, and so on.
B horizon	8–18 inches (20–46 cm)	Same texture as the preceding but with a minimum of 1% organic matter
C horizon	18+ inches (46+ cm)	May be existing soil material, but should not be finer than silty clay loam or coarser than sandy loam

Notes
Care should be exercised in choosing the appropriate mulch material for the planted shrubs; acidiphilous plants should have acid mulch, and so on.

Section 1D: Open Trees

This application is generally the open parklike setting with at least some open-soil area without an understory. The open-soil area is greater than the planting pit area, affording potential for root extension beyond the drip line as the tree establishes itself. The installation depends on the existing soil surrounding the tree pit and unamended backfill soil. The design is similar to the standard tree pit of Chapter 9 and is repeated here (Fig. 11.3).

Even when a group of trees or copse is employed in the landscape design, the underground design is simply a series of standard tree pits, since the soil surface is open, without an understory.

Section 1E: Open Trees with Understory

Subsection 1E1: Open Trees with Turf. Turf may be an understory for open individual or grouped trees. However, it should be recognized that there are incompatibility problems between some grass species and tree species (Watson, 1988). Part of the design must provide for a mulched surface around the trees. It is recommended that grass not be permitted to grow up against the tree trunk. The soil design follows that of the open tree as given in Section 1D.

Subsection 1E2: Open Trees with Plants or Ground Cover. The second situation is with bedding plants or ground cover or shrubs as the understory. There are several alternatives for handling the design of this element situation depending on the type of understory plants. Annual or biennial plants as understory for trees is not recommended. The annual disturbance to the tree

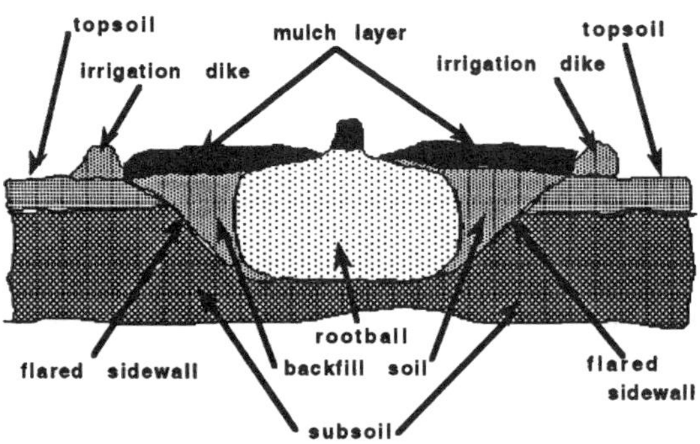

Fig. 11.3 The tree pit for open application.

root system is generally harmful, especially in a confined space. If annuals are to be used under trees, the best recommendation is to containerize them and place the containers under the tree. Permanent ground cover is generally beneficial to the protection of tree roots, provided there is compatibility between them, and in many instances is a pleasing design alternative (Fig. 11.4).

	The Profile Description	
A horizon	0–6 inches (0–15 cm)	Loam to sandy loam with 10% organic matter; thins to zero over tree framework roots or root ball
B horizon	6–12 inches (15–30 cm)	Loam to sandy loam with 1% or less organic matter
C horizon	12–24 inches (30–60 cm)	May be existing soil material but should not be finer than silt loam or coarser than sandy loam

Notes
Sandy loam should not be fine or very fine sandy loam. If the C horizon is not acceptable, then it should be modified or replaced to provide soil similar to the B horizon. The B and C horizons are primarily for tree rooting and stability against windthrow. It is sometimes appropriate to compact the soil immediately under each shrub root ball to create a stable pedestal. The process prevents settlement of the shrubs. Compaction must be done carefully by hand to prevent overcompaction. The technique works best in sandy soils.

ELEMENT 2: CURB

Pedestrian traffic impact is not the only impact associated with curbside locations. One is the impact of deicing salts in northern latitude cities. The plants must be salt-resistant and the soil must be well to excessively drained to prevent the buildup of salts. The soil surface (final grade) should be sloped

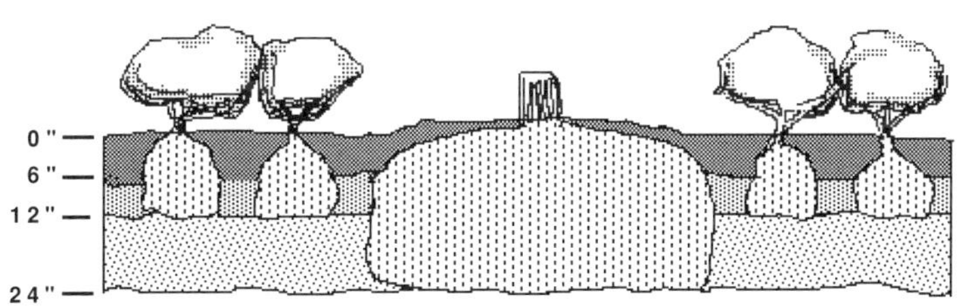

Fig. 11.4 The soil profile for trees with a shrub understory.

away from the plantings to prevent the accumulation of deicing salts in those areas where their use is frequent. Curbing surrounding the planting, either vertical or sloping away from the soil and at least 4 to 6 inches high, is very beneficial from the viewpoint of preventing pedestrian traffic over the beds and the interception of salt-laden water and roadside splash. The temporary placement of 36-inch wide black plastic fabric as a fence to prevent salt splash on soil and vegetation on median strips of Lakeshore Drive, Chicago, has been very successful in reducing salt damage (Pat Kelsey, private communication, 1997).

Another is consideration at curbside of the high degree and persistence of vibrational forces caused by vehicular traffic in the street. Therefore, the soil must be designed to be compaction-resistant in each case. Not to do so leads to serious compaction and all the ramifications of poor drainage and aeration and inhibition of root penetration (Craul and Klein, 1980).

Section 2A: Curb Turf

Because this situation occurs at curbside, it must be assumed that traffic is a major impact factor, unless the area is protected. The situation here is very similar to that of open turf with traffic in Subsection 1AB. Therefore, the three profiles described there and illustrated in Fig. 11.2 may be applied to this section. For northern applications, the deicing salt factor must be included, and the preference would be given to Profiles (a) or (b). Profile (c) with a finer texture may retain excessive amounts of salts and should probably be avoided, unless local experience indicates otherwise.

Section 2B: Curb Bedding

The placement of bedding plants generally occurs in median strips and as linear plantings between the curb and sidewalk. The location is precarious unless some protection from pedestrian traffic is afforded the plants. Another hazard is the delight exhibited by passersby in picking the blooms as they stroll along the street or hurry home from a hectic day at work. The recommended profile is similar to the open-bedding profile of Subsection 1B. The modification is primarily to increase the compaction resistance of the profile.

	The Profile Description	
A horizon	0–6 inches (0–15 cm)	Loam to silt loam with 10% organic matter and 20% fine structure amendment (such as PROFILE® or AXIS® fine)

B horizon	6–12 inches (15–30 cm)	Loam to sandy loam with 5% or less organic matter and 10% medium-structure amendment (PROFILE® or AXIS® Regular)
C horizon	12+ inches (30+ cm)	May be an existing soil material with good drainage or should be sandy loam

Notes
The A horizon thickness should match the average rooting depth of the plants; 6 to 8 inches is adequate for most annual plants (sometimes 4 inches is adequate for small annuals). The soil material below the C horizon should be well drained and not be subjected to a perched watertable at any season.

Section 2C: Curb Shrubs

It is not usual to find shrubs at curb side. It is hazardous to their health. If not protected, they can be easily trampled by pedestrian traffic. Salt splash in northern latitudes is another hazard. Further, curbside is a hot environment for these relatively small plants.

The Profile Description

A horizon	0–6 inches (0–15 cm)	Loam to medium sandy loam with 5% organic matter and mulched on the surface 2–4 inches (5–10 cm)
B horizon	6–12 inches (15–30 cm)	Medium to coarse sandy loam with 1% organic matter
C horizon	12+ inches (30+ cm)	Existing soil material well drained and free from rubble and a perched watertable

Notes
The soil is not disturbed annually as it is for curb bedding, so the inorganic amendment is not required. However, the soil must be sandy to resist compaction from vibrational forces. A mulch cover is required as well as protection from traffic. Curbing at least 4 inches high provides minimum protection. This arrangement will also prevent the incursion of salt-laden runoff into the

planting area. Greater rooting volume is provided by grouping the shrubs in larger areas rather than as individual plantings. Container plantings may be a viable alternative if proper maintenance is available.

Section 2D: Curb Trees

These are curb-side trees with open-soil surface. The situation is common in urban areas. In the past, there may have been some vegetative cover or mulch, but lack of maintenance may have caused the disappearance of the cover. The technique here is to protect the soil surface with a gravel mulch (Fig. 11.5).

The Profile Description

Mulch horizon	4–6 inches (10–15 cm)	Fine gravel, uniform size, 1/2- to 3/4-inch (7–19-mm) diameter
A horizon	0–12 inches (0–30 cm)	Medium sandy loam with 4% organic matter, unamended
B horizon	12–24 inches (30–60 cm)	Medium to coarse sandy loam
C horizon	24+ inches (60+ cm)	May be existing soil material but should be without rubble and well drained; otherwise install underdrainage

Notes

Gravel is recommended as the mulch layer for several reasons. It is able to withstand heavy loads, does not wash away in heavy rains, maintains a high infiltration rate, and reduces the potential for growth of weeds. Pea gravel is recommended by many practitioners, but the size can be scattered very easily by scuffing traffic. Angular rather than rounded gravel is preferred (see Fig. 11.5 for an exception) because of the interlocking effect and reduced chance of scatter. Limestone gravel is not recommended because of its high pH and continued contribution of calcium and magnesium on soil that probably already has elevated pH. Acid granitic gravel or sandstone gravel is recommended as well as brick chips. Marble gravel has the same disadvantage as limestone gravel. Organic mulch may be used as a soil protective layer but requires annual restoration and maintenance due to compaction and decomposition. It also has the disadvantage of washing away in heavy rains and loss by scuffing traffic.

Section 2E: Curb Trees with Understory

The understory may vary from turf to shrubs. Annuals or those species grown as annuals are not recommended because of the annual disturbance to the tree roots (Fig. 11.6).

Fig. 11.5 A compacted screened river gravel mulch along St. James Place, London, England.

	The Profile Decription	
A horizon	0–4 inches— (0–10 cm)	Loam to medium sandy loam with 5% organic matter, unamended

Fig. 11.6 Curb trees with understory along the "Golden Mile" of Michigan Avenue, Chicago.

B horizon	4–24 inches (10–60 cm)	Medium sandy loam to coarse sandy loam with 1% organic matter
C horizon	24+ inches (60+ cm)	May be existing soil material but should be without rubble and well drained; otherwise, install underdrainage

Notes
If shrubs are used as the understory, then organic mulch is recommended as soil cover. It should be no thicker than 4 inches (10 cm) and should be kept away from the base of the shrub stems and the tree trunks.

ELEMENT 3: COVERED

This element is the most common in central business districts and major urban plazas. Sidewalk and plaza plantings usually require a covered situation. Thus, impervious or jointed paving is the covering. Provision must be made for the gasous diffusion between the covered soil and the atmosphere even if the plants are irrigated—in fact, the more the reason for soil and root aeration as irrgation water under a cover evaporates or drains slowly. Covered turf, bedding, and shrubs practically do not exist and soil profiles are omitted here. In this case, the soil not only must be a favorable rooting medium for plants, but must also serve as a support system for the covering, as a sidewalk, paved plaza, and so on. The two functions are nearly contradictory, because a firm foundation for sidewalks that meet engineering standards may too dense for root growth, and, conversely, a soil with optimum rooting characteristics may be a poor foundation for a sidewalk. The information provided in the following sections meets both functions at the present stage of knowledge. Future research on "structured" soils may produce additional and more satisfactory solutions.

Section 3D: Covered Trees

There are two approaches to the solution of the tree soil profile under cover (Fig. 11.7). The second one is of recent development (Grabosky and Bassuk, 1995, 1996, Kristoffersen, 1998).

The Soil Profile Descriptions

Soil Profile (a)

Gravel layer	Minimum of 6 inches (15 cm)	May be coarse crusher-run, but No.2 is preferred
A horizon	0–24 inches (0–60 cm)	Very coarse to coarse loamy sand with no more than 3% organic matter

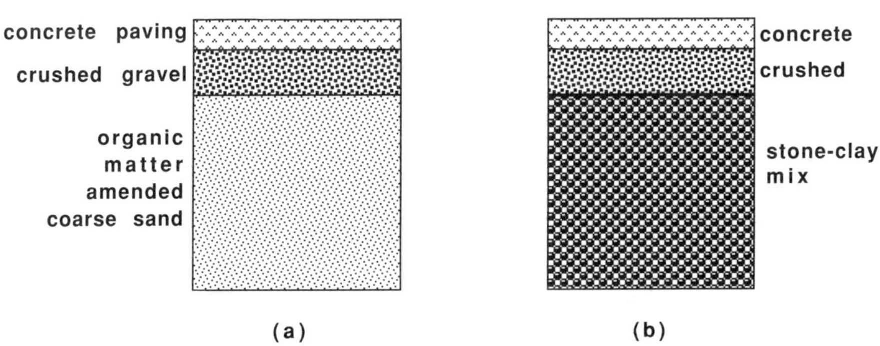

Fig. 11.7 Two soil profiles for covered trees: (a) the gravel and sand approach, and (b) the stone matrix approach of Grabosky and Bassuk (1995, 1996). The crushed gravel layer in (b) is assumed. Crushed gravel below concrete is required in most city engineering specifications.

B horizon	24+ inches (60+ cm)	May be existing soil material, but should be without rubble and well drained; otherwise, install underdrainage

Notes
This homogeneous profile makes use of the large pore spaces within the very coarse to coarse loamy sand as the channels for root extension. The large sand particles are in contact by the 12-point tetrahedral packing arrangement (Deresiewicz, 1958), which provides great strength for pavement support. The sand may be compacted to 95% Proctor and still the large pore spaces of the matrix are maintained by the strength of the packing arrangement. Because the soil is excessively drained, its favorable aeration under the pavement is assured if a pathway to the air atmosphere is provided. The distinct disadvantage of the coarse loamy sand is the low fertility, which may supplemented by organic matter amendment, and low waterholding capacity, which requires irrigation. The gravel layer below the covering is necessary to meet engineering requirements and provides a pathway from the open pit and extending under the pavement for aeration of the roots and soil. In some cases, the pavement is suspended over the soil (Chapter 9). The technique has been successfully applied to several projects (Chapter 12).

Soil Profile (b)

Gravel layer	Minimum of 6 inches (15 cm)	May be coarse crusher-run, but No. 2 is preferred
A horizon	0–24 inches (0–60 cm)	A mixture of No. 2 stone and clay loam soil in various ratios (tested: 3.82:1 to 5.98:1)
B horizon	24+ inches (60+ cm)	May be existing soil material, but should be without rubble and well drained; otherwise, install underdrainage

Notes
The use of large gravel provides the high-strength support for the pavement. The very large interstitial spaces between the gravel contains a precise amount of soil that has pore space. Roots extend through the soil. The presence of the soil provides fertility and waterholding capacity that sand does not provide in Profile (a). The disadvantage of the technique is the soil amount added to the gravel must be precise and the mixing must be highly controlled; it should not exceed 25% of the total weight. Further, the gravel itself does not provide nutrients and waterholding capacity, so the questions arise: What is the vol-

ume occupied by the gravel and is this compensated by the included soil, all compared to the coarse sand of Profile (a)? A somewhat similar technique using English "mud-balls" has been installed in the restored dock areas of London with apparent success (Olin Partnership, personal communication, 1994).

Section 3E: Covered Trees with Understory

This situation is a combination of Section 2E, "Curb Trees with Understory," and Section 3D, "Covered Trees." The major difference is that the open-soil area is confined to the dimensions of the tree pit itself. The cover extends toward the trunk within the drip line, even at planting. Thus, most lateral root extension needs to occur under the covering, whereas in the case of Section 2E, a major portion of lateral root extension occurs in open soil. The only difference is that the tree root ball must be planted deeper than in the two previous cases in order to provide sufficient rooting depth for the understory. Topsoil is applied only to the open-tree pit area; otherwise, the soil profile is the same as Section 3D. The danger here is that the tree root collar is planted too deep for many species (especially the maples) and the tree subsequently dies. The problem not withstanding, many covered street-side trees are found in this condition. The author does not recommend this system (Fig. 11.8).

ELEMENT 4: ROOF

Rooftop plantings are unique, as described in Chapter 6. Though the subject is covered in detail with respect to trees in that chapter, some further discussion is necessary to illustrate various types of planting and the soil design here, and to better complete the element-plant palette matrix with the associated soil profiles.

Since weight limitations are usually encountered in rooftop applications, the designed soil must be lightened, as described in Chapter 6. The degree of lightening is obviously dependent on the severity of the constraint and the provision for a soil profile (depth) on the rooftop. In cases in which there are no significant constraints and the the minimum soil depth for the plant palette is permitted (Table 11.2), the soil profiles for the appropriate application previously described in the matrix can be applied. If severe constraints exist and/or the minimum soil depth is not possible, the following soil profiles are given.

It should be noted that rooftop soils are not covered with paving.

Section 4A: Roof Turf

Subsection 4AA: Roof Turf without Traffic. If sufficient depth is provided and no weight limitation exists, the soil profile from Subsection 1AA may be used. If the full depth is permitted by planned space, the A and B horizons

Fig. 11.8 New York City side street.

may require further structure amendment amounts to meet the loading requirements, as calculated in Chapter 6. If minimum depth is available (8–12 inches or 20–30 cm), then a homogeneous topsoil mix (A horizon) should be used. If it requires lightening at minimum depth, the amount of structure amendment to lighten to the limit is calculated and the amendment is applied accordingly. Irrigation is required in this instance because the minimum soil

TABLE 11.2 Minimum Soil Depths for Various Types of Plants

Planting	Minimum soil depth, ft (cm)
Turf	8 to 12 (20 to 30)
Flowers and ground cover	10 to 12 (26 to 30)
Shrubs	24 to 30 (60 to 75)
Trees	30 to 42 (75 to 120)

Source: R. W. Harris, personal communication, 1992

depth is inadequate to provide sufficient waterholding capacity to maintain the turf between rainfall events. It is likely that the irrigation will need to be frequent and thorough.

Subsection 4AB: Roof Turf with Traffic. The soil profile from Subsection 1AB may be applied here with sufficient soil depth and no loading constraints. If depth is minimal, the soil is the homogeneous topsoil (A horizon) with the stated structure amendment as minimum for compaction-resistance purposes. The amount of amendment needs to be adjusted if the weight of the amended soil exceeds the prescribed loading. Normally, the structure amendment should not exceed 50% by volume as the soil becomes too dry.

Section 4B: Roof Bedding

The applicable profile here is Section 1B, "Open Bedding." Since the A horizon is already specified with 15% fine-structure amendment, it may meet load limitations as is. Amendment of the B horizon would further lighten the load. Coarser amendment may be used in the B horizon.

Section 4C: Roof Shrubs

The soil profile described for Section 1C, "Open Shrubs," is applicable here. The profile will require structure amendment if loading limits are set below the soil profile weight. The amended material should be medium to fine for the A horizon. If the B horizon is lightened, a coarse-structure amendment would be a more effective lightener.

Section 4D: Roof Trees

Rooftop trees should have at least minimum soil depth for windthrow resistance. If space cannot be provided in the rooftop itself, the trees should be containerized as suggested in Chapter 6. If weight is the only constraint, then styrofoam may be used to reduce the soil volume. Irrigation may be necessary due to the reduced waterholding capacity.

Section 4E: Roof Trees with Understory

In most cases, trees with understory on a rooftop are treated in the same way as in Section 1E, "Open Trees with Understory." Many rooftops have been designed to support the total weight when this planting alternative is used. It is unwise to force the alternative when the rooftop loading limits the soil depth. In this case it is better to simply plant understory plants in containerized trees as a miniature adaptation to the larger-scale design.

CONCLUSIONS

Thirteen soil profiles and their modifications are described for application to the more common situations encountered in the urban environment. The details for the profiles should be considered as starting points in soil design. Project conditions together with the experience of the practitioner may dictate modifications and revisions to these details. Simplification may often occur with little detrimental effect to the outcome. The initial level of detail is provided to assist the design process as it is more difficult to move from simplification or homogeneity to complexity or heterogeneity rather than progressing from complexity to simplicity.

12

CASE STUDIES

INTRODUCTION

These case studies are provided as guidelines in the preparation of other soil specifications. Readers should not use these specifications directly as they may not be appropriate and adequate for numerous purposes. Each practitioner should develop an individual approach and style to the writing of soil specifications using the knowledge of fundamental soil science and the practical experience described here.

The case studies are presented in the chronological order in which the projects were executed. The reason for this order is to show the evolution of the written soil specifications and how the details are expanded as our knowledge in soil design progressed over time. Also, the experience of working with various contractors indicated the need to be more specific in what is to be done on the project and how it is to be carried out. It may be difficult for many people not familar with "working with soil" that there are right ways and very wrong ways of handling soil and its treament after placement. Some projects potentially may fail if there is omission of these details along with the lack of very close supervision. Some further details as numbered "Notes" are added for explanatary reasons in each project that were not part of the original documents.

The major lesson learned over more than 25 years of involvement with landscape projects is that close supervision by the landscape architect or a direct agent during the installation of soil and planting stock is *absolutely necessary*. When the project budget is developed, provision must be made to fund this supervision.

The original specifications have been reformatted into that of the remainder of the book to maintain a consistent style. Thus, they do not appear as the same as specifications would normally be organized with numbered sections and paragraphs. The details and information remain unchanged.

BATTERY PARK CITY SOUTH COVE, NEW YORK CITY

This was an xerophytic landscape design to represent wooded beach conditions by Susan Child Associates, Boston, Massachusetts. Wave-erosion resistance and self-flushing soil for salt removal are two criteria for the soil design. The specifications feature simultaneous planting stock and soil installation.

The project specifications were completed in 1985. Contrasted to those for the Getty Center in 1994 and the Allegheny Riverfront Park in 1995–1996, progression of greater specification detail is observed as stated in the Introduction. Experience dictated this progression to ensure quality material and specification conformance. The Battery Park South Cove project benefited from the craftmanship of a very competent contractor and soil-mix supplier though the specifications were probably inadequate compared to present standards.

Soil Specifications

Planting Soil Mix

A. The planting soil mix is to be specified pending analysis. Topsoil, granular fill, and other amendments as required (pending analysis) shall be used to create the planting soil mix.

The planting soil mix shall have the following mechanical properties:

Textural Class	Weight	% of Total Average
Sand (0.05—2.0-mm diameter range)	45 to 75	70*
Silt (0.002—0.05-mm diameter range)	15 to 35	15
Clay (less than 0.002-mm diameter range)	5 to 25	15

*With 60% medium sand (0.25 mm) or larger.

1. Planting soil mix for the island shall be 80% sand with 70% medium (0.25 mm) or larger.
2. Peat moss shall *not* be used.

B. Planting soil shall have a pH value range of 5.5 to 6.5 without adjustment.

C. The planting soil mix shall have the following chemical properties:

Nitrogen Phosphorus Potassium Calcium Soluble salts	*Note 1:* These values have been lost from the original specifications and are not recoverable. They were probably routine values.

Note 2: A sand/soil mix meeting the specifications was eventually provided by a topsoil contractor. The significant feature in this project is that the base soil was brought to rough grade, and then the specified planting soil was placed simultaneously with the planting stock, eliminating traffic over final grade; thus, subsequent compaction was avoided.

Figure 12.1 (from Craul, 1992) shows the section through an area planting. Notice that the compacted fill is carried up under the sidewalk. The 6-inch (15-cm) gravel layer permits roots to extend under the walkway. The planting stock, with varying root ball heights, is placed on compacted pedestals to prevent settlement.

BOSTON COMMONS PARADE GARAGE

A rooftop application with weight limitations coupled with the requirement of resistance to heavy all-weather traffic for Sasaki Associates, Watertown,

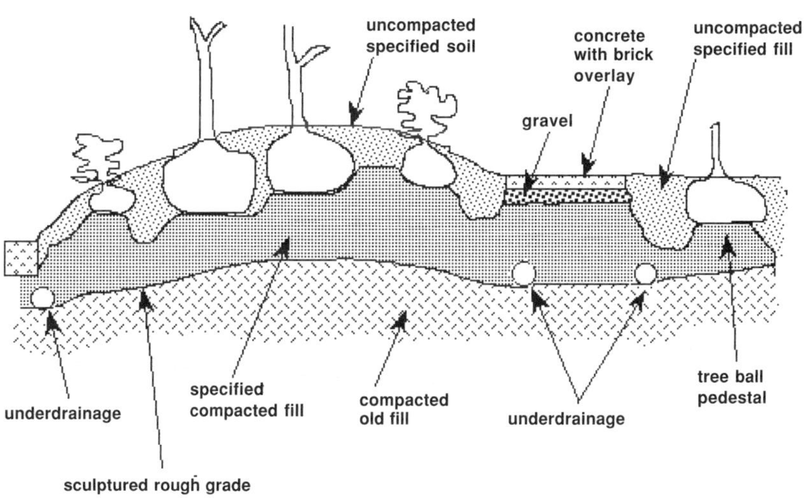

Fig. 12.1 Soil cross-section of Battery Park City South Cove, New York City. The compacted specified fill is actually the same as the uncompacted specified soil, the difference only being the former is mechanically compacted to 95% Proctor when placed.

Massachusetts. The plant palette was restricted to turf. Many events are held on this site throughout the year, including the erection of tents and other shelters. The resulting soil design is almost a prepared athletic turf (PAT).

Soil and Turf Specifications

Introduction. The soil over the roof of the parking garage of the Boston Commons acts as a counterbalance for the bouyancy of the garage below the watertable. Therefore, not only must the soil meet specifications as a planting medium with minimum maintenance under heavy use, but must meet engineering weight specifications.

Sasaki Associates proposes to compensate for the removal of the roof vents by replacement with molded/cast styrofoam blocks ("Liteguard"), the volumes based on the vent volumes. The new kiosk design requires an increase in final grade elevation for better access walk and landscape gradients. This increases the soil volume and weight. The increase can be offset by increasing the amount of styrofoam placed on the roof. Soil weight should not exceed 120 pounds per cubic foot. Because of the complexity of the loading configuration by section of roof cells, no prior soil thickness (profile depth) across the roof can be given. Adjustments are to be made later by Sasaki Associates after soil thickness and the weight of the planting medium are provided. Weight calculations and the conformance to specifications are critical.

Soil Requirements and Constraints

Loading Constraints. The soil weight limit of 120 lb/ft^3 means that the soil density may not exceed 1.5 Mg/m^3 with a total pore space of 43.4% or a void ratio (e) of 0.77. Saturated (100%) weight equals 120.75 lb/ft^3, and it must be assumed that the soil may have this moisture content at some period during the year, especially the lower portion of the soil profile, unless the subsoil or fill allows this layer to readily drain. The upper portion would generally have lesser moisture content and thus weigh less than 120 lb/ft^3. For loading evaluation, it is important to estimate the moisture content distribution within the soil profile in addition to the dry density and weight.

Resistance to Compaction. The Parade is a heavily used area of the Commons with several annual events having large concentrated impact. The present soil appears to have moderate to low resistance to compaction and spotty turf cover.

Medium- to fine-textured soils (heavy loams, silt loams to clays) are susceptible to compaction (Figure 12.2a). Sandy soils with large proportions of coarse and very coarse sand fractions are resistant to compaction (Figure 12.2b). Unfortunately, these latter soils have limited waterholding capacity and fertility. Those sandy soils with dominant fine and very fine fractions are susceptible to compaction.

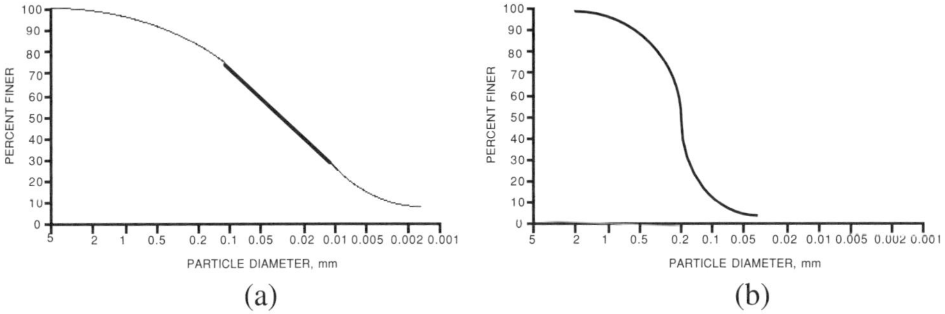

Fig. 12.2 Particle size distribution for (a) a well-graded versus (b) a poorly graded soil.

Wear-Resistant Turf and Minimum Maintenance. Certain turfgrass mixtures are more wear-resistant than others, so they are recommended for play and athletic fields. Other less resistant mixtures may be used, but they require moderate to high levels of maintenance input for sustained coverage and appearance characteristics. For this application, heavy use and low-level maintenance is anticipated.

Soil Specifications

The Substratum or Fill. The substratum has the purpose of filling the void between the roof insulation and/or the styrofoam surface and the lower boundary of the planting medium. Plant roots are not expected to penetrate into the material. The material may be any porous sandy fill free of stones, building rubble, wood, plastic, metal, glass, asphalt, or other anthropeic materials. It shall not contain more than 5% silt plus clay, and no more than 5% organic matter. Water movement through the material shall not be impeded, with drainage occurring freely to the roof surface. The *in situ* density shall not exceed 1.5 Mg/m^3. This material shall be excavated from well-drained upland areas. No material from bogs, swamps, or waste sites is to be used.

The Planting Medium. The planting medium shall be material from well-drained upland sites or specially mixed sand prepared for this project. No material from bogs, swamps, or waste sites is to be used.

Texture

The planting medium is a coarse to medium sandy loam (USDA texture class), with a particle size distribution that fits into the envelope, as given in Fig. 12.3. The material is SP in the Unified Soil Classification System. The range in particle size distribution allows some flexibility in sand material procurement as it is usually difficult to meet exact specifications.

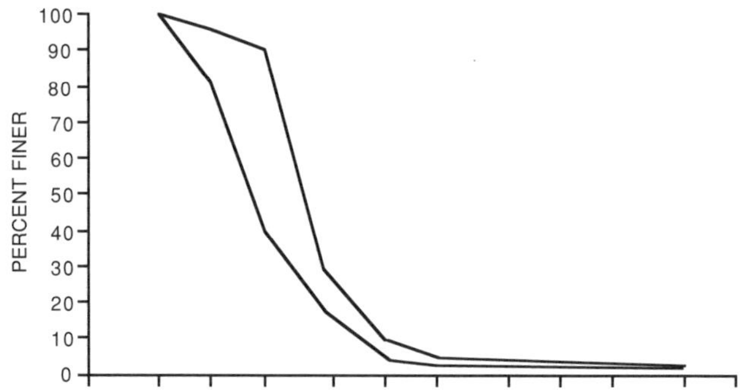

Fig. 12.3 Specified soil particle size distribution envelope.

This envelope permits a range in the coefficient of uniformity (C_u) from about 4.0 to 4.7 and the coefficient of gradation (C_c) from about 1.21 to 1.56.

As a guide, the fractionation of the sands and the soil separates are given in Table 12.1. Deviation from these values shall be limited to 10% or less for very coarse, coarse, and medium sand, and to less than 5% for fine and very fine sand, and less than 2% for silt and clay.

Density

The density of the planting medium shall not exceed 1.4 Mg/m^3 to prevent overload for the garage roof. This allows for 47.2% total pore space or a void ratio (e) of 0.89. The saturated weight is approximately 116.8 lb/ft^3 and field capacity weight (saturated and allowed to drain for 24 hours) is approximately 98.53 lb/ft^3. This bulk density value is below the threshold

TABLE 12.1 Particle Size Classes and Their Percentages Passing and Contents

Particle Class	Diameter, mm	Sieve, No. Passing	Percent, Separate or Fraction	Percent
Fine gravel	12.7 to 2.00	10	100.0	—
Very coarse sand	1.00 to 2.00	18	65.9	17.7
Coarse sand	0.50 to 1.00	35	32.5	41.7
Medium sand	0.25 to 0.50	60	14.3	22.7
Fine sand	0.10 to 0.25	140	7.1	9.0
Very fine sand	0.05 to 0.10	270	5.3	2.2
Silt	0.002 to 0.05	—	—	1.0
Clay	< 0.002	—	—	5.6

value of 1.75 Mg/m^3, which inhibits root penetration in sandy soil and greatly reduces gaseous diffusion within the soil profile.

Water is differentially distributed within a soil profile at any time depending mainly on the physical properties of texture, structure, and density, together with the drainage condition present at the base of the profile. In this case, drainage from the bottom of the planting medium is expected. The lower portion of the profile will tend to contain a higher moisture content than the upper portion during the growing season, whereas moisture content will be high throughout most of the profile during the dormant season. However, it is expected that only a portion of the soil profile will be saturated and only for short periods of time. Thus, maximum density is assumed with saturation, but since it may occur infrequently, this acts as a safety feature concerning roof loading. Profile weight will be reduced during the growing season and maximum weight will occur during the dormant season.

Waterholding Capacity

Turf grass requires 1/4 to a 1/3 of an inch of water per day. The specified soil supplies the estimated available water in Table 12.2.

Optimistically, the soil profile is able to sustain the turfgrass for 9 days without precipitation or watering. However, this estimate does not include water loss through evaporation and makes the assumption that the turf roots occupy total soil profile depth. Most turf roots occur 12 inches in the surface; the preceding soil profile then sustains the turf approximately 6½ days with a potential reserve supply of 2½ days. The daily growing season (June–September) potential evapotranspiration rate amounts to about 0.20 inch (Logan Airport and Chestnut Hill weather data), which may further reduce the sustaining period.

The amount of available water can be significantly increased by specifying a finer texture, but this modification increases the compaction potential and may lead to the unsuitable conditions of poor aeration, water drainage, and root penetration. Increasing the organic matter component

TABLE 12.2 Estimated Soil Profile Waterholding Capacity

Horizon, in.	Field Capacity, %	Available Water, in./ft	Available Water, %
0–6	17.9	2.15	17.9
6–12	17.9	2.15	17.9
12–18	8.8	1.16	7.7
18–24	6.7	0.79	6.6

Profile total 3.125

also increases waterholding capacity but increases the settlement potential of the soil material.

Organic Matter Amendment

Because the specified texture is relatively coarse with limited waterholding capacity and fertility, it is necessary to amend the material with organic matter to increase waterholding capacity and improve the nutrient retention of at least the upper 18 inches of the planting medium. Therefore, 10% by volume of peat moss or 15% by volume of compost shall be mixed with the specified sand. The peat is coarser and decomposes slowly, lasting longer than the finer compost.

If compost is to be used, certification shall be made that it contains no pesticides or other toxic substances in excess of EPA or State of Massachusetts guidelines, and that the pH of the compost is within the range of 5.5 to 7.5. Separate batches of compost must be tested and the results submitted to the landscape architect for approval.

Soil Reaction (pH) and Chemical Composition

The initial soil pH shall be in the range of 5.0 to 7.0. The specified soil shall not contain excess soluble salts, alumimum, pesticides, or other toxic substances. The contents of nitrogen, phosphorus, and potassium should meet at least the minimum limit or low range as published by the University of Massachusetts Soil Testing Laboratory. Analyses results shall be submitted to the landscape architect for approval.

Soil Profile Construction

Placement of Soil Materials. The importance of careful placement of the soil materials and the sequencing of operations cannot be overemphasized for sucessful installation and the sustainability of the design.

Spreading and Rolling. It is imperative that spreading and rolling be done when the material is dry and during dry weather. Work during rainy weather shall be prohibited. Wet material leads to excessive compaction.

Fill

The fill material shall be placed in 6-inch lifts and rolled to meet density specifications of 1.5 Mg/m^3 or a dry density of 93.6 lb/ft^3. Rolling force shall not be excessive. Density of each lift shall be checked to determine conformance. The rough grade surface shall be parallel (equal to 2 feet below final grade) to the final grade surface except at the walk locations where the fill shall be sloped upward to subgrade elevation for the walks

(Fig. 12.4). Thus, the turf areas between the walks and project edges become basins 24 inches deep to be filled with planting medium material. Compaction caused by vehicles shall be minimized by backfilling over previously rolled fill.

Planting Medium

The planting medium shall be laid in lifts of varying thickness (Fig. 12.5) with interspersed mixing and rolling. The first lift of soil is spread to a thickness of 6 inches and rolled to a density not to exceed 1.4 Mg/m^3 or a dry density of 87.4 lb/ft^3. For the planting medium, the roller weight shall be 75 to 100 pounds per foot of roller width. The second lift is started by spreading soil to a thickness of 3 inches without rolling, and then the first layer of organic amendment is spread to a thickness of 1 inch or at a rate of 3 $yd^3/1000\ ft^2$. Then, 3 inches of the soil is spread. The three layers are now tilled or disked to thoroughly mix the organic amendment with the soil to provide an approximate 6-inch planting medium lift. The lift is then carefully rolled. Another 3-inch layer of soil is spread, then covered with a 2-inch layer of organic amendment at a rate of 6 $yd^3/1000\ ft^2$, followed by another 3-inch layer of soil. This lift is thoroughly tilled then rolled. The last lift is constructed in the same manner: 3 inches of soil, 2 inches of organic amendment, and the final 3 inches of soil, thoroughly tilled and carefully rolled. The final lift shall conform to final grade. If low areas need to be filled, a sufficient amount of planting medium shall be mechanically mixed to match the final lift. This surface becomes the turf seedbed. Starter fertilizer and lime are applied prior to final tilling.

Continunal inspection shall be exercised throughout the soil profile phase to make certain that lifts are placed as specified and that the tilling results in thorough mixing of the organic amendment and that the rolling does not overly compact the soil but is sufficient to prevent excessive soil settlement.

Checking *In Situ* Density

It is imperative that fill and planting medium density be checked periodically during the construction of the soil profile so that the final profile density does not exceed the loading constraints. The density of each lift

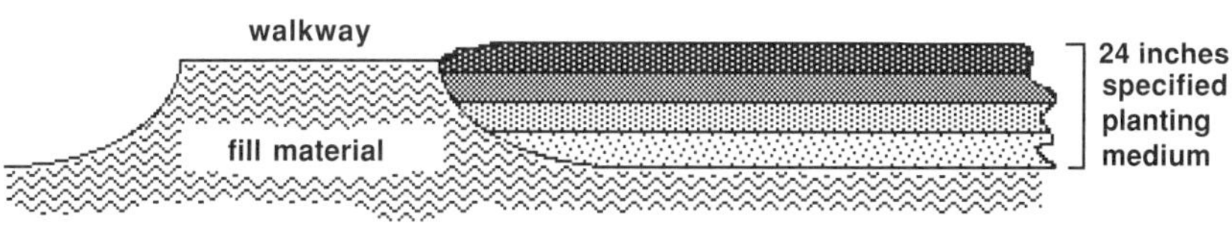

Fig. 12.4 The spreading of fill and upward sloping to a walk subgrade.

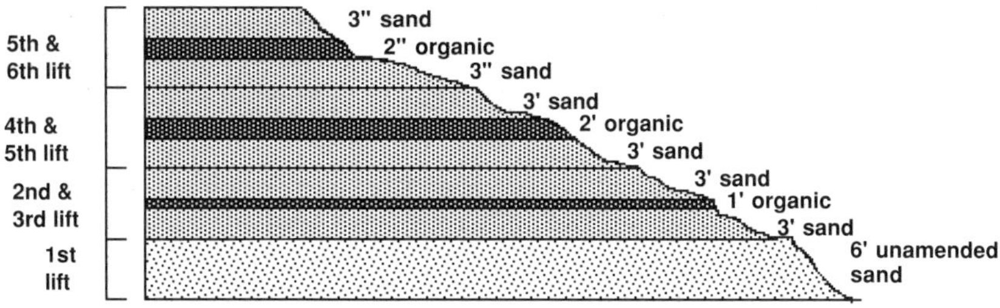

Fig. 12.5 The spreading of the planting medium in varying lifts.

shall be checked utilizing standard density tests by sand excavation, penetrometer, or nuclear methods, as recommended by Sasaki Associates engineers.

Rototilling or Rotovating Interface and Organic Matter into Planting Medium

The tilling for mixing the organic amendment and soil may be accomplished by heavy rotovator or farm disk. The depth of tilling shall be at least 6 inches for each lift for thorough mixing. Several passes shall be made in one direction and then at a right angle to the first direction. A lightweight rototiller with standard tines is not adequate for thorough mixing.

The second layer of soil is placed over the organic amendment before tilling so that the heavier soil may fall into the organic material and bring about more thorough mixing.

Sequencing of Construction Operations. The sequencing of the construction operations is as important to the successful installation as is the design and materials used.

All of the fill shall be placed first and built up to the walk subgrade below walk locations and rough grade to 24 inches below the planting medium final grade. Since the fill is to be built up to the walk subgrades, it is then appropriate for the placement, tilling, and rolling of the planting medium to be done from the walk subgrades and the edges of the construction site. This procedure eliminates the danger of excessive traffic and its compaction on the planting medium after placement. See "Seedbed Preparation."

Turf Specifications

Turf Mixture and Application Rate. For playing fields and areas receiving heavy pedestrian use, a turf mixture of at least three varieties of Kentucky bluegrass with one or two varieties of perennial ryegrasses shall be used.

Each bluegrass variety shall be at least 20% of the blend. Table 12.3 gives the recommended mixes and application rates.

All of these varieties are recommended for heavy-use areas, have good winter color and cold and drought resistance, good disease resistance, and tolerate low soil fertility. Other varieties may be substituted if the preceding are not available. However, the substitutes shall have the same adaptation and disease-tolerance characteristics as those listed. All proposed substitutes shall be submitted to the landscape architect for approval.

Seedbed Preparation and Planting

Seedbed Preparation. The tilling and rolling of the final lift shall be adequate for seedbed preparation. On completion of this step, no further vehicle traffic shall be permitted on the prepared seedbed area. All subsequent operations shall be carried out from the walkways. See the sections on "Planting Medium" and "Sequencing of Construction Operations."

Starter Fertilization and Liming. A starter fertilizer with a ratio of 2:1:1 at a rate of 3 to 4 lb of nitrogen per 1000 ft^2. when the fertilizer contains at least 35% slowly available nitrogen (ureaform, sulfur-coated urea) should be incorporated into the top 3 inches of soil on tilling of the final lift.

Soil pH may adjusted by the application of lime during the final tilling operation. The amount to be applied depends on the initial soil pH. Standard tables for application rates are available from the state agricultural experiment station or extension service at the University of Massachusetts. For best results, the final pH should be in the range of 6.5 to 7.5.

Planting and Mulching. The seed may added to the hydromulch, which is recommended for this project. The slurry (wood-fiber products) shall be spread from the walkways and the edges of the garage limits. The application rate may range from 1000 to 1500 pounds per acre. The chosen rate must be adequate to thoroughly cover the entire area uniformily and must apply the

TABLE 12.3 Turfgrass Mixes and Application Rates

Formula	Percentage of Mixture	Rate of Seeding
1. Kentucky bluegrass variety blend		
America	20	
Bonnieblue	20	
Princeton	20	
2. Perennial ryegrass		
Allstar or Yorktown II	40	4 lb per 1000 ft^2

correct rate of grass seed. Areas that are too heavy to allow seedlings to grow through must be thinned by hand. The final hydromulch-seed formulation of the landscape contractor must be approved by Sasaki Associates.

Maintenance

Soil Aeration. Aeration is necessary to sustain the turf. Aeration of the soil with a hollow-tube aerator is recommended at least twice to three times a year. This is necessary to reduce the surface compaction that is inevitable with heavy use of the Parade. The aerator shall be heavy enough to penetrate 2 to 3 inches into the surface soil and the hollow opening shall be 3/4 to 1 inch in diameter. A spike aerator shall not be used. Heavy aeration (six to eight times over the area) in the spring prior to fertilization is recommended, followed by light aeration (one to three times over the area) in late summer or early fall. A chainlink fence section or a chain shall be dragged over the area after aeration to crumble and spread the plugs for the early aeration treatment. The plug holes should be left open during the winter so that freezing and thawing may penetrate the soil and help reduce compaction.

The spring aeration may be followed by topdressing of compost spread to a depth of about 1/3 to 1/2 inch at an approximate rate of 300 to 700 lb/ 1000 ft^2 depending on its moisture content. The compost will fill the aerator holes, aiding the incorporation of organic matter into the surface soil.

Fertilization Program. An annual fertilization program, coupled with aeration, is necessary for the maintenance of a strong turf. The following program is recommended on the assumption that the recommended soil material has minimal to adequate levels of phosphorus and potassium. A soil test should be perfomed to determine if modifications should be made to the program. The following program (Table 12.4) is based on a fertilizer ratio of 2:1:1 and on application rate of approximately 3 to 4 lb of nitrogen per 1000 ft^2.

TABLE 12.4 Turfgrass Fertilizer Program

	Time of Application			
Nitrogen Source	Late Spring	Early Summer	Late Summer	Early Fall
25% to 49% of the N derived from slowly available source[1]	pounds of N–P_20_5–K_20 per acre applied			
	60–30–30		50–25–25	50–25–25

[1] Ureaform or sulfur-coated urea. Do not confuse with urea.

Special Considerations

There are some issues in the landscaping effects on soil drainage that require some special attention. These issues need to be considered as the design and installation process occurs.

Soil Surrounding Kiosks and Its Drainage. The presence of the kiosk plenums close below (24 inches) the soil surface pose some drainage problems. The sandy soil material as recommended should provide adequate drainage within the soil profile, provided that runoff from the kiosk roof is not drained into the shallow soil. If the plenum surface slopes away from the kiosk, this may prevent water from collecting in the soil near the kiosk roof eves.

Later in the project planning, it may be suitable to reexamine the soil material specification around the kiosk. If the design shows that water accumulation may be a problem, it might be necessary to respecify the soil around the kiosks to a coarser texture together with a deliberate underdrainage design.

Observation of Soil Drainage on Excavation of Existing Soil. Some observations may be made during the process of existing soil excavation that can be very helpful for obtaining some indication of present soil-water drainage. This information may be used to confirm the present design or to suggest modification to improve the present design.

From the Upslope Into the Garage Area. It would be well to know how much, if any, water flows downslope from the hillside into the soil of the garage roof area. When the soil is trenched or excavated along the uphill side or on the line from Kiosk 3 to Kiosk 4, the soil profile should be examined for wet or very moist segments or areas that may indicate lateral flow. If significant flow is detected, it is recommended that a longitudinal curtain drain be installed to intercept water flow from upslope. This interception would reduce the danger of excess water accumulation within the rooftop area.

Drainage within Existing Roof Soil Profile. As the soil is removed from the roof, drainage within the soil profile should be observed to determine if wet spots or areas exist and to note their locations. These observations indicate roof areas that may present some drainage problems and the fill material may be modified in placement or specification if it appears that a wet area may recur.

Note

The project was succesfully installed in 1994–1995. Unfortunately, the subsoil grade was not fully protected from traffic during construction because some other activities on the site were not accomplished in sequence. A major disturbance occurred after final grading when it was discovered that a conduit had not been placed by the electrical contractor. A major trench was opened across one end of the project. The effort to segregate the planting medium and the subsoil and replace them in proper sequence was not successful, and so a line was formed where the soil profile is now nonexistent. The line is detectable by the differential response of the turf. Fortunately, the success of the project was assured in a major way by the City Authority complying with the need to leave the planting undisturbed (protected by an 8-foot high cyclone fence) for two growing seasons. This allowed the turf roots to become fully established and more resistant to the effects of future traffic.

J. PAUL GETTY FINE ARTS CENTER, LOS ANGELES

Hanna/Olin, Ltd. (now Olin Partnership, Philadelphia) requested that a thorough introduction be prepared and that the entire soil specification process should be considered as an educational endeavor for the client. The situation was fortuitous because it enabled the author to pursue every detail to an appropriate degree.

The landscape project involved the design of soil for 67 different planters, two-thirds of which were on-structure and the remaining ones on-grade. The soil weight limit was more constrained than for the Boston Commons Parade garage roof; hence, the need for the cast styrofoam blocks in the deeper profiles. The plant palettes ranged from tropical tree ferns to xeric species as cactii, etc.

The author's activities in this project ceased due to impending serious health reasons when the specifications had been prepared and sent out for bid. He was not involved in the eventual installation and finishing of the project. As best as can be determined at the time of publication, the specifications were followed with minor modifications.

Introduction

The J. Paul Getty Fine Arts Center is located in the Santa Monica Mountains that experiences a Mediterranean type of climate with a dry growing season and moist to wet dormant season. Soil moisture retention during the growing season and active or passive irrigation become primary concerns in a design situation that involves a diverse plant palette as proposed for the Center. And yet, there must be conservation of the valuable water resource. The dormant season presents a condition of irregular rainfall distribution and intensity that requires good soil drainage and a high degree of stability (resistance to raindrop dispersion). Most of the plantings are to be placed in a "planter" configuration on the structure as a rooftop application; hence, rooting (soil) volume is limited, placing further constraints on soil design. Urban (1989) and others (Lindsey and Bassuk, 1991) have found that reasonable tree growth requires minimum rooting volumes of 300 cubic feet under common urban soil conditions. With the variation of available soil depth ranging from 14 inches (11.5 inches in one instance) on structure to 6 feet on grade, available rooting volume and depth are moderately to severely restricted for trees and woody shrubs in the shallow depths. Volume limitations can be overcome to a certain extent by soil design, use of shared rooting space, and sound irrigation design. All of these factors are present in the J. Paul Getty Fine Arts Center project.

The Soil Design Philosophy

The soil design philosophy for this project is to meet the needs of the landscape design and its plant palette with soil specifications that emulate appropriate natural soils as closely as possible and that they respond well to the irrigation design. The soil is a three-phase system of solid, liquid, and gas. Most soil specifications concentrate on the solid component by specifying the texture and organic matter content, with pH, soluble salts, and nutrient content assuming almost a secondary role in many cases. Waterholding capacity, density, and degree of drainage are almost always neglected or omitted. For a soil to be properly designed, especially for the unique applications, all these properties and characteristics must be entered into the design criteria list. A further concern is the propensity of soils with a high organic matter content for settling and compacting.

The Soil Design

The soil solid phase or component, which includes both inorganic and organic materials, has profound influence on other properties and characteristics, so it more than forms the matrix of the soil: It determines inherent fertility and, partially, waterholding capacity, degree of drainage, and density. Texture is the soil science term that refers to the proportion of sand, silt, and clay particles (also called the particle size distribution) in the soil material. The other soil property that influences these latter characteristics is structure, which is defined as the degree and pattern of aggregation of the solid particles. Since structure is usually destroyed or greatly reduced in disturbed soil or may not exist in a soil mix, some compensating factors must be applied. Compensation is gained by additions of organic matter or artificial amendments such as sintered fly ash, expanded clay, pumice, and so on.

The soil design best begins with specifying the texture. With the variation in soil depth and volume relations coupled with the plant palette diversity, it would be unwise to specify a single texture for all plantings. It is easiest to start with a single texture as a basic matrix and then modify it with amendments to fit the various soil requirements of the plantings in the construction of the planter soil profile leading to several different layers or horizons. This may serve well for moderately deep and uniform soil profile depths, but is not especially appropriate for the Getty Center.

The intention is to emulate the "topsoil," "subsoil," and "parent material" of the natural soil with changes in the basic matrix together with different amendments. Textural changes are accomplished with additions of coarse sand or clay (or zeolite) materials, depending on the direction of the textural change. Waterholding capacity, bulk density, and drainage are modified accordingly. The use of sintered flyash, pumice, ceramic products such as ISOLITE®, AXIS®, or TURFACE® are used to further the modification of the latter soil characteristics with or without texture change.

Soil amendment with organic matter of some type will also provide for the modifications described before in addition to contributions to the fertility status. However, the addition of excessive amounts of organic matter may be required to attain an effective modification. Excessive amount of organic matter causes several problems: (1) the soil is prone to settlement; (2) waterholding capacity may be increased to the point where the soil remains wet for prolonged periods after rainfall or irrigation, decreasing the movement of gases into and out of the soil and creating unfavorable aeration conditions; and, (3) decomposition by soil organisms may be actually inhibited. Many natural forest soils have only 4 to 5% by weight of organic matter content coupled with a well-developed structure. The addition of organic matter does increase the fertility of the urban soil and over time enhances the development of natural structure that is initially destroyed to varying degrees in these disturbed soils, but an excess amount is not benefical. To overcome the problem of excessive organic matter content, inorganic materials that lighten the soil are employed as amendments to the basic matrix.

Fertility and soil reaction (pH) of the layers will be adjusted during the mixing process to provide a broad range of nutrient availability. Local adjustment for each planter can be made after installation and applied according to the needs of each plant palette either by the contractor or the client in routine maintenance.

The Standard Specifications

Because the project is located in Los Angeles, useful knowledge of the Standard Specifications for Public Works Construction of 1991 (hereafter, STAND SPECS) is appropriate. Section 212-1, paragraphs 212-1.1.1 through 212.1.1.2 provide the following topsoil specification. However, these specifications are not totally satisfactory.

212-1.1 Topsoil

212-1.1.1 General. Topsoil shall be designated as Class A (imported), Class B (selected), or Class C (unclassified) as specified herein. The engineer shall determine the suitability of topsoil prior to use. Topsoil shall be transported from the source to its position unless stockpiling is specified.

212-1.1.2 Class A Topsoil. Class A topsoil shall be from a source outside the limits of the project selected by the contractor and in compliance with the requirements specified herein. The engineer may make such inspections and perform such tests as deemed necessary to determine that the material meets the requirements.

At least 15 days before scheduled use, the proposed source of topsoil must be submitted to the engineer for approval. The contractor shall submit a written request for approval which shall be accompanied by a written report of a testing agency registered by the state for agricultural soil evaluation which states that the proposed source complies with these specifications. Class A topsoil shall have the same relative composition and structure, a friable sandy loam character, and be free of roots, clods, and stones larger than 1 inch in greatest dimension, pockets of coarse sand, noxious weeds, sticks, brush, and other litter. It shall not be infested with nematodes or other undesirable insects and plant disease organisms.

Class A topsoil shall meet the following additional requirements:

1. *Gradation Limits.* Sand, 50–80 percent, clay 20 percent maximum, and silt, 30 percent maximum. The sand, clay and silt gradation limits shall be as defined in ASTM D 422.
2. *Permeability Rate.* Not less than 0.5 inch per hour nor more than 2 inches per hour when tested in accordance with ASTM D 2434, Calif. Test 220, or other approved methods.
3. *Agricultural Suitability.* The topsoil shall be suitable to sustain the growth of the plants specified.

These specifications are not fully satisfactory for application to this project for the following reasons. The gradation limits are not sufficiently specific as "friable sandy loam character." There is a very wide range in the particle sizes of the five fractions (USDA Classification) within the sand separate as well as in the sandy loam texture class. The fractions, ranging from very coarse to very fine (2 to 0.05 mm) have significant effect on rooting and plant growth and on susceptibility to passive or active compaction on installation. The specification must be more specific in the particle size distribution of the soil, especially in the sand fractions; hence, the sieve sizes and percentage passing each sieve must be specified for each mix.

Soil structure is one of the physical soil properties that strongly influences its characteristics of porosity and degree of aeration, drainage, bulk density, and waterholding capacity. In-place, undisturbed soils will have this natural structure to a greater or lesser degree. Disturbed soil will almost always have greatly reduced structure. Sandy soils do not have inherently a high degree of structure, but the large average pore sizes tend to compensate for the loss of the large pores that exist in a well-developed natural structure. Therefore, sandy soils are nearly always specified to compensate for the loss of soil structure in the excavation, loading, hauling, stockpiling, and spreading processes involved in all soil construction operations. The addition of organic matter and inorganic porous materials such as agricultural pumice also provide additional porosity and "lightens" the soil.

The STAND SPECS as written do not ensure uniformity as well as the specifications provided herein. A "friable sandy loam" can embrace a wide range of soil properties and conditions that may not be satisfactory for use in a particular project. The term "friable" specifically refers to the physical condition of a soil at a given moisture content, usually at field capacity. Unfortunately, friability varies widely among different soils, even within sandy loams.

The specifications provided herein will meet and/or exceed those provided in the STAND SPECS and will embrace more stringent requirements. This is necessary for the planting not only to be able to survive and thrive for this timeless institution but also for the desire for diversity.

Two alternative sets of specifications are submitted to provide some degree of flexibility to Hanna/Olin and the client for use in the final design and decision process that may have some impact on costs and installation operations. Alternative I is an attempt to match the planter soil mix and the designed profile to each planter condition (vertical dimensions) and to its plant palette; therefore, it is composed of four distinct mixes. Alternative II has only two mixes, a "master mix" used in all planters except for turf planters that have a separate specification in both alternatives. This is necessary since these planters may receive pedestrian traffic use. The other areas do not have traffic use. It must be borne in mind that the second alternative as represented by the "master mix" is a compromise for most of the planters and does not necessarily meet the requirements of the individual planters as well as the mixes of Alternative I together with the designed profiles. Modifications after installation may be required for individual planters.

A major constraint on the design of the specifications lies in the limit of 85 lb/ft^3 (1.36 Mg/m^3) moist weight for the specified soils in both alternatives. This requires a relatively "light" soil that may have the disadvantage of greater potential for settlement over time, resulting in increased bulk density with its attendant decrease in air porosity (increase in anaerobic conditions) and decreased permeability. High organic matter content soils experience compaction, through settlement, which makes the organic nutrients unavailable to the plants. Thus, a balance must be attained between "lightness" of soil and resistance to settlement and compaction, as well as maintaining a suitable waterholding capacity, fertility level, and favorable physical rooting environment for the plants. More details will be given within each specified mix.

Soil Mix Specifications

Alternative I

Topsoil Mix (S1)

This is a mixture of specified sand and specified organic matter in a volume ratio of 7:3.

Sand: USDA loamy sand with a coefficient of uniformity (C_u) = 4.00 to 6.00.

Sieve No.	Percent Passing
8	100
10	90–100
18	65–70
35	35–40
60	15–20
140	5–10
300	2–5

Organic matter: Kellogg's NITROHUMUS (Kellogg Supply Co.) with the following specifications:

(a) pH of 6.5 to 7.5

(b) soluble salts less than 5000 ppm (EC < 7.8) and SAR (sodium adsorption ratio) < 7

(c) 1–2% total nitrogen

(d) 0.5–1.5% phosphorus

(e) 0.2–1.5% potassium

(f) the compost must not contain any PCBs or Dioxins, and must meet EPA Type 1 organic waste material and STAND SPECS Type 2 organic amendment.

(g) Must have 100% passing 1/2-inch or smaller screen.

The resultant uncompacted dry bulk density of this mix is approx. 1.21 Mg/m^3, or 75.5 lb/ft^3.

(a) field capacity weight is approx. 84.9 lbs/cuft. at 15% moisture content.

Subsoil Mix (S2)

This is a mixture of specified sand, organic matter and an inorganic amendment in a volume ratio of 7:2:1.

Sand: USDA loamy sand with a coefficient of uniformity (C_u) = 2.86.

Sieve No.	Percent Passing
8	100
10	95–100
18	90–95
35	80–85
60	45–50
140	10–15
300	1–5

Organic matter: Kellogg's NITROHUMUS (Kellogg Supply Co.) with the following specifications:

(a) pH of 6.5 to 7.5
(b) soluble salts less than 5000 ppm (EC $<$ 7.8) and SAR (sodium adsorption ratio) $<$ 7
(c) 1–2% total nitrogen
(d) 0.5–1.5% phosphorus
(e) 0.2–1.5% potassium
(f) the compost must not contain any PCBs or Dioxins, and must meet EPA Type 1 organic waste material and STAND SPECS Type 2 organic amendment
(g) Must have 100% passing 1/2 inch or smaller screen

Inorganic amendment: A stable, noncrushable calcined diatomaceous earth: AXIS® AXT-2 manufactured by Eagle-Picher Industries, Reno, Nevada:

(a) porosity of at least 74%
(b) pH of 7.0
(c) apparent dry density of about 34 lb/ft^2.
(d) pore size range of 0.1 to 1 micron
(e) bulk density of 0.51 g/cc
(f) cation-exchange capacity of 27 meq/100 g
(g) 2 mm in diameter

The resultant uncompacted dry bulk density of this mix is approx. 1.25 Mg/m^3, or 77.9 lb/ft^3.

(a) field capacity weight is approx. 84.1 lb/ft^3. at 10% moisture content

Base Material (*S3*)

This is a mixture of specified sand without amendment. It is used only when a lower layer may be required to raise the soil profile to finished grade. It is used only for on-grade applications, not for on-structure applications. The coarse sandy texture is to provide for a gradation from the specified subsoil to the gravel drainage layer rather than an abrupt boundary, as may be the case elsewhere. This reduces the amount of perched water above the gravel layer.

Sand: USDA sand with a coefficient of uniformity (C_u) = 1.84.

Sieve No.	Percent Passing
18	100
35	75
60	10
140	1

The resultant uncompacted dry bulk density of this mix is approx. 1.45 Mg/m^3, or 90.5 lb/ft^3. Since this material will be used only for on-grade applications, weight is not a roof loading factor.

Turf Topsoil Mix (S1A) (On-Structure)

This is a mixture of specified sand, organic matter and an inorganic amendment in a volume ratio of 6:2:2.

Sand: USDA sandy loam with a coefficient of uniformity (C_u) = 3.08.

Sieve No.	Percent Passing
10	100
18	95–100
35	70–75
60	25–30
140	5–10
300	0–5

Organic matter: Kellogg's NITROHUMUS (Kellogg Supply Co.) with the following specifications:

(a) pH of 6.5 to 7.5
(b) soluble salts less than 5000 ppm (EC $<$ 7.8) and SAR (sodium adsorption ratio) $<$ 7
(c) 1–2% total nitrogen
(d) 0.5–1.5% phosphorus
(e) 0.2–1.5% potassium
(f) the compost must not contain any PCBs or Dioxins, and must meet EPA Type 1 organic waste material and STAND SPECS Type 2 organic amendment
(g) must have 100% passing 1/2 inch or smaller screen

Inorganic amendment: A stable, noncrushable calcined diatomaceous earth: AXIS® REGULAR manufactured by Eagle-Picher Industries, Reno, Nevada:

(a) porosity of at least 74%

(b) pH of 7.0
(c) apparent dry density of about 25 lb/ft^3
(d) pore size range of 0.1 to 1 micron
(e) bulk density of 0.40 g/cc
(f) cation-exchange capacity of 27 meq/100 g
(g) diameter distribution of

6 mesh	5%
10 mesh	54%
20 mesh	87%
80 mesh	98.8%
minus 80 mesh	0.2%

The resultant uncompacted dry bulk density of this mix is approximately 1.13 Mg/m^3, or 70.5 lb/ft^3.

(a) field capacity weight is approx. 79.9 lb/ft^3 at 15% moisture content

Alternative II

Master Mix (SM)

This is a mixture of specified sand, organic matter, and an inorganic amendment in a volume ratio of 6:1:3.

Sand: USDA loamy sand with a coefficient of uniformity (C_u) = 2.50.

Sieve No.	Percent Passing
8	100
10	95–100
18	95–100
35	90–95
60	25–30
140	5–10
300	0–5

Organic matter: Kellogg's NITROHUMUS (Kellogg Supply Co.) with the following specifications:

(a) pH of 6.5 to 7.5
(b) soluble salts less than 5000 ppm (EC < 7.8) and SAR (sodium adsorption ratio) < 7
(c) 1–2% total nitrogen
(d) 0.5–1.5% phosphorus
(e) 0.2–1.5% potassium

(f) the compost must not contain any PCBs or Dioxins, and must meet EPA Type 1 organic waste material and STAND SPECS Type 2 organic amendment
(g) must have 100% passing 1/2 inch or smaller screen

Inorganic amendment: A stable, noncrushable calcined diatomaceous earth: AXIS® COARSE (AXC) manufactured by Eagle-Picher Industries, Reno, Nevada:

(a) porosity of at least 74%
(b) pH of 7.0
(c) apparent dry density of about 24 lb/ft^3.
(d) pore size range of 0.1 to 1 micron
(e) bulk density of about 0.51 g/cc
(f) cation-exchange capacity of 27 meq/100 g
(g) about 4 mm in diameter.

The resultant uncompacted dry bulk density of this mix is approx. 1.16 Mg/m^3, or 72.4 lb/ft^3.

(a) field capacity weight is approx. 81.8 lb/ft^3. at 15% moisture content

Turf Topsoil Mix (S1A) (On-Structure)

This is a mixture of specified sand, organic matter and an inorganic amendment in a volume ratio of 6:2:2.

Sand: USDA sandy loam with a coefficient of uniformity (C_u) = 3.08.

Sieve No.	Percent Passing
10	100
18	95–100
35	70–75
60	25–30
140	5–10
300	0–5

Organic matter: Kellogg's NITROHUMUS (Kellogg Supply Co.) with the following specifications:

(a) pH of 6.5 to 7.5
(b) soluble salts less than 5000 ppm (EC $<$ 7.8) and SAR (sodium adsorption ratio) $<$ 7
(c) 1–2% total nitrogen
(d) 0.5–1.5% phosphorus
(e) 0.2–1.5% potassium

(f) the compost must not contain any PCBs or Dioxins, and must meet EPA Type 1 organic waste material and STAND SPECS Type 2 organic amendment
(g) must have 100% passing 1/2 inch or smaller screen

Inorganic amendment: A stable, noncrushable calcined diatomaceous earth: AXIS® REGULAR manufactured by Eagle-Picher Industries, Reno, Nevada:

(a) porosity of at least 74%
(b) pH of 7.0
(c) apparent dry density of about 25 lb/ft^3
(d) pore size range of 0.1 to 1 micron
(e) bulk density of 0.40 g/cc
(f) cation-exchange capacity of 27 meq/100 g
(g) diameter distribution of

6 mesh	5%
10 mesh	54%
20 mesh	87%
80 mesh	98.8%
minus 80 mesh	0.2%

The resultant uncompacted dry bulk density of this mix is approximately 1.13 Mg/m^3, or 70.5 lb/ft^3

(a) field capacity weight is approximately 79.4 lb/ft^3 at 15% moisture content

Discussion

Though there may be apparent similarities among the specified sands for the various mixes in both alternatives, there are significant differences concerning the resulting characteristics when mixed with the organic and inorganic amendments. The blending of the various components and the resultant interactions must be taken into account in order to ensure that the final mix will have the desired characteristics to match the requirements of the plant palette and the planting conditions. The importance of this consideration becomes greater when constructing a soil profile.

Soil weight is an important consideration for on-structure applications. The field capacity weights provided before are for the individual layers as separate entities, not as a component of a constructed soil profile. Weight will be discussed further in the "Soil Profile Types" section. It is easier to control and predict soil weights by construction of soil profiles composed of different mixes than one homogeneous mix; hence, a distinct advantage to Alternative I.

Soil Profile Types

In the case of Alternative I, one to three layers compose the soil profiles installed on the project. Therefore, it is necessary to construct a design sequence for four potential profiles to meet conditions in the range of planter depths and plant palettes.

Type P1: Total Planter Depth of 18 Inches or Less On-Structure. This profile consists of only the topsoil mix (S1) and designed for the shallow ground cover and shrub planters or shallow portions of variable-depth planters, either on-grade or on-structure. It is to be filled to the 18-inch depth in three lifts of 6 inches each. See the "Placement and Restrictions" section for further details. It is not for the turf planters that will receive pedestrian traffic.

Type P1A: Turf Planter 18 Inches or Less On-Structure. This profile is used for pedestrian traffic areas. It is to be filled to the 18-inch depth in three lifts of 6 inches each.

Type P2: Planters Greater Than 18 Inches Deep On-Structure. The upper 18-inch layer is of the topsoil mix (S1) and the second or lower layer 18 inches thick is the subsoil mix (S2). The remainder of the planter depth is filled with cast Styrofoam. See the next section for discussion of the use and placement of the Styrofoam.

Use of Cast Styrofoam in the Profiles. To eliminate an excess weight problem due to unknown and unpredictable moisture contents of the profiles, it is necessary to lower the total profile weights for the deeper profile depths where the safety factors are potentially lessened. Soil moisture tends to accumulate in the bottom layers of the soil profile under heavy rainfall or irrigation conditions. This condition will frequently occur during the winter rainy season in California. It is difficult to predict the moisture contents and their distribution within the profile until after construction. However, some guidelines exist based on some research at the University of California, Davis (Craul, 1992, p. 336; Harris, 1991). A containerized sandy soil (which exists in this project for all practical purposes) saturated and then allowed to drain by gravity will have a 55 to 60% moisture content in the lowest 6-inch (0–6 inches) layer 35% in the next higher 6-inch (6–12 inches) layer, 30% in the 12–18-inch layer, 28% in the 18–24-inch layer, and 27% in the 30–36-inch layer. A 6-inch layer alone would have a moisture content of 58%. Under these conditions, the lower layers may have an estimated density exceeding 100 lb/ft^3, approaching 110 lb/ft^3. Obviously, these moisture contents would vary depending on the soil physical properties of each layer. For this reason, the various layers specified herein have been designed to vary in their physical properties and characteristics to offset the large amount of moisture held in the lower layers. But, this design is not sufficient from a safety factor standpoint.

Lightening of the overall soil load can be accomplished by the placement of cast Styrofoam in the layer portions of the deeper profiles and aside the tree root boxes. The placement of cast Styrofoam in the lower portion of the deeper profiles will not interfere with tree rooting since most tree roots will occur primarily in the upper 18 to 24 inches of the profile. Profile drainage is not impeded if the blocks are spaced to allow water to reach the drainage gravel. Even with the placement of the Styrofoam, a two-layer profile can be installed in most on-structure planters.

Each deep or variable depth on-structure planter will need to be examined for the potential employment and placement of the cast Styrofoam blocks. These details can be considered later in the specification development.

Styrofoam will need to be employed for either soil mix alternative.

Type P3: Planters On-Grade and Depending On Depth. The profile is composed of three layers as follows:

1. 18 inches of the topsoil mix (S1)
2. 18 inches of the subsoil mix (S2)
3. the remaining profile depth filled with the base material (S3)

Installation follows the same procedure as described in the "Placement and Restrictions" section.

Discussion of Alternative I versus Alternative II

Alternative I. The design of soil profiles deeper than 18 inches (46 cm) composed of layers more nearly emulates the natural soil profile. The natural profile usually exhibits vertical gradients in those properties and characteristics that influence liquid and gas movement, root penetration, and especially plant growth through the supply of water and nutrients. The most favorable environment is in the upper portion of the soil generally called the topsoil. It is characterized by a high organic matter content and thus high fertility, relatively high oxygen, and low carbon dioxide concentrations, a large and diverse microorganism population and loose soil that is easy to work. With increase in depth, these favorable conditions are decreased. Organic matter decreases, and if it does exist at depth or is placed there, it decomposes very slowly and may decrease the oxygen supply through anaerobic decomposition. At a depth of about 36 to 48 inches (91 to 122 cm), very few roots occur even in forest soils. But this depth does contribute to the drainage of the entire profile. To overcome the disadvantages of a homogeneous profile, it is necessary to reduce the proportion of organic matter and increase the structural matrix of the mix to prevent the settlement and its accompanying decrease in porosity. This process also reduces the inherent fertility of the mix and may also reduce its waterholding capacity, increasing the irrigation requirements. For these reasons, a layered profile, with its decreasing organic matter content but contrasting physical properties with those of the topsoil, provides a better plant growing medium than one that is uniform or isotropically homogeneous.

Alternative II. A specification for a homogeneous soil with high organic matter content for placement below 24 inches (61 cm) or more may create more unfavorable rather than favorable rooting conditions. The favorable characteristics created in a homogeneous mix by specifying a greater organic matter content may be wasted in the lower portions of a deep profile. This in turn increases the potential for settlement over time after installation, leading to compaction and decreased favorable aeration and water drainage in the profile. In order to overcome the disadvantages of a homogeneous profile, it is necessary to reduce the proportion of organic matter and increase the structural matrix of the mix to prevent the settlement and its accompanying decrease in porosity. This process also reduces the inherent fertility of the mix and may also reduce its waterholding capacity, increasing the irrigation requirements. These factors were considered in the development of the master mix (SM). If a homogeneous mix is used, the organic matter of the surface can be increased over time by postinstallation maintenance where organic matter is incorporated into the soil surface and gradually becomes intermixed with the soil by organism and root activity.

Soil Amendments

Amendments to the soil consist of liming or acidifying materials for the adjustment of pH, fertilizers, mulches, and top dressings. The last two items are usually covered in the planting specifications as well as in the specific fertilization for the plant palettes for each planter. The pH (soil reaction) adjustment and fertilization discussed here is for the purpose of preparing the chemical properties of the soil mixes for general plant growth.

Soil reaction adjustment materials:

(a) ground agricultural limestone if pH must be raised
(b) elemental sulfur, 96%, to lower pH

Commercial fertilizers:

(a) 15–10–10 complete fertilizer with micronutrients, including iron, boron, copper, and manganese

Other materials may become necessary subsequent to sampling and analysis of the mixes prior to installation.

Soluble salt content (EC) of amendments must be carefully evaluated to prevent a significant increase in the resultant soluble salt content of the final soil material. Final EC should not exceed 3.0 dS/m.

Placement Procedures and Restrictions

Soil will be placed by 6-inch (15 cm) lifts within each layer and compacted to 85% Proctor (Holtz and Kovacs, 1981) or not to exceed 1.3 to 1.4 Mg/m^3 bulk density. The compaction is best accomplished by hand tamping or lightweight rolling. Vibration tamping is not recommended. The density is to be verified by cone penetrometer readings (187 to 250 psi or 1.29×10^6 to 1.73×10^6 Pa) for each lift. However, some loosening of each lift surface will be required to prevent the formation of an interface or discontinuity between the layers, again to emulate the natural soil profile. Even the shallow profiles should be placed in lifts to provide for uniformity of packing in the entire volume. In the case of the Master Mix, this also should be placed in 6-inch (15 cm) lifts to provide for uniformity of compaction.

Soil mixes shall not be handled, hauled, or placed during rain or wet weather or when wet near or above field capacity. This restriction shall also apply to the plant installation operation as planting in wet weather can also have detrimental effects on the soil.

Careful sequencing of installation of the soil is just as important as the development of the mix. The lower soil layers are placed until the grade of plant rootball bottom is reached, then the plants placed, and backfilled with the appropriate mix for the sequence of layers up to the surface. It is not recommended that the entire profile be placed and then excavated for plant placement; this causes the mixing of the layers upon backfilling. This consideration is less important for Alternate II and shallow rootballs.

Stockpiling. The stockpiling on-site should be restricted to no more than the needs of several days. Stockpiles of mixes S1, S1A, and S2 should be no higher than 6 feet to prevent anaerobic conditions within the piles. The stockpiles should be sheltered from the elements to prevent excessive water absorption and blowing by high winds. These restrictions should apply to off-site or source stockpiling as well as on-site stockpiling.

Testing and Inspection

The soil testing and analysis process falls into three major phases:

Testing of the materials for specification conformance prior to placement.

Physical properties:

(a) particle size distribution according to ASTM D422-63
(b) dry density and saturated density according to Soil and Plant Laboratory, Inc., Tests A06, or *Methods of Soil Analysis*, Agronomy Monograph No. 9, Part 1, American Society of Agronomy
(c) permeability according to ASTM D2434
(d) moisture retention curve to determine water availability (Hillel, 1980). Necessary for irrigation evaluation.

Chemical properties:

(a) pH
(b) fertilizer element content, N, P, and K
(c) calcium, magnesium, sodium and chlorides
(d) organic matter content
(e) micronutrient content such as boron, manganese, and iron
(f) soluble salts

Tests A01, A06, A07, A10, A11, and A20-00-N2,3 of the Soil and Plant Laboratory would be the most appropriate array of analyses for the preceding testing. Duplicate samples of each mix must be tested. Those mixes that do not meet the requirements of a well-drained soil with moderate moisture retention and dry density less than 85 lb/ft^3 (1.36 Mg/m^3), appropriate pH range of approximately 6.5 to 7.5, and soluble salts less than 200 ppm should be rejected.

Sampling and testing during delivery and installation. Duplicate samples from each batch and/or truckload should be obtained for conformance testing. Major emphasis should be on

(a) particle size distribution or texture
(b) organic matter content and its dispersion in the mix
(c) inorganic amendment content and its dispersion in the mix
(d) soil pH
(e) soluble salts

Post-testing after installation for soil performance. In-place testing involves density tests during installation and after placement, permeability or drainage of the entire profile, and general conditions. This testing will be more of inspection and observation by on-site personnel than a large amount of laboratory testing.

In-place testing of infiltration and drainage can be accomplished in conjunction with testing the irrigation system. Probing of the soil across the planter to refusal depth and examining moisture content should suffice. One or two deeper holes should be augered to the soil profile bottom to check for saturated or perched condition above the gravel layer after heavy irrigation or precipitation.

Postinstallation Maintenance

Routine postinstallation maintenance must include some constraints. No maintenance work on the planter soils should be done during or immediately following rain or irrigation. Puddling and clodding of the soil results and amelioration efforts are difficult and drawn out.

No maintenance vehicles, except for light wheelbarrows and other lightweight manually propelled equipment, shall be permitted on the planter soils. Too many instances arise when wheeled vehicles are driven over pavements or walks and the wheels override the edges onto planter soils. This does harm to the soil and plants and causes unnecessary maintenance effort. Equipment used on the turf planters should be restricted to lightweight wheeled vehicles as approved by the structural engineer, if any, even though the soil is designed for pedestrian use.

ALLEGHENY RIVERFRONT PARK, PITTSBURGH

The Allegheny Riverfront Park is a linear public park along the Allegheny River designed by Van Valkenburgh Associates, Cambridge, Massachusetts. The lower portion is subject to annual flooding and potential ice erosion. Flooding during the winter of 1997–1998 deposited a layer of silt on the site, which appears to reduce the infiltration of water. The problem is yet to be solved at the time of publication. The use of a biodegradable blanket and tackified mulch on a steep slope is one of the features. Well-draining soil specifications are required due to the frequent flooding. Boulders mixed with river birch as a floodplain approach is part of the design.

General

Scope of Work

A. This section specifies administrative and procedural requirements for planting soil including but not limited to the following:
   1. Planting soil material acquisition
   2. Testing and analysis for specification conformance
   3. Preparation of mixes and testing for conformance
   4. Installation and placement of soils
   5. Final in-place testing

B. References to other sections are given that would duplicate provisions in this section.

Submittals

A. Certificates: Provide certificates required by authorities having juristiction, espcially for any composted materials containing sewage sludge. Approval as EPA Type 1 is required.
B. Test Reports: Submit certified reports for tests required. Particle size analysis reports will use the USDA sieve cluster which is #4, #10, #35, #60, #140, and #300. Percent clay at 0.002 mm shall be reported separately in addition to silt.
C. Samples: Prior to ordering the following listed materials, submit representative samples to the landscape architect for selection and approval. Do not order materials until the landscape architect's approval has been obtained. Delivered materials shall closely match the approved samples.
   1. Organic amendment: duplicate samples of 1 quart
   2. Loamy sand (USDA): duplicate samples of 1 pint
   3. Sandy loam (USDA): duplicate samples of 1 pint
   4. Fine concrete aggregate (ASTM): duplicate samples of 1 pint

Delivery, Storage and Handling

A. Refer to Section 02900, 1.07, page 3, and 3.11A1, page 12, for overall material handling requirements.
B. In addition, the following provision is established: Material should not be handled or hauled, placed or compacted when it is wet as when after a heavy rainfall. Soil should be handled only when the moisture content is less than field capacity.

Products

The Soil Layers. The planting soils are comprised of several layers (*horizons* in pedological terms) that when combined form soil profiles. The profiles are described later.

The Drainage Layer (*S3*). The drainage layer is a gravel aggregate (ASTM fine concrete aggregate as recommended by Interflue, Inc.) that provides a high rate of water flow from the bottom of the soil profile to the underdrainage system. This feature is extremely essential for the survival and maintenance of the trees of the Lower Park that may be subjected to frequent flooding. The aggregate size envelope is given in Table 12.5.

TABLE 12.5 Range in Percent Passing for the Drainage Layer (S3)

Sieve Size	Percent Finer
3/8 inch	100
No. 4	95–100
No. 8	80–100
No. 16	50–85
No. 30	25–60
No. 50	10–30
No. 100	2–10

The Lower Park Subsoil (*S2A*). The subsoil is designed so that a sufficient amount of moisture retention provides trees with water from one precipitation event to another, resists self-compaction, and yet has adequate drainage to remove excess water during the dormant season or to remove surface ponded water and excess water retained within the profile during river flooding events. This layer is a coarse loamy sand (USDA classification). The particle size envelope is given in Table 12.6 and illustrated in Fig. 12.6.

The Upper Park Subsoil (*S2B*). This soil has a finer texture than the Lower Park subsoil (Table 12.7 and Fig. 12.7). This change in texture is necessary to provide greater moisture retention and a relative increase in potential fertility. The Upper Park is not susceptible to flooding as the Lower Park making safe a finer texture. Underdrainage is included at all locations to ensure uniform drainage across the planting areas. Geotechnical exploration data have indicated a high degree of variability in material over the area. It is not necessary to depend on the drainage characteristics of the underlying material if underdrainage is provided.

TABLE 12.6 Range in Percent Passing Sieve Sizes for Lower Park Subsoil

Sieve Size	Percent Finer
No. 10	100
No. 18	88–100
No. 35	70–80
No. 60	25–35
No. 140	13–23
No. 300	10–20
Silt	8–18
Clay	2–12

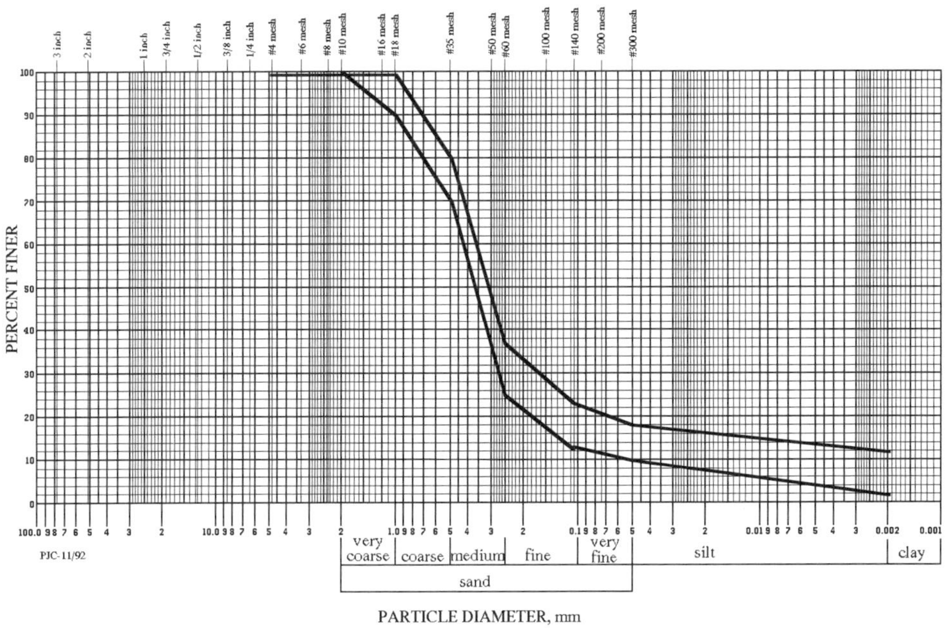

Fig. 12.6 The Lower Park Subsoil particle size distribution.

The Topsoil (*S1*). The topsoil for either the Lower or Upper Park is the organic matter amended subsoil. The organic amendment recommended to be composted is organic matter that meets or exceeds EPA Type 1 composted material as well as Commonwealth of Pennsylvania specifications and regulations. This includes material such as "Allgro" brewer's waste or composted sewage sludge without the woodchips or "Earthlife" from Philadelphia. The Washington/Maryland composted sewege sludge (especially the Blue Plains

TABLE 12.7 Range in Percent Passing Sieve Sizes for Upper Park Subsoil

Sieve Size	Percent Finer
No. 10	100
No. 18	88–100
No. 35	70–80
No. 60	40–50
No. 140	29–39
No. 300	25–35
Silt	10–30
Clay	5–15

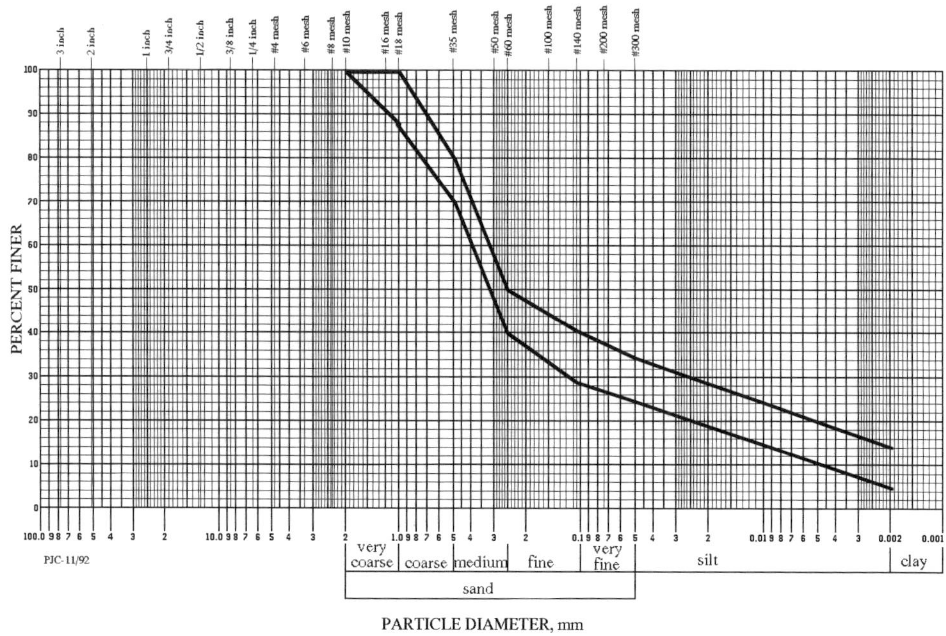

Fig. 12.7 The Upper Park Subsoil (S2B) particle size distribution.

source) is another organic matter alternative. It should be thoroughly cured material after composted before delivery to the site. All forms of composted material requires prior approval of the landscape architect.

The Soil Profiles. The soil profiles are made up of different combinations of soil layers to provide appropriate physical and chemical characteristics for satisfactory plant root growth and extension for the different planting applications encountered in the project. The features of each are described under each heading.

The Riverwalk Bioreinforced Slope Profile (*P1*). This profile is applied to only the soil slope in the Lower Park, or Planting Area A2 (Fig. 12.8). It is comprised of a topsoil 6 inches thick, which is simply the loamy sand subsoil amended with 4% (w/w or 20% volume) organic matter. A biodegradable blanket is placed 3 inches (8 cm) below the surface or at the midpoint of the topsoil. The subsoil is the loamy sand subsoil S2A as specified earlier. It is 24 inches (61 cm) thick. The drainage layer, S3, is 6 inches (15 cm) thick. Filter fabric is wrapped around the pipe itself or a 12 in. × 12 in. (30 × 30 cm) gravel encasement about the pipe.

The Riverwalk Profile without Boulders (*P2A*) (*River Birch*). This profile is applied to Planting Areas B1 and C2, which are planted to River birch without

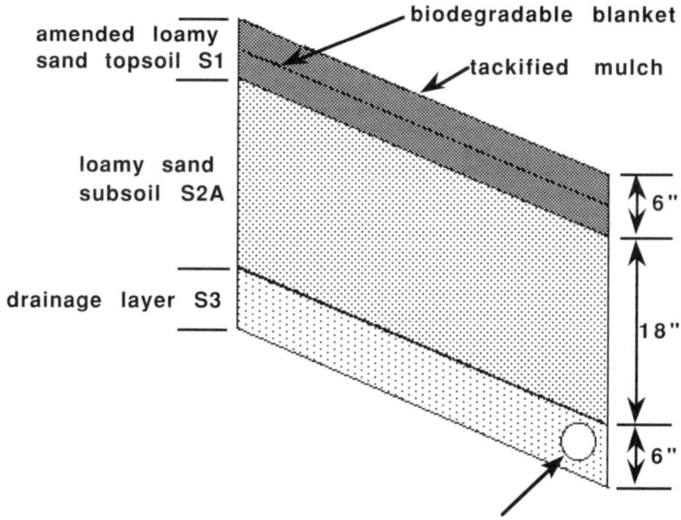

Fig. 12.8 The riverwalk bioreinforced slope profile (P1).

boulders on the soil surface. This profile (Fig. 12.9) is essentially the same as Profile P1 but with a level surface and without the biodegradable blanket and the tackified mulch.

The Riverwalk Profile with Boulders (P2B) (River Birch). This profile is applied only to the planting areas where surface boulders are to be placed with the River birch (Fig. 12.10). These are Planting Areas B2 and C1.

The Sidewalk Profile (P3). This profile is applied to both the Lower and Upper Parks for plantings in pits below the sidewalk (Fig. 12.11). In the

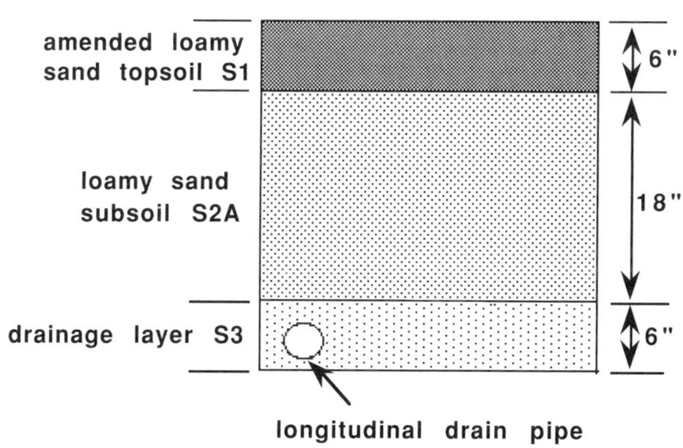

Fig. 12.9 The riverwalk profile without boulders (P2A).

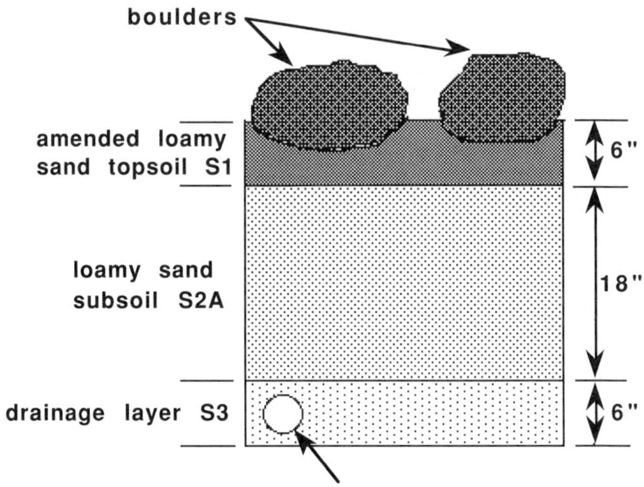

Fig. 12.10 The riverwalk profile with boulders (P2B).

Lower Park, these plantings are located at an elevation generally higher than the annual flood level. Thus, they have the same specifications as the Upper Park sidewalk profile. This profile applies to Planting Areas B4, (B6?), B8–12, C4, (C6?), and D6-9.

The Sloping Lawn Profile (*P4*). This profile is applied to Planting Areas B5 and C5, which are the sloping lawns located between Seventh and Ninth streets, and Sixth and Seventh streets, respectively (Fig. 12.12). Since these areas are to receive heavy leisure use, a thicker (12-inch) (30 cm) topsoil is

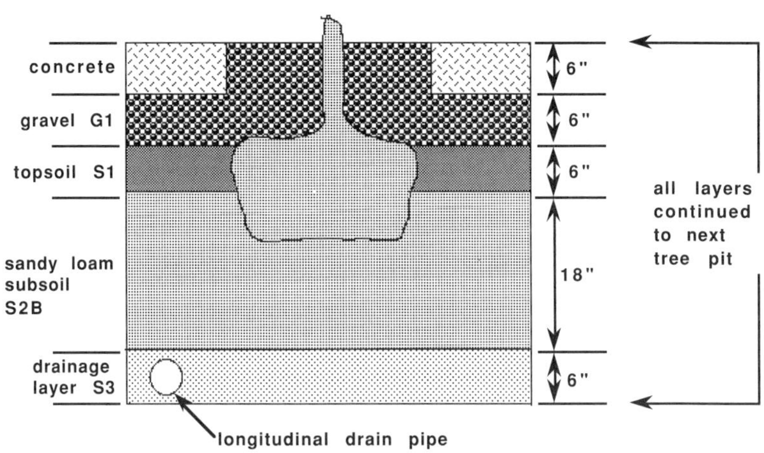

Fig. 12.11 The sidewalk soil profile (P3).

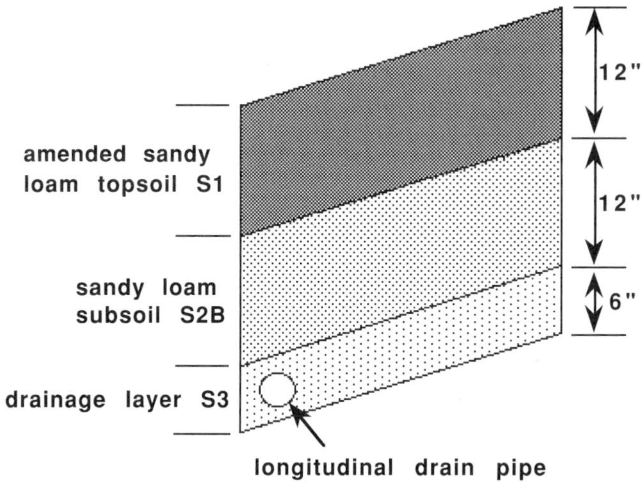

Fig. 12.12 The sloping lawn profile (P4).

specified. The subsoil is reduced accordingly. The thick topsoil provides more waterholding capacity and a greater nutrient supply to aid the concept of minimum maintenance. Underdrainage is also provided as elsewhere.

Median Profile (*P5*) (*London Plane Tree*). The median soils in Planting Areas A4, B7, C7, D4, and D5, will probably have little internal pedestrian traffic (Fig. 12.13). The soil is to be covered by grass. This profile is applied primarily to Planting Area A3 and possibly B6, C6, and D3, as well as the Ft. Duquesne Blvd. median plantings of London plane trees.

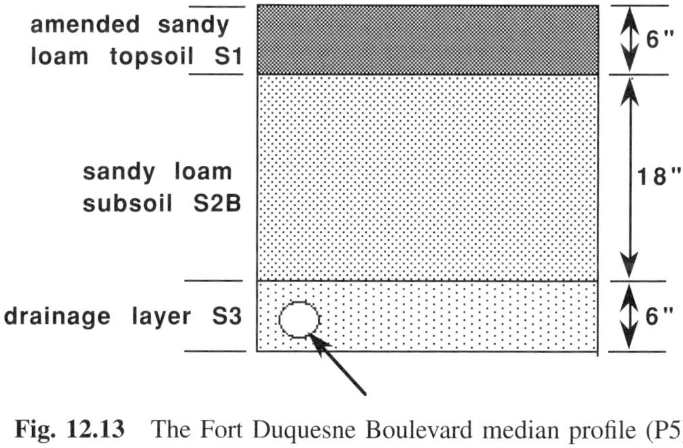

Fig. 12.13 The Fort Duquesne Boulevard median profile (P5).

Execution

Coordination. Coordinate activities with other project contractors so that there is no or only minimum amount of light soil disturbance from traffic or other construction activities subsequent to placement.

NEW SCHOOL FOR SOCIAL RESEARCH COURTYARD, NEW YORK CITY

This is a diffuse-lighted open courtyard requiring sheltered subsurface irrigation of sharply sloping soil with highly variable depth designed by Van Valkenburgh Associates, Cambridge, Massachusetts. The irrigation system must be located below the soil surface and be a low-volume system to prevent soil erosion onto to the terrace surface.

The Soil Mix

This is a mixture of specified sand, organic matter, and an inorganic amendment at porportions of 75%, 10%, and 15%, respectively. (On a volume basis, the ratio is approximately 1.68:1:1.78.)

Sand: USDA medium loamy sand with an approximate coefficient of uniformity (C_u) = 2.4 to 3.0 (Table 12.8).

Organic matter: Composted brewer's waste, (ALLGRO), composte sewage sludge, or other composted organic waste that meets the following requirements:

(a) pH of 6.5 to 7.5
(b) soluble salts less than 600 ppm
(c) 1–2% total nitrogen
(d) 0.5–1.5% phosphorus

TABLE 12.8 Range in Percent Passing Sieve Sizes for Sand

Separate Class	Sieve No.	Range in Percent Passing
Gravel	10	100
Very coarse sand	18	94–99
Coarse sand	35	90–99
Medium sand	60	25–35
Fine sand	140	1–10
Very fine sand	300	1–5
Silt	Hydrometer	1
Clay	Hydrometer	2–5

(e) 0.2–1.5% potassium

(f) the compost must not contain any PCBs or Dioxins, and must meet EPA Type 1 organic waste material

(g) must have 100% passing 1/2 inch (10 mm) or smaller screen

Inorganic amendment: A stable, noncrushable calcined diatomaceous earth, such as ISOLITE® CG-2 or an equivalent material (TURFACE®, AXIS®):

(a) porosity of at least 74%

(b) pH of 7.0

(c) apparent dry density of about 34 lb/ft^3

(d) pore size range of 0.1 to 1 micron

(e) bulk density of 0.60 g/cc

(f) cation-exchange capacity of 27 meq/100 g

(g) 2 mm in diameter

The resultant uncompacted dry bulk density of this mix is approximately 0.88 Mg/m^3, or 54.9 lb/ft^3. Saturated weight is approximately 96.6 lb/ft^3.

The organic matter content at a volume of approximately 44.6% is somewhat high and may cause the soil to settle excessively.

Placement of Soil Material

The soil should be placed in lifts not to exceed 6 inches in thickness and only moderately hand-tamped not to exceed 90% Proctor (ASTM 180) with the following exception:

> In those cases where the final soil profile depth exceeds the height of the plant rootball, the entire soil depth beneath the rootball should be compacted to 100% Proctor to create a pedestal to prevent settlement of the rootball (Fig. 12.14). If stock is to be planted bare-rooted, a compacted pedestal is not necessary. Ordinary compaction of the soil around the bare roots remains appropriate and necessary.

Placement of soil and planting of stock should be carried out simultaneously to prevent excessive traffic over soil lifts and the final grade and create undesirable soil compaction. This can be accomplished by beginning the work in corners or against walls, or the center of isolated beds, and progressing outward toward the borders.

Testing

The testing program is comprised of three phases:

Component testing: The sand is tested for proper gradation by ASTM D-422-63 (1972), particle size analysis.

The organic matter should be tested for

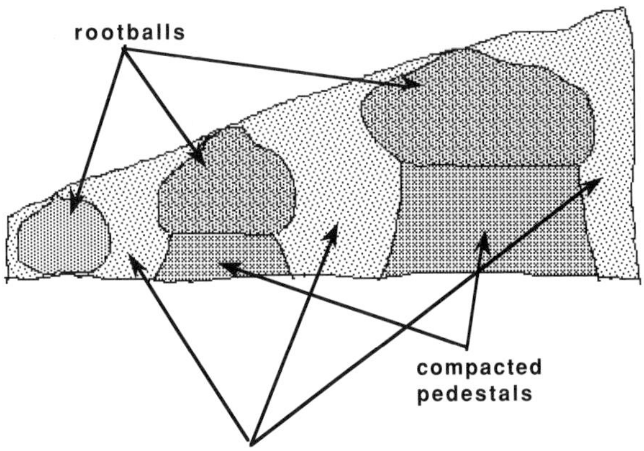

Fig. 12.14 Compacted rootball pedestals.

(a) pH
(b) soluble salts
(c) nitrogen, phosphorus, potassium, calcium, magnesium, and aluminum
(d) certification that it meets EPA Type 1 standards if composted sewage sludge

The inorganic amendment should be approved by the landscape architect on the basis of submitted samples and accompanying manufacturer's specifications.

Mix testing: Samples of the specified mix should be submitted for compliance prior to installation. In addition, separate mix samples should be tested for compaction density and infiltration. Density should not exceed 90% Proctor and infiltration should be more than 1 inch per hour nor exceed 4 inches per hour (American Society of Agronomy, *Methods of Soil Analysis*, Part 1, 1986 or ASTM D-2434-68 [1974]).

Postinstallation testing:

In-place density (e.g., AASHTO 180) and drainage (ASTM D-3385-75 or satisfactory substitute approved by the landscape contractor) should be tested at selected points within each planting bed.

Results of all tests and analyses should be submitted to the landscape architect for approval.

APPENDIX

STANDARD TESTS FOR ANALYSES OF DESIGNED SOIL

A.1 ORGANIC COMPONENT

A.1.1 Stability

Dewar Self-Heating Test. Based on the measurement of the maximum temperature rise observed in moist compost under controlled conditions (Dewar flask). State of Illinois adopted standards are maximum rise less than 10°C is classified as *very stable,* and maximum rise between 10° and 20°C is classified as *stable.* Data from the Woods End Institute show the classes in Table A.1.

Respiration Test. The measurement of the evolution of carbon dioxide from microbial decomposition of the compost under closed, controlled conditions. It is measured in micrograms of CO_2 per unit of dry compost weight, or sometimes reported on a unit volume basis.

A.1.2 Maturity

Growth Screening Test. Compost is placed in a replicated series of pots and planted to wheat in one series and cress in another. Other plant seeds are sometimes used, but wheat and cress are considered standards. Failure for seed to germinate or evidence of inhibited growth of seedlings is a positive test for phytotoxicity and rating as *immature.* Complete growth without deficiencies indicates the compost is *mature.*

A.1.3 Pathogens

As specified in Part 503, EPA regulations.

TABLE A.1 Compost Stability Classes

Temperature Rise	Stability Class	Stability Description
0–10°C	V	Very mature compost
10–20°C	IV	Maturing, curing compost
20–30°C	III	Material still decomposing
30–40°C	II	Immature, active compost
40–50°C	I	Fresh, very raw compost

A.1.4 Organic Content

Weight loss on ignition. ASTM or American Society of Agronomy, *Methods of Soil Analysis,* Part 2, 1986.

A.1.5 Nutrient Content

Analyses may be those used by the University of Massachusetts Soil Testing Laboratory, or an equivalent such as the Association of Official Agricultural Chemists.

A.1.6 Reaction (pH)

A 1:1 soil–distilled water suspension using a glass electrode pH meter. American Society of Agronomy, *Methods of Soil Analysis,* Part 2, 1986 or ASTM D-4972 with water.

A.1.7 Soluble Salts

Electrical conductivity measured with a glass electrode calibrated in dS/m and may be converted to ppm. American Society of Agronomy, *Methods of Soil Analysis,* Part 2, 1986. The readings are classified in Table A.2 (U.S.

TABLE A.2 Soluble Salt Ratings for Plant Tolerance

Reading		
dS/m	ppm	Interpretation
0–2	0–1280	Low, few plants affected
2–4	1280–2560	Very slightly saline, sensitive plants restricted
4–8	2560–5120	Moderately saline, many plants will be restricted
8–16	5120–10,240	Strongly saline, only tolerant plants grow
> 16	> 10,240	Very strongly saline, very few plants grow

Source: U.S. Soil Salinity Laboratory, 1954.

Salinity Laboratory, 1954). The effects of soluble salts are also detected in the maturity test, Section B.1.2.

A.2 MINERAL COMPONENT

A.2.1 Particle Size Analysis

Dry sieving for stones, gravels (ASTM C-136), and sands (2 to 0.05 mm) and hydrometer for silt and clay (< 0.002 mm) ASTM D-422-63.

A.2.2 Organic Matter Content

Weight loss on ignition. ASTM or American Society of Agronomy, *Methods of Soil Analysis,* Part 2, 1986.

A.2.3 Reaction (pH)

A 1:1 soil–distilled water suspension using a glass electrode pH meter. American Society of Agronomy, *Methods of Soil Analysis,* Part 2, 1986 or ASTM D-4972 with water.

A.2.4 Nutrient Content

Analyses may be those used by the University of Massachusetts Soil Testing Laboratory, or an equivalent such as the Association of Official Agricultural Chemists.

A.2.5 Soluble Salt Content

Electrical conductivity measured with a glass electrode calibrated in dS/m and may be converted to ppm. American Society of Agronomy, *Methods of Soil Analysis,* Part 2, 1986. The readings are classified as in Table A.2 (U.S. Salinity Laboratory, 1954).

A.3 TESTS ON MIXES

A.3.1 Bulk Density

American Society of Agronomy, *Methods of Soil Analysis,* Part 1, 1986. Density should be less than 1.65 Mg/m^3 (or g/cc), with 1.40 to 1.50 Mg/m^3 optimum.

A.3.2 Infiltration

ASTM D-3385-75. The observed infiltration rate should be between 1 and 3 inches per hour. Other values are unacceptable.

A.3.3 Organic Matter Content

Weight loss on ignition. ASTM or American Society of Agronomy, *Methods of Soil Analysis,* Part 2, 1986.

A.3.4 Particle Size Analysis

ASTM D-422-63, dry sieving for a sands (2 to 0.05 mm) and hydrometer for silt and clay ($<$ 0.002 mm).

A.4 POSTINSTALLATION TESTS

A.4.1 Infiltration

ASTM D-3385-75. The observed infiltration rate should be between 1 and 3 inches per hour. Other values are unacceptable.

A.4.2 In-Situ Density

ASTM D-2937-83 or AASHTO 180. Density should be less than 1.65 Mg/m^3 (or g/cc), with 1.40 to 1.50 Mg/m^3 optimum.

A.4.3 Thickness or Depth

Thickness of horiozns should be checked as installed. The thickness should be within the specified value by ± 5% for horizons 6 inches thick or less. For horizons 12 to 24 inches, the tolerance should be ± 10%. For horizons greater than 24 inches, the tolerance may be ± 15%.

A.4.4 Percolation or Permeability

Methods of Soil Analysis, Part 1. Hydraulic conductivity of saturated soils—field methods. Section 29-3.2, pp. 758–763. American Society of Agronomy, 1986 or ASTM D-2434.

GLOSSARY

Definitions of soil science terms are taken from the list in *Glossary of Soil Science Terms* published by the Soil Science Society of America, 1987, with some modification for clarification by the author. Those relating to soil classification are taken from *Soil Taxonomy,* published by the U.S. Dept. of Agriculture, 1976. Other definitions are by the author or taken from appropriate references given in the text.

***Actinomycetes* spp.** A group of soil-inhabiting microorganisms intermediate between the bacteria and the true fungi that usually produce a characteristic branched mycelium. They are the organisms that give the newly plowed soil a unique but pleasant odor.

A horizon. *See* Topsoil.

Albedo of the surface below the object. The fraction of incident shortwave radiation reflected from a surface. The surface below the object usually refers to the ground.

AXIS®. A commercial product derived from diamotaceous earth (fossil diatoms). Porous and light in weight. Used as an amendment to lighten soil weight and provide structure that is resistant to compaction.

B horizon. A soil horizon in which the dominant features include illuvial concentration of silicate clay, iron, aluminum, or humus, alone, or in combination with other compounds, coatings of sesquioxides, and usually has blocky, subangular blocky, prismatic, columnar or platy structure.

Brewer's waste. A commercial product of composted waste from beer making. It is the grain residue left after the malting process. Somewhat similar to other composted organic products and compares favorably on pH and nutrient contents. Generally more fertile than peat moss. Has a faint brewery odor when first spread, but quickly dissipates. Very uniform with good consistency in quality.

Bulk density. The weight of a unit volume of oven-dry soil material expressed in units of megagrams per cubic meter (Mg/m^3). A bulk density of 1.33 Mg/m^3 is considered nearly ideal. A value in excess of about 1.60 Mg/m^3 for most soil materials inhibits plant root penetration.

Carbon:nitrogen ratio. The ratio of total carbon to total nitrogen in a carbonaceous material. In the soil, the humus of the A horizon stabilizes at about 10:1 to 12:1. In forested soils, the ratio may be 20:1 and relatively undecomposed organic matter may be much higher.

C horizon. A mineral horizon or layer, exluding bedrock, that is either like or unlike the material from which the solum is derived. It is little affected by pedogenic processes.

Clay. A primary particle of aluminosilicate composition and crystalline structure in soil being less than 0.002 mm in diameter and having properties of surface charges for nutrient, water, and gas adsorption, and potential for swelling and shrinking. Contributes to the fertility status of soil materials. Colloidal clay particles (exhibit Brownian movement) are smaller than 0.001 mm in diameter.

Coefficient of gradation (curvature). Given by the formula $C_c = (D_{30})^2/(D_{60} \times D_{10})$, where D_{60} is the particle diameter at which 60% of the particles are smaller. D_{10} is the particle diameter at which 10% of the particles are smaller, and D_{30} is the particle diameter at which 30% of the particles are smaller.

Coefficient of permeability. Expresses the rate of water flow through the soil based on soil physical parameters. In soil physics literature, it is referred to as the hydraulic conductivity and is the term k in Darcy's law of saturated flow where $q = k(d\Psi/dx)$.

Coefficient of uniformity. Given by the formula $C_u = D_{60}/D_{10}$.

Complete fertilizer. A fertilizer material that contains sources of all three major fertilizer nutrients of nitrogen, phosphorus, and potassium. The nutrients may be present in various ratios and grades.

Composted sewage sludge (biosolids). Tertiary sewage sludge that has been composted with wood chips (Beltsville method) and aerated, or treated in aerated vats for 21 days, and then allowed to cure to stages of maturity and stability. The material is odor-free and nontoxic if meeting EPA Grade A requirements. Useful as an organic matter component in a soil mix.

Conduction. The flow of heat within a material by molecular kinetic energy. Flow occurs from a region of high (hot) temperature to a region of low (cool) temperature.

Convection. The exchange of heat by a moving fluid, usually water or air. The heat is first transferred to the fluid by conduction, and then is removed as the fluid moves away from the surface.

Drainage coefficient. The excess water that must be removed from a site or land area by a drainage system to prevent crop or structure damage. The total amount of water considered for determination of the coefficient includes water applied by irrigation.

EPA Grade A composted biosolids (exceptional quality). Composted biosolids that meet the minimum standards for limits on heavy metal concentration, odor, vector attraction, and so on, under 40 CFR, Part 503 rule (EPA CFR, Part 503 Regulations, Table 3, p. 9392, Vol. 58, No. 32, Friday, February 19, 1993).

Friable. The consistency of the soil when it is moist, not wet, and workable in the hand. Does not become hard when dry and can be broken or crumbled easily. When wet, it is not plastic nor stricky and does not flow.

Geographic information system(s). The input, storage, analysis, and output of geographic data in digital form within a computer system. The data are rectified to specific geographic points, making it useful for spatial analysis and interpretation.

Horizon. A genetic layer of soil material within a soil profile differentiated on the basis of physical, chemical, and biological properties such as color, texture, structure type, consistence, pH, roots, and organisms. Usually designated by capital letters (the major horizons) as A (topsoil), B (subsoil), and C (parent material). Prefixes and suffixes are used to designate subdivisions.

Humus. The highly decomposed amorphous residue remaining after nearly complete decomposition of organic matter deposited on and incorporated into the surface of the soil. It is the organic matter primarily of the A horizon but may be contained elsewhere in the profile at lower concentrations .

ISOLITE®. A commercial ceramic material produced from the crushing of diatomaceous earth, firing at high temperature and recrushing to prescribed sizes. Has internal porosity and light density. Used as a soil amendment to increase waterholding capacity and decrease the bulk density of the mix.

Lift. A layer of placed unconsolidated material in filling operations. Its thickness is usually prescribed to enhance treatment such as compaction with rollers. Thus, a fill is built up in a series of lifts.

Macropores. The pores created by the voids between structural peds, usually called as interpedal pores. The water contained in these pores is drained by gravitational force and are air-filled at field capacity. Their minimum diameter is about 0.03 mm.

Maturity. The stage of decomposition of composted organic matter that does not inhibit plant growth.

Mean Radiant Temperature. The uniform temperature of black surroundings that will give the same radiant heat loss from an object (a plant–the tree).

Micropores. These are very small pores that exist primarily within structural peds or at points of contact between peds. They retain water against gravitational force and contain plant-available water to wilting point and beyond. Their maximum diameter is about 0.03 mm.

Mollisols. The soil order represented by those soils with a highly developed topsoil, the mollic epipedon with a base saturation of greater than 50%, usually formed in nature under native grassland vegetation. It is quite fertile in most cases.

Municipal yard waste. A mixture of plant waste including leaves, needles, twigs, bark, branches, limbs, wood, garden and house plant debris, and sometimes various types of garbage, processed by shredding or chipping, then allowed to compost for a period. Used as a soil amendment or mulch. Greatly varies in quality depending on the available feedstock.

Mycorrhiza. The symbiotic association of fungi with the roots of seed plants. They invade the roots and increase the capacity of the roots to absorb nutrients and moisture, whereas the mycorrhiza derive growth substances directly from the plants.

Organic horizon. The horizon(s) found as the upper layers on relatively undisturbed mineral soils and consist primarily of organic debris in varying stages of decomposition.

Oi. The litter layer composed of undecomposed remains: leaves, twigs, bark, etc.

Oe. Partially physically disintegrated and chemically decomposed organic debris.

Oa. Wholly decomposed dark brown to black organic material, homogeneous and comprised of lignin, celluloses, and so on. Sometimes called the humus layer.

Organic matter. The general term applied to the organic component of a soil material. May be of natural humus, leaf mold, peat moss, sawdust, wood chips, brewer's waste, or any other organically derived product.

Orthophoto. An aerial photograph that has been processed by computer to eliminate parallax errors and is geographically rectified to a known base. It shows true shape and direction.

Parent material. The unconsolidated, partially weathered geological material from which the topsoil and subsoil have been derived. It is usually designated the C horizon in pedological nomenclature. In transported soils, the C horizon may not be the parent material for the A and B horizons. It is then designated as IIC.

Particle size distribution. The distribution of the primary soil particles of gravel, sand, silt, and clay of a soil material among the array of diameters ranging from several inches to less than 0.002 mm. Usually described by the percentage finer passing an array of sieve sizes, or represented by a curve drawn on five-cycle semi-log graph paper. The shape of the curve indicates its relative degree of resistance or susceptibility to compaction. Laboratory analysis is by the ASTM D-422-63 standard.

Well-graded. A soil with a broad representation in particle size classes, from gravels to clay. A nonuniform soil.

Poorly graded. A soil with a narrow range of particle size classes, sometimes just several are present. A uniform soil.

Gap-graded. A soil that has one or more classes missing within the common range of particle size classes.

Peat moss. Partially decomposed organic remains of sphagnum moss obtained from bog deposits. Lightens the soil mix and greatly increases its waterholding capacity. Must be shreaded and well mixed with other components to be totally effective. Does decompose over time.

Ped. An individual aggregate formed by natural processes of sand, silt, clay, and organic matter having some describable geometric pattern such as granular, blocky, and prismatic.

Pedology. The study of soil based on its genesis, morphology, scientific classification, and relation to other earth sciences.

Pedosphere. The weathered upper portion of the earth's crust where pedogenesis occurs. It is the locus of all the soil-forming processes, including the influence of plant roots, soil organisms, water movement, and gaseous diffusion. In a designed soil, it includes all the foregoing in addition to any urban infrastructure such as pipe traces, underdrainage, curbing, roof or vault ceilings, cable lines, and so on, that may compete with roots for soil space. The term "landscape below ground" is similar in meaning.

Perched water. Water that is retained within a soil layer above a boundary between two distinctly different materials in texture and/or structure and density. Most often occurs in the soil above a gravel layer or at the bottom of a container.

Percolation. The downward movement of water within the soil, especially in saturated or nearly saturated soil under very small hydraulic gradients.

Permeability. The ease (rate) of water and gas movement within the soil (in any direction) as determined by soil physical properties of texture, structure, density, and pore size distribution. It is also influenced by antecedent moisture content. Sometimes it is defined to include the ease of root penetration. The term is misapplied to the infiltration of precipitation or irrigation water into the soil.

Piping. The migration of soil fines from one point to another within the soil profile. Soil is most susceptible to piping when the coefficient of permeability is approximately 10^{-4} cm/sec.

Pore size distribution. The volume of various sizes of pores in the soil. Expressed as percentages of the dry bulk volume. Similar to particle size distribution, but emphasis is on the voids. Usually obtained from moisture retention curves determined in the laboratory.

Puddled soil. Dense, massive soil artificially compacted when wet and having no aggregate structure. The condition commonly results from the tillage of a clayey soil when it is wet, but the effect is not necessarily restricted to clayey soils. Can be caused by foot or vehicle traffic on saturated or nearly saturated soils.

Reaction, soil. *See* Soil Reaction.

Restoration site. A previously occupied or developed site, usually derelict in nature, which is to undergo new construction or landscaping or both. Sometimes called a rehabilitation site. The site frequently contains contaminated soil.

***Rhizobium* spp.** Bacteria capable of living symbiotically on the roots of higher plants, usually legumes, from which they receive their energy. In exchange, they fix atmospheric nitrogen for their own use as well as for the host plant.

Sand. A primary soil particle derived from fragmented and weathered rock, composed mainly of quartz and having a diameter of 2.00 to 0.05 mm (USDA classification). Includes the fractions of *very coarse, coarse, medium, fine,* and *very fine* sand. If present in high proportion tends to cause soil material to be well to excessively drained, low waterholding capacity, and low fertility.

Silt. A primary soil particle derived from fragmented and weathered rock, composed mainly of quartz and having a diameter of 0.05 to 0.002 mm (USDA classification). If present in high proportion tends to enhance compaction susceptibility of the soil material but increases waterholding capacity.

Slumping. The failure of a mass of soil on a slope in response to gravity. It is usually triggered by an external force, such as heavy rains saturating the soil, earthquakes, or vibration set up by high winds rocking tree roots, or a combination of several of these factors.

Soil profile. The vertical construct of natural or designed soil comprised of layers designated as horizons. It usually consists of a topsoil horizon, subsoil horizon(s), and a parent material horizon. It exhibits the vertical gradient of pedogenesis for a given point on the earth's surface.

Soil reaction (pH). The hydrogen-ion concentration of the soil solution. It is expressed in the form of the reciprocal of the log of the hydrogen-ion number. A neutral soil has a pH value of 7.0. An acid soil has a pH less than 7.0 and an alkaline soil has a pH value greater than 7.0.

Stability. In reference to compost, it indicates the relative state of decomposition of the compost. An unstable compost is one that is only somewhat decomposed and will over a short time continue the process with evolution of large amounts of water, heat, carbon dioxide, and other by-products that may be temporarily harmful to plants.

Structure (soil). The aggregation of the primary soil particles of sand, silt, and clay into geometric forms or patterns as *single-grained, granular, sub-*

angular blocky, blocky, platy, columnar, prismatic, and *massive.* Results as the the soil forms under natural conditions. Disturbance or manipulation of the soil generally tends to reduce or even destroy the natural structure. This loss tends to cause compaction, reduced water infiltration, decreased gaseous diffusion, and increased resistance to root penetration.

Subsoil. The second layer or set of horizons lying below the topsoil. May have bright colors, an accumulation of clay and other translocated materials, low organic matter content, and few but large roots.

Surficial geology. The geologic material or strata that is exposed at earth's surface or lies immediately below the soil and extends no more than tens of feet in depth. It usually includes glacial, coastal and fluvial deposits, and so on.

Sustainable soil. Sustainable soil is comprised entirely of recyclable products alone or in a mixture with derelict soil material, the latter of which is useless without supplementation. It contains little, if any, nonrenewable resources.

Texture. The proportion of sand, silt, and clay in the soil material, usually expressed on a percentage basis by weight, and classified into one of 12 classes as defined by the USDA. Not directly interpolated into the Unified Soil Classification system.

Thixotrophy. The loss of cohesion and even liquifaction of saturated or nearly saturated soils under vibratory stress. May be observed in all but the coarsest, gravelly soils.

Tilth. An old Anglo-Saxon term that describes the physical condition of the soil as related to its ease of tillage, fitness as a seedbed, and impedance to seedling emergence and root penetration. A soil with good tilth is very desirous.

Topsoil mix. A mix of several components designed to emulate the natural topsoil. It may be comprised of inorganic and organic materials. An example is one-third sand, one-third peat moss, and one-third natural soil.

Topsoil. The surface or top horizon of a soil profile usually exhibiting dark color from high organic matter content, favorable structure, and containing many soil organisms and roots. Defined as the A horizon in profile descriptions. Generally, moderate to high fertility. It is also used as a term in soil specifications but may not always meet the strict definition.

TURFACE®. A commerical product derived from calcined clay applied to soil to enhance drainage and aeration and resistance to compaction. It will reduce soil density.

Vermiculite. A natural clay material found in many soils as a silicate clay mineral and occurs in natural deposits where it is mined, processed by heating to high temperatures into an expanded form. It is used as insulation material and is applied to soil to enhance structure and tilth.

REFERENCES

Aandahl, A. R. 1958. Soil survey interpretation theory and purpose. *Soil Sci. Soc. Amer. Proc.* 22: 152–154.

Anon. 1990. Root barriers and root pruning. *Arbor Age* 10(6): 38–45.

Anon. 1995. Root barrier update. *Arbor Age* 15(9): 22, 24, 27.

Appleton, B. L. 1994. Elimination of ciricling tree roots during nursery production. In G. W. Watson and D. Neely, eds., *The Landscape Below Ground,* pp. 93–97. Proceedings of an International Workshop on Tree Root Development in Urban Soils. International Society of Arboriculture Savoy, IL.

Arnold, M. A., and D. F. Welsh. 1995. Effects of planting hole configuration and soil type on transplant establishment of container-grown live oak. *J. Arboric.* 21(4): 213–218.

Bassuk, N., and T. Whitlow. 1985. Evaluating street tree microclimates in New York City. *METRIA: 5, Proceedings of the Fifth Conference of the Metropolitan Tree Improvement Alliance,* pp. 18–27. Pennsylvania State University Press, University Park.

Beasley, R. P. 1978. *Sediment and Erosion Control.* Iowa State University Press, Ames.

Bernatzky, A. 1978. *Tree Ecology and Preservation.* Elsevier North-Holland, New York.

Berrang, P., D. F. Karnosky, and B. J. Stanton. 1985. Environmental factors affecting tree health in New York City. *J. Arboric.* 11(6): 185–189.

Brady, N. C. 1990. *The Nature and Properties of Soils,* 10th ed. Macmillan, New York.

Bridel, R., B. L. Appleton, and C. E. Whitcomb. 1983. Planting techniques for tree spade-dug trees. *J. Arboric.* 9: 282–284.

Bridges, E. M. 1987. *Surveying Derelict Land.* Clarendon Press, Oxford.

Buol, S. W., F. D. Hole, and R. J. McCracken. 1989. *Soil Genesis and Classification,* 3rd ed. Iowa State University Press, Ames.

Burger, D. W., G. W. Forister, and P. A. Kiehl. 1996. Height, caliper growth, and biomass response of ten shade tree species to tree shelters. *J. Arboric.* 22(4): 161–166.

Campbell, G. S. 1977. *An Introduction to Environmental Biophysics*. Springer-Verlag, Heidelberg.

Carlson, Chris R. 1997. Survey of curbside planting width in 116 cities. Poster report, International Society of Arboriculture Annual Convention, Salt Lake City.

Chaney, W. R. 1997. How wind affects trees. *Arborist News* 6(5): 31–38.

Clark, D. B., ed. 1994. *Xerigation: Landscape Low Volume Irrigation Design Manual.* Rain Bird Sales, Inc., Contractor Division. Glendora, CA.

Coalition of Northeast Governors (CONEG) Source Reduction Task Force. 1996. Model procurement guidelines for source-separated compost. Washington, DC.

Composting Council. 1996. *Draft Field Guide to Compost Use.* The Compost Council, Alexandra, VA.

Cool, R. A. 1976. Tree spade vs. bare root planting. *J. Arboric.* 2: 92–95.

Corley, W. L. 1984a. Re-evaluating the value of amending planting holes with organic material. *Amer. Nurseryman* 159(6): 113–116.

Corley, W. L. 1984b. Soil amendments at planting. *Environ. Hort.* 2: 27–30.

Costello, L. R., C. L. Elmore, and S. Steinmaus. 1997. Tree root response to circling root barriers. *J. Arboric.* 23(6): 211–218.

Couenberg, E. A. M. 1994. Amsterdam tree soil. In G. W. Watson and D. Neely, eds., *The Landscape Below Ground,* pp. 24–33. Proceedings of the International Workshop on Tree Root Development in Urban Soils. Morton Arboretum, Lisle, IL. International Society Arboriculture, Savoy, IL.

Coutts, M. P. 1983. Root archtecture and tree stability. *Plant and Soil* 71: 171–188.

Cox, L. D. 1916. A street tree system for New York City, Borough of Manhattan. *The New York State College of Forestry, Syracuse, Bulletin.*

Craul, P. J. 1964. An analysis of site index variation and certain soil characteristics of DeKalb very stony sandy loam. Ph.D. diss. Pennsylvania State University, University Park.

Craul, P. J. 1985. A description of urban soils and their desired characteristics. *J. Arboric.* 11(11): 330–339.

Craul, P. J. 1988. Assessing soil for urban tree survival. *Grounds Maint.* 23(9): 28, 30, 84, 86–87, 89.

Craul, P. J. 1992. *Urban Soil in Landscape Design.* John Wiley, New York.

Craul, P. J. 1994. Urban soils: An overview and their future. In G. W. Watson and D. Neely, eds., *The Landscape Below Ground: Proceedings of an International Workshop on Tree Root Development in Urban Soils,* pp. 115–125. International Society of Arboriculture, Savoy, IL.

Craul, P. J. 1997. Designing sustainable soil. In Margarita Hill, ed., *Proceedings Opportunities in Sustainable Development: Strategies for the Chesapeake Basin,* pp. 47–51. American Society of Landscape Architects, Washington, DC.

Craul, P. J., and C. J. Klein. 1980. Characterization of streetside soils of Syracuse, New York. In *METRIA* 3: *Proceedings of the Third Conference of the Metropolitan Tree Improvement Alliance,* pp. 88–101. Pennsylvania State University Press, University Park.

Craul, P. J., and M. S. Switzenbaum. 1996. Developing biosolids compost specifications. *BioCycle* 37(12): 44–47.

Das, B. M. 1979. *Introduction to Soil Mechanics*. Iowa State Univ. Press, Ames.

Davis, W. B., and others. 1964. *Landscaping in Containers without Natural Drainage.* Agricultural Science Leaflet 2577. University of California, Davis.

deBertoldi, M. 1987. *Compost: Production, Quality and Use.* Elsevier Applied Science, London.

Deresiewicz, H. 1958. Mechanics of granular matter. *Adv. Appl. Mech.* 5: 233–306.

Estes, J. E., ed. 1983. *Manual of Remote Sensing,* 2nd. ed., Vol. 2, pp. 1233–2440. American Society of Photogrammetry, Falls Church, VA.

Evans, J. and M. J. Potter. 1985. Treeshelters—A new aid to tree establishment. *Plasticulture* 68: 7–20.

Fanger, P. O. 1970. *Thermal Comfort: Analysis and Applications in Environmental Engineering*. McGraw-Hill, New York.

Fazio, J. R., ed. 1993. *How to Landscape to Save Water.* Tree City USA Bulletin No. 17. The National Arbor Day Foundation, Nebraska City, NE.

Federer, C. A. 1970. *Measuring Forest Evapotranspiration—Theory and Problems.* USDA Forest Service Research Paper NE-165. USDA, Washington, DC.

Field, R., H. Masters, and M. Singer. 1982. Status of porous pavement research. *Water Res.* 16: 849–858.

Flint, H. R. 1997. *Landscape Plants for Eastern North America.* John Wiley, New York.

Fortier, J. H. 1994. Arbor irrigation: creating separate stations. *Arbor Age* 14(5): 43–45.

Fraedrich, B. R. 1995. Solutions to tree-sidewalk conflicts. In *Tree Topics, Winter Issue,* pp. 3–6. F. A. Barlett Tree Expert Co., Stamford, CT.

Frearson, K., and H. D. Weiss. 1987. Improved growth rates with treeshelters. *Quart. J. For.* 81(3): 184–187.

Gilman, E. F. 1996. Root barriers affect root distribution. *J. Arboric.* 22(3): 151–154.

Good, G. L. 1989. Fertilizing shade trees. *Landscape Contractors* (May): 20–21.

Goodwin, C., and G. Lumis. 1992. Embedded wire in tree roots; implications for tree growth and root function. *J. Arboric.* 18(3): 115–123.

Grabosky, J., and Nina Bassuk. 1995. A new urban tree soil to safely increase rooting volumes under sidewalks. *J. Arboric.* 21(4): 187–201.

Grabosky, J., and Nina Bassuk. 1996. Testing of structural urban tree soil materials for use under pavement to increase street tree rooting volumes. *J. Arboric.* 22(6): 255–263.

Greenly, K. M., and D. A. Rakow. 1995. The effect of wood mulch type and depth on weed and tree growth and certain soil parameters. *J. Arboric.* 21(5): 225–232.

Guinaudeau, Claude. 1987. *Planter aujourd'hui, batir demain: le preverdissement.* Institut pour le developpement forestier, Nancy, France.

Haines, R. 1981. *Contaminated sites in the West Midlands Country. A prospective survey.* JURUE, University of Aston, Birmingham.

Harada, Y., A. Inoko, M. Tadaki, and T. Izawa. 1980. The measurement of the cation-exchange-capacity of composts for the estimation of the degree of maturity. *Soil Sci. Plant Nutr.* 26: 126.

Hardin, G. 1968. The tragedy of the commons. *Science* 162: 1243–1248.

Harris, R. W. 1983. *Arboriculture: Care of Trees, Shrubs, and Vines in the Landscape.* Prentice Hall, Englewood Cliffs, NJ.

Harris, R. W. 1991. *Arboriculture: Integrated Management of Landscape Trees, Shrubs, and Vines,* 2nd ed. Prentice Hall, Englewood Cliffs, NJ.

Hartin, J., and D. Pittenger. 1992. Irrigation benefits from new technology. *Arbor Age* 12(6): 27–28.

Hensley, D. L., and C. F. Fackler. 1984. Do water-holding compounds help in transplanting? *Amer. Nurseryman* 159(3): 93.

Hensley, D. L., R. E. McNeil, and R. Sundheim. 1988. Management influences on growth of transplanted Magnolia grandiflora. *J. Arboric.* 14(8): 204–207.

Hillel, D. 1980. *Fundamentals of Soil Physics.* John Wiley, New York.

Himelick, E. B. 1981. Mechanical tree digging and planting with a tree spade. In *Tree and Shrub Transplanting Manual,* pp. 29–32. International Society of Arboriculture, Urbana, IL.

Himelick, E. B., and G. W. Watson. 1990. Reduction of oak chlorosis with wood chip mulch treatments. *J. Arboric.* 16(10): 275–278.

Hirai, M. F., V. Chanyasak, and H. Kubota. 1983. A standard measurement for compost maturity. *BioCycle* 24(6): 54–56.

Holtz, R. D., and W. D. Kovacs. 1981. *An Introduction to Geotechnical Engineering.* Prentice Hall, Englewood Cliffs, NJ.

Hornick, S. B., J. J. Murray, R. L. Chaney, L. J. Sikora, J. F. Parr, W. D. Burge, G. B. Willson, and C. F. Tester. 1979. Use of sewage sludge compost for soil improvement and plant growth. *USDA-Sci. and Educ. Adm. Ag. Rev. and Man.* ARM-NE-6, Washington, DC.

Ingram, D. L., and W. Burbage. 1985. Effetcs of antitranspirants and a water-absorbing polymer on the establishment of transplanted live oaks. *South. Nur. Assoc. Nur. Res. Conf. Proc.* 30: 114–115.

Jewell, Linda. 1981. Construction: Planting trees in city soils. *Landscape Architect.* 71(3): 387–389.

Jones, R. H., A. H. Chappelka, and D. H. West. 1995. Use of plastic shelters for low-cost establishment of street trees. *South. J. Appl. For.* 20(2): 85–89.

Jones, R. K., G. Pierpont, G. M. Wickware, J. K. Jeglum, R. W. Arnup, and J. M. Bowles. 1983. *Field Guide to Forest Ecosystem Classification for the Clay Belt, Site Region 3e.* Ontario Ministry of Natural Resouces, Agriculture Canada/Environment Canada, Ottawa.

Juncker, P. H., and J. J. Madison. 1967. Soil moisture characteristics of sand-peat mixes. *Soil Sci. Soc. Amer. Proc.* 31: 5–7.

Kays, B. L. 1982. Methodology for on-site soil analysis. In P. J. Craul, ed., *Urban Forest Soils: A Reference Workbook.* SUNY-College Environmental Science and Forestry, Syracuse.

Kays, B. L., and J. C. Patterson. 1982. Soil drainage and infiltration. In P. J. Craul, ed. *Urban Forest Soils: A Reference Workbook.* SUNY-Coll. *Environ. Sci. and Forestry,* Syracuse, NY.

Knowles, R. N. 1981. *Sun Rhythm Form.* The MIT Press, Cambridge, MA.

Kopinga, J. 1985. Research on street tree planting practices in the Netherlands. In *METRIA 5: Proceedings of the Fifth Conference on the Metropolitan Tree Improvement Alliance,* pp. 72–84. Pennsylvania State University Press, University Park.

Kourik, R. 1992. *Drip Irrigation for Every Landscape and All Climates.* Metamorphic Press, Occidental, CA.

Kozlowski, T. T. 1986. Soil aeration, flooding and tree growth. *J. Arboric.* 11(3): 85–96.

Kramer, P. J. 1987. The role of water stress in tree growth. *J. Arboric.* 13(2): 33–38.

Kramer, P. J., and T. T. Kozlowski. 1979. *Physiology of Woody Plants.* Academic Press, New York.

Kristoffersen, P. 1998. Designing urban pavement sub-bases to support trees. *J. Arboric.* 24(3): 121–126.

Krumbein, W. C., and F. J. Pettijohn. 1938. *Manual of Sedimentary Petrography.* Appleton-Century, New York.

Kuuluvainen, T., and T. Pukkala. 1987. Effect of crown shape and tree distribution on the spatial distribution of shade. *Agric. For. Meteorol.* 40: 215–231.

Lake, D. W., Jr., ed. 1989. *New York Guidelines for Urban Erosion and Sediment Control.* New York State Soil and Water Conservation Committee, Albany.

Leopold, L. B. 1968. *Hydrology for Urban Land Planning. A Guidebook on the Hydrologic Effects of Urban Land Use.* Circular 554. USGS, Washington, DC.

Lindsey, P., and N. Bassuk. 1991. Specifying soil volumes to meet the water needs of mature urban street trees and trees in containers. *J. Arboric.* 17(6): 141–149.

Lindsey, P., and R. Waters. 1995. Use of recycled water in the urban landscape. *California Landscaping* May: 12–15.

Luthin, J. N., ed. 1957. *The Drainage of Agricultural Lands.* American Society of Agronomy, Madison, WI.

Lyford, W. H. 1975. Rhizography of nonwoody roots of trees in the forest floor. In J. G. Torrey and D. T. Clarkson, eds., *The Development and Function of Roots,* pp. 179–196. Academic Press, New York.

Lyford, W. H. 1980. *Development of the root system of northern red oak (Quercus rubra, L.).* Harvard Forest Paper No. 21. Harvard University, Petersham, MA.

Marsh, W. M. 1991. *Landscape Planning: Environmental Applications,* 2nd. ed. John Wiley, New York.

Marx, D. H. 1991. The practical significance of ectomycorrhizae in forest establishment. In *Ecophysiology of Ectomycorrhizae of Forest Trees,* pp. 54–90. The Marcus Wallenberg Foundation Symposium Proc. 7, Stockholm.

Marx, D. H. 1997. *Plant health care benefits of mycorrhizae.* Paper presented at Annual Convention, International Society of Arboriculture, Salt Lake City.

Marx, D. H., M. Murphy, T. Parrish, S. Marx, D. Haigler, and D. Eckard. 1997. Root response of mature live oaks in coastal South Carolina to root zone inoculations with ectomycorrhizal fungal inoculants. *J. Arboric.* 23(6): 257–263.

Mazria, E. 1979. *The Passive Solar Energy Book.* Rodale Press, Emmaus, PA.

McAdams, W. H. 1954. *Heat Transmission,* 3rd ed. McGraw-Hill, New York.

McCammon, R. B. 1962. Efficiencies of percentile measures for describing mean particle size and sorting of sedimentary particles. *J. Geol.* 70: 453–465.

McCoy, E. L. 1992. Quantitative physical assessment of organic materials used in sports turf rootzone mixes. *Agron. J.* 84: 375–381.

McGavin, C. T., ed. 1974. *Ecological Parking Lots.* Transportation Terminal Techniques, Bulletin No. 158. Applied Parking Techniques, Inc., Reston, VA.

McHarg, Ian. 1969. *Design with Nature.* Natural History Press, Falcon Press, Philadelphia.

McHarg, Ian. 1992. Green the earth, heal the earth. *J. Soil Water Cons.* 47(1): 39–46.

Nicoll, B. C., and D. Ray. 1996. Adaptive growth of tree root systems in response to wind action and site conditions. *Tree Physiology* 16: 891–898.

Olson, G. W. 1976. *Criteria for Making and Interpreting a Soil Profile Description.* Bulletin No. 212. University of Kansas Publications, Lawrence.

Olson, G. W. 1981. *Soils and the Environment: A Guide to Soil Surveys and Their Applications.* Chapman and Hall, New York.

Patterson, J. C. 1976. Soil compaction and its effects upon urban vegetation. In *Better Trees for Metropolitan Landscapes Symposium Proceedings.* General Technical Report NE-22. USDA-Forest Service, Washington, DC.

Patterson, J. C. 1985. Creative site planning—alternatives. In *METRIA 5: Proceedings. of the Fifth Conference of the Metropolitan Tree Improvement Alliance,* pp. 85–91. Pennsylvania State University Press, University Park.

Patterson, J. C., and D. L. Mader. 1982. Soil compaction: causes and control. In P. J. Craul, ed., *Urban Forest Soils: A Reference Workbook.* SUNY-College of Environmental Science and Forestry, Syracuse.

Peck, T. R. 1981. What to expect from soil tests. *J. Arboric.* 7(1): 11–12.

Pellet, H. 1971. Effect of soil amendments on growth of landscape plants. *Amer. Nurseryman* 134(12): 103–106.

Perry, R. C. 1992. *Landscape Plants for Western Regions.* Land Design Publishing, Claremont, CA.

Perry, T. O. 1982. The ecology of tree roots and the practical significance thereof. *J. Arboric.* 8(8): 197–211.

Potter, K. N., F. S. Carter, and E. C. Doll. 1988. Physical properties of constructed and undisturbed soils. *Soil Sci. Soc. Amer. J.* 52: 1435–1438.

Potter, M. J. 1988. Treeshelters improve survival and increase early growth rates. *J. For.* 86(6): 39–41.

Pritchett, W. L., and R. F. Fisher. 1987. *Properties and Management of Forest Soils,* 2nd. ed. John Wiley, New York.

Rawls, R. J. 1983. Estimating soil bulk density from particle size analysis and organic matter content. *Soil Sci.* 135(2): 123–125.

Reifsnyder, W. E., and H. W. Lull. 1965. *Radient Energy in Relation to Forests.* USDA-Forest Service Technical Bulletin No. 1344. U.S. Government Printing Office, Washington, DC.

Riffaldi, R., R. Levi-Minzi, A. Pera, and M. deBartoldi. 1986. Evaluation of compost maturity by means of chemical and microbial analyses. *Waste Manag. Res.* 4: 387.

Rose, C. W. 1966. *Agricultural Physics.* Pergamon Press, Oxford.

Saviozzi, A., R. Levi-Minzi, and R. Riffaldi. 1989. Maturity evaluation of organic waste. In staff of *BioCycle,* eds., *The BioCycle Guide to Composting Municipal Wastes.* IG Press, Emmaus, PA.

Smalley, T. J., and C. B. Wood. 1995. Effect of backfill amendment on growth of red maple. *J. Arboric.* 21(5): 247–249.

Smeltz, A. 1998. Urban shadows. *Research/Penn State* 19(1): 12.

Smiley, E. T., D. H. Marx, and B. R. Fraedrich. 1997. Ectomycorrhizal fungus inoculation of established residential trees. *J. Arboric.* 23: 113–115.

Smiley, E. T. 1997. *Mycorrhizae: the latest and greatest.* Paper presented at Annual Convention, International Society of Arboriculture, Salt Lake City.

Smith, A. M., and D. A. Rakow. 1992. Strategies for reducing water input in woody landscape plantings. *J. Arboric.* 18(4): 165–170.

Smith, S. W. 1997. *Landscape Irrigation: Design and Management.* John Wiley, New York.

Smith, W. H. 1985. Forest quality and air quality. *J. For.* 83(2): 82–92.

Soil Conservation Service. 1973. Drainage of Agricultural Land. In *National Engineering Handbook,* Section 16. Water Information Center, Port Washington, New York.

Soil Survey Staff. 1992. *Draft National Soils Handbook.* Part 620. USDA-Natural Resources Conservation Service. U.S. Government Printing Service, Washington, DC.

Soil Survey Staff. 1993. *Soil Survey Manual.* Handbook No. 18. U.S. Department of Agriculture. U.S. Government Printing Office, Washington, DC.

Solotaroff, W. 1911. *Shade Trees in Towns and Cities.* John Wiley, New York.

Sopper, W. E., and L. T. Kardos, eds. 1973. *Recycling Treated Municipal Wastewater and Sludge through Forest and Cropland.* Pennsylvania State University Press, University Park.

Spirn, A. W. 1984. *The Granite Garden.* Basic Books, New York.

Spirn, A. W. 1992. Designing with the land. *J. Soil Water Cons.* 47(1): 35–38.

Spomer, L. A. 1983. Physical amendment of landscape soils. *J. Environ. Hort.* 1(3): 77–80.

Stark, T. F., and D. R. Miller, 1977. Effect of synthetic surfaces and vegetation in urban areas on human energy balance and comfort. In *Proceedings of the Conference on Metropolitan Physical Environment,* pp. 139–151. USDA-Forest Service Northeast Forest Experimental Station General Technical Report NE-25.

Stone, H. M. 1993. Root collar inspection: Detecting and correcting hidden damage. *Arbor Age* 13(2): 8–13.

Stuller, J. 1997. Golf gets back to nature, inviting everyone to play. *Smithsonian* 28(1): 56–58, 60, 62–64, 66–67.

Svihra, P., D. W. Burger, and R. Harris. 1994. The effect of treeshelters on shoot/root development of some ornamentals grown in containers or in the landscape. In G. W. Watson and D. Neely, eds., *The Landscape Below Ground.* Proceedings of an International Workshop on Tree Root Development in Urban Soils. International Society Arboriculture Savoy, IL.

Svihra, P., D. W. Burger and R. Harris. 1996. Treeshelter effect on root development of redwood trees. *J. Arboric.* 22(4): 174–179.

Thorpe, M. R., R. Saugier, A. Auger, A. Berger, and M. Methy. 1978. Photosynthesis and transpiration of an isolated tree. *Plant, Cell Environ.* 1: 269–277.

Tomlinson, P. B. 1982. Chance and design in the construction of plants. In R. Sattler, ed., *Axioms and Principles of Plant Construction,* pp. 162–183. Acta Biotheoretica 31A. Junk, The Hague.

Townsend, A. M., and W. F. Kwolek. 1987. Relative susceptibility of thirteen pine species to sodium chloride spray. *J. Arboric.* 13(9): 225–228.

Urban, J. 1989. New techniques in urban tree plantings. *J. Arboric.* 15(11): 281–284.

Urban, J. R. 1992. Bringing order to the technical dysfunction within the urban forest. *J. Arboric.* 18(2): 85–90.

Urban J., and P. J. Craul. 1996. Success with soils in urban landscapes. *Arbor Age* 16(7): 18, 20, 22.

U.S. Department of Agriculture. 1982. *A National Program for Soil and Water Conservation.* Final Program Report and Environmental Impact Statement. U.S. Government. Printing Office, Washington, DC.

U.S. Salinity Laboratory Staff. 1954. L. A. Richards, ed. *Diagnosis and Improvement of Saline and Alkali Soils.* US Department of Agriculture Handbook No. 60. Government Printing Office, Washington.

Van der Ryn, S., and S. Cowan. 1996. *Ecological Design.* Island Press, Washington, DC.

Veihmeyer, F. J., and A. H. Hendrickson. 1948. Soil density and root penetration. *Soil Sci.* 65: 487–493.

Vepraskas, M. J. 1988. Bulk density values diagnostic of restricted root growth in coarse-textured soils. *Soil Sci. Soc. Amer. J.* 52: 1117–1121.

Vogel, S. 1996. Blowing in the wind: Storm-resisting features of the design of trees. *J. Arboric.* 22(2): 92–98.

Vrecenak, A. J., and L. P. Herrington. 1984. Estimation of water use of landscape trees. *J. Arboric.* 10(12): 313–319.

Waggoner, P. E., and W. E. Reifsnyder, 1961. Difference between net radiation and water use caused by radiation from the soil surface. *Soil Sci.* 91: 246–250.

Wargo, P. M. 1975. *Estimating Starch Content in Roots of Deciduous Trees. A Visual Technique.* USDA-Forest Service Research Paper NE-313. Washington, DC.

Wargo, P. M. 1988. *Judging Vigor of Deciduous Hardwoods.* USDA Agricultural Information Bulletin 418. U.S. Government Printing Office, Washington, DC.

Waring, R. H., and B. D. Cleary. 1967. Plant moisture stress: Evaluation by pressure bomb. *Science* 155: 1248.

Warren, S. L. 1993. Growth and nutrient concentration in flowering dogwood after nitrogen fertilization and dormant root pruning. *J. Arboric.* 19(2): 57–63.

Warrick, A. W., D. F. Myers, and D. R. Nielsen. 1986. Geostatistical methods applied to soil science. In A. Klute, ed., *Methods of Soil Analysis, Part 1. Physical and Mineralogical Methods,* pp. 53–82. American Society of Agronomy, Madison, WI.

Waters, R., and P. Lindsey. 1995. Irrigation technology and recycled water use. *California Landscaping* May: 16–19.

Watson, G. W. 1988. Organic mulch and grass competition influence tree root development. *J. Arboric.* 14(8): 200–203.

Watson, G. W., S. Clark, and K. Johnson. 1990. Formation of girdling roots. *J. Arboric.* 16: 197–202.

Watson, G. W., and E. B. Himelick. 1997. *The Principles and Practice of Planting Trees and Shrubs*. International Society of Arboriculture, Savoy, IL.

Watson, G. W., P. Kelsey, and K. Woodli. 1996. Replacing soil in the root zone of mature trees for better growth. *J. Arboric.* 22(4): 167–173.

Watson, G. W., G. Kupkowski, and K. G. von der Heide-Spravka. 1992. The effect of backfill soil texture and planting hole shape on root regeneration of transplanted green ash. *J. Arboric.* 18(3): 130–135.

Watson, G. W., and G. Kupkowski. 1991. Effects of a deep layer of mulch on the soil environment and tree root growth. *J. Arboric.* 17(9): 242–245.

Way, D. S. 1973. *Terrain Analysis: A Guide to Site Selection Using Aerial Photographic Interpretation.* Dowden, Hutchinson and Ross, Stroudsburg, PA.

Whitcomb, C. E. 1975. Effects of soil amendments on growth of silver maple trees in the landscape. *South. Nurs. Assoc. Nurs. Res. Conf. Proc.* 20: 49–50.

Whitcomb, C. E. 1987. *Establishment and Maintenance of Landscape Trees*. Lacebark Publications, Stillwater, OK.

Wilde, S. A. 1958. *Forest Soils*. Ronald Press, New York.

Wilson, B. F. 1975. Distribution of secondary thickening in tree root systems. In J. G. Torrey and D. T. Clarkson, eds., *The Development and Function of Roots*. Academic Press, London.

Wischmeier, W. H., and D. D. Smith. 1978. *Predicting rainfall erosion losses—A Guide to Conservation Planning.* Agricultural Handbook No. 537. U.S. Dept. of Agriculture, Washington, D.C.

Wise, M. E. 1952. Dense random packing of unequal spheres. *Philips Res. Rep.* 7: 321–343.

Wolman, M. G., and A. P. Schick. 1967. Effects of construction on fluvial sediment, urban and suburban areas of Maryland. *Water Resources Res.* 3(2): 451–464.

Yelonosky, G. 1964. Tolerance of trees to deficiencies of soil aeration. *Intern. Shade Tree Conf. Proc.* 40: 127–147.

Yorke, T. H., and W. J. Herb. 1978. Effects of urbanization on streamflow and sediment transport in the Rock Creek and Anacostia River Basins, Montgomery County, Maryland, 1962-1974. U.S. Geological Survey Professional Paper No. 1003. US Government Printing Office, Washington, D.C.

Zucconi, F., M. Forte, A. Monaco, and M. deBartoldi. 1981. Biological evaluation of compost maturity. *BioCycle* 22(4): 27–29.

Zucconi, F., and M. deBartoldi. 1989. Specifications for solid waste compost. In The Staff of BioCycle, eds. *The BioCycle Guide to Composting Municipal Wastes.* IG Press, Emmaus, PA.

AUTHOR INDEX

SUBJECT INDEX